AF353031

Biochemical and Thermochemical Conversion Processes of Lignicellulosic Biomass Fractionated Streams

Biochemical and Thermochemical Conversion Processes of Lignicellulosic Biomass Fractionated Streams

Editors

Leonidas Matsakas
Anna Trubetskaya

MDPI • Basel • Beijing • Wuhan • Barcelona • Belgrade • Manchester • Tokyo • Cluj • Tianjin

Editors
Leonidas Matsakas
Luleå University of Technology
Sweden

Anna Trubetskaya
University of Limerick
Ireland

Editorial Office
MDPI
St. Alban-Anlage 66
4052 Basel, Switzerland

This is a reprint of articles from the Special Issue published online in the open access journal *Processes* (ISSN 2227-9717) (available at: https://www.mdpi.com/journal/processes/special_issues/lignicellulosic_conversion).

For citation purposes, cite each article independently as indicated on the article page online and as indicated below:

LastName, A.A.; LastName, B.B.; LastName, C.C. Article Title. *Journal Name* **Year**, *Volume Number*, Page Range.

ISBN 978-3-0365-1942-5 (Hbk)
ISBN 978-3-0365-1943-2 (PDF)

Contents

processes

Editorial

Special Issue: Biochemical and Thermochemical Conversion Processes of Lignocellulosic Biomass Fractionated Streams

Anna Trubetskaya [1],* and Leonidas Matsakas [2],*

1   Department of Sciences and Engineering, University of Limerick, Castletroy, V94 T9PX Limerick, Ireland
2   Biochemical Process Engineering, Division of Chemical Engineering, Department of Civil, Environmental and Natural Resources Engineering, Luleå University of Technology, 97187 Luleå, Sweden
*   Correspondence: anna.trubetskaya@ul.ie (A.T.); leonidas.matsakas@ltu.se (L.M.)

Citation: Trubetskaya, A.; Matsakas, L. Special Issue: Biochemical and Thermochemical Conversion Processes of Lignocellulosic Biomass Fractionated Streams. *Processes* **2021**, *9*, 969. https://doi.org/10.3390/pr9060969

Received: 17 May 2021
Accepted: 26 May 2021
Published: 31 May 2021

Publisher's Note: MDPI stays neutral with regard to jurisdictional claims in published maps and institutional affiliations.

Global consumption of materials such as forest resources, fossil fuels, earth metals and minerals are expected to double in the next 30 years, while annual waste production is estimated to increase by approximately 70% by 2050 [1]. Keeping the resource consumption within planetary boundaries, we strive to minimize the carbon and environmental footprint and concurrently double the waste material use in the coming decades. Preventing food waste from being generated could have a major impact on waste collection systems and on the capacity of bio-waste management facilities worldwide [2]. Therefore, sustainable food waste management is a key part of any green business strategy to convert food waste into green fuels.

Thermochemical and biochemical conversion utilizes biomass and waste in an efficient and sustainable way for a wide variety of applications, such as heat, electricity, biofuels, chemicals and biomaterials. This Special Issue aims to explore the most advanced solutions in biomass and waste pre-treatment techniques. Moreover, we have looked into the woody and herbaceous biomass conversion using thermochemical and biochemical processes. This Special Issue has also considered plant species originating from all around the globe. The production of solid or liquid fuels from low-cost lignocellulosic biomass, i.e., forestry residue, agricultural or pulp waste, very often includes a pre-treatment, followed by enzymatic saccharification of the carbohydrates and/or microbial conversion of the sugars to biofuel [3]. Pre-treatment aims to efficiently separate hemicellulose and lignin from cellulose [4]. Removal of lignin from cellulose is particularly important as lignin has a negative influence on enzymatic saccharification due to the irreversible adsorption of cellulolytic enzymes onto lignin and their inhibition from soluble lignin-derived molecules [5]. Hydrothermal pre-treatment is one of the most common methods to effectively degrade hemicelluloses without using chemicals and increase the biomass porosity [6]. The common challenge of hydrothermal pretreatment is related to the fact that the direct separation of lignin is limited; hence, lignin partly tends to rearrange on the surface of the lignocellulosic biomass, causing an inhibitory effect on the saccharification process [7,8]. A recent study has shown that steam explosion is not a suitable pretreatment for acid hydrolysis of hardwood lignocellulosic biomass [9]. This Special Issue provides the experimental results on the organosolv separation of agricultural waste and forestry residues [10]. Organosolv pretreatment is known as an effective method to fractionate biomass into three lignocellulosic compounds, i.e., cellulose, hemicellulose and lignin, by using aqueous-organic solvent mixtures, with high solvent concentration (30–70%) for temperatures ranging from 100 to 220 °C in the presence or absence of catalysts [11,12]. Thus, the organosolv process provides high-quality cellulose and lignin [13]. A recent study has demonstrated that using the vanadate–hydrogen peroxide system on acetosolv pine lignin, vanillin and isovanillin can be generated as main products with depolymerization yields of 31% [14]. In another study, the etherification of both organosolv and Kraft lignin with alkyl halides led to the lignin product with the low glass transition temperature and improved thermal stability that can be used as thermoplastic [15]. With regards to the conversion of the cellulose to products via microbial fermentation (such as bio-ethanol), two approaches are commonly used. These involve separate hydrolysis and fermentation (SHF) and simultaneous saccharification and fermentation (SSF), and recently, it was shown that SHF is advantageous for bioethanol production from pretreated Napier grass [16]. Apart from glucose, lignocellulosic biomass has other types of C6 and C5 sugars. From the different available sugars, processes that use glucose (or other C6 sugars) are more established compared to processes using C5 sugars (such as xylose). Xylose can serve as a carbon source for the cultivation of oleaginous yeasts aiming to produce microbial lipids that can serve as feedstock for biodiesel production [17,18]. A recent study demonstrated the potential of the bioconversion of abundant xylose-containing wastewaters into yeast biomass and lipid with satisfactory productions and conversion yields using various yeasts [19].

Knowledge of the composition of raw and pre-treated products is important to better understand the pre-treatment process as well as further refinement of value-added products. The analysis of pre-treated products is tedious due to the complexity of mixtures containing compounds of low and high polarity, which are not always possible to resolve with conventional techniques [20]. The pyrolysis product, i.e., bio-oil, has its challenges, such as the high-water content, high viscosity, low pH, instability, presence of solids, high oxygen content and low calorific value [21]. However, the comparison of bio-oil properties from pinewood and acacia indicated that the yield of bio-oil can be predicted using standard analysis methods such as elemental analysis and volatile matter characterization. Structural and physicochemical characteristics of five different lignins were elucidated using established analysis methods, i.e., gel permeation chromatography for molecular mass features, quantitative ^{31}P NMR and comparative two-dimensional ^{1}H-^{13}C HSQC analyses for more detailed structural aspects [10]. The selection of analytical methods depends on the complexity of the pre-treatment products and on the final application of the value-added products within the circular economy.

The circular economy plays a dominant role in tackling the climate crisis by redefining the product value with a focus on both the social and environmental benefits. Circular economy principles and strategies using life cycle assessment can be used for analyses of food systems, storage, transportation, feedstock pretreatment, pollutions, etc. [22]. Life cycle assessment within a circular economy can illustrate how the products can be transformed in a way to reduce greenhouse emissions. Thus, the lignocellulosic composition of biomass has a strong impact on the final thermochemical conversion product, leading to consideration of compositional differences in life cycle assessment [23]. The review article on the life cycle assessment of biochar production and utilization for metallurgical processes has been discussed from industrial perspectives. This Special Issue points at several possibilities to integrate the production of bio-based reductants in ferroalloy industries with bio-refineries to lower the cost and increase the total efficiency. Despite challenges related to energy-efficient charcoal production and formation of air pollutions in classical biochar kilns, the potential of bio-based reductant usage in ferroalloy reduction process was underlined as a sustainable pathway to convert forestry to value-added products in metallurgical industries [24]. In addition, the mechanical durability of biochar slightly increased after heat treatment, whereas coal and semi-coke-based reductants showed a decrease in durability. The results indicate that biochar can be used as an efficient carbon source for electric arc furnaces [25]. Overall, more research has to be carried out to identify potential feedstock mixtures for the optimization of the biochar properties.

A promising pre-treatment route of low-value wood refers to various biorefinery processes developed to produce green chemicals. Forestry residues are an excellent potential source of a plethora of renewable chemicals with applications ranging from the energy field to the chemical industry and the potential to replace fossil resources. Therefore, new pre-treatment concepts for low quality wood processing within the circular economy are required to meet the guidelines, as described in the EU green deal [1]. Extraction, microbial fermentation, pyrolysis and organosolv pre-treatments are some of the main biomass processing technologies that can be utilized for the conversion of low-quality wood into value-added products. The upscale of such processes will strongly depend on the development of analytical methods that will allow to appropriately evaluate the end products and allow the correlation with the different stages of the process, which in turn will enable to tune the whole process for high yield production of high-quality green chemicals and materials.

Funding: This research received no external funding.

Conflicts of Interest: The authors declare no conflict of interest.

References

1. European Commission. Communications from the Commission to the European Parliament, the European Council, The Council, The European Economic and Social Committee and the Committee of the Regions. *Eur. Green Deal Rep.* **2019**, *640*, 24.
2. Kalmykova, Y.; Sadagopan, M.; Rosado, L. Circular economy—From review of theories and practices to development of implementation tools. *Resour. Conserv. Recyc.* **2018**, *135*, 190–201. [CrossRef]
3. Kalogiannis, K.G.; Matsakas, L.; Aspden, J.; Lappas, A.A.; Rova, U.; Christakopoulos, P. Acid Assisted Organosolv Delignification of Beechwood and Pulp Conversion towards High Concentrated Cellulosic Ethanol via High Gravity Enzymatic Hydrolysis and Fermentation. *Energies* **2018**, *23*, 1647. [CrossRef] [PubMed]
4. Zhao, X.; Cheng, K.; Liu, D. Organosolv pretreatment of lignocellulosic biomass for enzymatic hydrolysis. *Appl. Microbiol. Biotechnol.* **2009**, *82*, 815–827. [CrossRef]
5. Matsakas, L.; Nitsos, C.; Raghavendran, V.; Yakimenko, O.; Persson, G.; Christakopoulos, P. A novel hybrid organosolv: Steam explosion method for the efficient fractionation and pretreatment of birch biomass. *Biotechnol. Biofuels* **2018**, *11*, 1–14. [CrossRef] [PubMed]
6. Zhang, J.; Tang, M.; Viikari, L. Xylans inhibit enzymatic hydrolysis of lignocellulosic materials by cellulases. *Bioresour. Technol.* **2012**, *121*, 8–12. [CrossRef] [PubMed]
7. Kristensen, J.B.; Thygesen, L.G.; Felby, C.; Jorgensen, H.; Elder, T. Cell wall structural changes in wheat straw pretreated for bioethanol production. *Biotechnol. Biofuels* **2008**, *1*, 1–9. [CrossRef]
8. Sipponen, M.H.; Rahikainen, J.; Leskinen, T.; Pihlajaniemi, V.; Mattinen, M.L.; Lange, H. Strutural changes of lignin in biorefinery pretreatments and consequences to enzyme-lignin interactions. *Nordic Pulp Pap. Res. J.* **2018**, *32*, 550–571. [CrossRef]
9. Steinbach, D.; Kruse, A.; Sauer, J.; Storz, J. Is Steam Explosion a Promising Pretreatment for Acid Hydrolysis of Lignocellulosic Biomass? *Processes* **2020**, *8*, 1626. [CrossRef]
10. Trubetskaya, A.; Lange, H.; Wittgens, B.; Brunsvik, A.; Crestini, C.; Rova, U.; Christakopoulos, P.; Leahy, J.; Matsakas, L. Structural and Thermal Characterization of Novel Organosolv Lignins from Wood and Herbaceous Sources. *Processes* **2020**, *8*, 860. [CrossRef]
11. Kalogiannis, K.G.; Matsakas, L.; Lappas, A.A.; Rova, U.; Christakopoulos, P. Aromatics from Beechwood Organosolv Lignin through Thermal and Catalytic Pyrolysis. *Energies* **2019**, *12*, 1606. [CrossRef]
12. Wyman, C.E.; Dale, B.E.; Elander, R.T.; Holtzapple, M.; Ladisch, M.R.; Lee, Y.Y. Coordinated development of leading biomass pretreatment technologies. *Bioresour. Technol.* **2005**, *96*, 1959–1966. [CrossRef]
13. Constant, S.; Wienk, H.L.J.; Frissen, A.E.; de Peinder, P.; Boelens, R.; van Es, D. New insights into the structure and composition of technical lignins: A comparative characterisation study. *Green Chem.* **2016**, *18*, 2651–2665. [CrossRef]
14. Penín, L.; Gigli, M.; Sabuzi, F.; Santos, V.; Galloni, P.; Conte, V.; Parajó, J.; Lange, H.; Crestini, C. Biomimetic Vanadate and Molybdate Systems for Oxidative Upgrading of Iono- and Organosolv Hard- and Softwood Lignins. *Processes* **2020**, *8*, 1161. [CrossRef]
15. Bhattacharyya, S.; Matsakas, L.; Rova, U.; Christakopoulos, P. Melt Stable Functionalized Organosolv and Kraft Lignin Thermoplastic. *Processes* **2020**, *8*, 1108. [CrossRef]
16. Kongkeitkajorn, M.; Sae-Kuay, C.; Reungsang, A. Evaluation of Napier Grass for Bioethanol Production through a Fermentation Process. *Processes* **2020**, *8*, 567. [CrossRef]
17. Patel, A.; Arora, N.; Sartaj, K.; Pruthi, V.; Pruthi, P.A. Sustainable biodiesel production from oleaginous yeasts utilizing hydrolysates of various non-edible lignocellulosic biomasses. *Renew. Sustain. Energy Rev.* **2016**, *62*, 836–855. [CrossRef]
18. Patel, A.; Mikes, F.; Bühler, S.; Matsakas, L. Valorization of brewers' spent grain for the production of lipids by oleaginous yeast. *Molecules* **2018**, *23*, 3052. [CrossRef] [PubMed]
19. Xenopoulos, E.; Giannikakis, I.; Chatzifragkou, A.; Koutinas, A.; Papanikolaou, S. Lipid Production by Yeasts Growing on Commercial Xylose in Submerged Cultures with Process Water Being Partially Replaced by Olive Mill Wastewaters. *Processes* **2020**, *8*, 819. [CrossRef]
20. Trubetskaya, A.; Johnson, R.; Monaghan, R.F.D.; Ramos, A.S.; Brunsvik, A.; Wittgens, B.; Han, Y.; Pisano, I.; Leahy, J.J.; Budarin, V. Combined analytical strategies for chemical and physical characterization of tar from torrefaction of olive stone. *Fuel* **2021**, *291*, 120086. [CrossRef]
21. Charis, G.; Danha, G.; Muzenda, E. Optimizing Yield and Quality of Bio-Oil: A Comparative Study of Acacia tortilis and Pine Dust. *Processes* **2020**, *8*, 551. [CrossRef]

22. Fassio, F.; Tecco, N. Circular Economy for Food: A Systemic Interpretation of 40 Case Histories in the Food System in Their Relationships with SDGs. *Systems* **2019**, *7*, 43. [CrossRef]
23. Surup, G.; Trubetskaya, A.; Tangstad, M. Life Cycle Assessment of Renewable Reductants in the Ferromanganese Alloy Production: A Review. *Processes* **2021**, *9*, 185. [CrossRef]
24. Surup, G.; Trubetskaya, A.; Tangstad, M. Charcoal as an Alternative Reductant in Ferroalloy Production: A Review. *Processes* **2020**, *8*, 1432. [CrossRef]
25. Surup, G.; Pedersen, T.; Chaldien, A.; Beukes, J.; Tangstad, M. Electrical Resistivity of Carbonaceous Bed Material at High Temperature. *Processes* **2020**, *8*, 933. [CrossRef]

 *processes*

Article

Is Steam Explosion a Promising Pretreatment for Acid Hydrolysis of Lignocellulosic Biomass?

David Steinbach [1], Andrea Kruse [2], Jörg Sauer [1,* and Jonas Storz [1]

[1]   Karlsruhe Institute of Technology (KIT), Institute for Catalysis Research and Technology,
    Hermann-von-Helmholtz-Platz 1, 76344 Eggenstein-Leopoldshafen, Germany;
    David.Steinbach@partner.kit.edu (D.S.); jonas.storz@ocbc.uni-freiburg.de (J.S.)
[2]   Institute of Agricultural Engineering, University of Hohenheim, Garbenstraße 9, 70599 Stuttgart, Germany;
    Andrea_Kruse@uni-hohenheim.de
*   Correspondence: j.sauer@kit.edu

Received: 13 November 2020; Accepted: 8 December 2020; Published: 10 December 2020

Abstract: For the production of sugars and biobased platform chemicals from lignocellulosic biomass, the hydrolysis of cellulose and hemicelluloses to water-soluble sugars is a crucial step. As the complex structure of lignocellulosic biomass hinders an efficient hydrolysis via acid hydrolysis, a suitable pretreatment strategy is of special importance. The pretreatment steam explosion was intended to increase the accessibility of the cellulose fibers so that the subsequent acid hydrolysis of the cellulose to glucose would take place in a shorter time. Steam explosion pretreatment was performed with beech wood chips at varying severities with different reaction times (25–34 min) and maximum temperatures (186–223 °C). However, the subsequent acid hydrolysis step of steam-exploded residue was performed at constant settings at 180 °C with diluted sulfuric acid. The concentration profiles of the main water-soluble hydrolysis products were recorded. We showed in this study that the defibration of the macrofibrils in the lignocellulose structure during steam explosion does not lead to an increased rate of cellulose hydrolysis. So, steam explosion is not a suitable pretreatment for acid hydrolysis of hardwood lignocellulosic biomass.

Keywords: glucose; xylose; 2nd generation sugars; lignocellulose; hydrolyzate; biorefinery; furfural; hydroxymethylfurfural; bioeconomy

---

1. Introduction

In view of a bioeconomy, promising products from lignocellulosic biomass can be value-added platform chemicals produced in a biorefinery [1,2]. As lignocellulosic biomass consists mainly of the polymeric constituents cellulose, hemicelluloses and lignin, the fractionation of feedstock is the first step for maximizing the value of these materials. Cellulose is composed of glucose building blocks linked together by glycosidic bonds forming a linear polymer. Between adjacent cellulose chains, intermolecular hydrogen bonds are formed, which result in a water-insoluble and highly ordered configuration that makes cellulose partly crystalline [3]. Hemicelluloses hydrolyze faster than cellulose, because they are group of amorphous heteropolymers which are considerably shorter than cellulose. Hemicelluloses provide a linkage between cellulose and lignin in the lignocellulosic fiber structure. Lignin is constructed of phenylpropane units which form a complex three-dimensional macromolecule.

One method for the fractionation of lignocellulosic biomass is acid hydrolysis whereby cellulose and hemicelluloses are hydrolyzed to water-soluble sugars which are partly dehydrated to furfurals. In particular, furfural and hydroxymethylfurfural can be produced in a lignocellulose biorefinery. These furfurals were named as one of the top 10 value-added biobased chemicals [4].

However, the complex structure of lignocellulosic biomass generally causes a low hydrolysis rate, especially of cellulose during acid hydrolysis [5]. Hemicelluloses and lignin surrounding the cellulose form a physical barrier against the permeation for the hydrolysis catalyst [6]. Additionally, different hydrolysis rates for crystalline and amorphous cellulose regions exist. This, combined with the shielding matrix of lignin and hemicelluloses, results in a gradual sugar release [5].

A pretreatment is necessary to increase the accessibility of cellulose for the subsequent hydrolysis step [5]. As a general estimation, the pretreatment step accounts for about 40% of the processing cost in a lignocellulose biorefinery [7]. Therefore, it is important to choose pretreatment methods and conditions carefully. A relatively inexpensive pretreatment is steam explosion because no addition of an external catalyst is needed [8]. However, the energy consumption for steam explosion was estimated as 1.8 MJ/kg$_{wood}$ and is therefore considerable [9]. Steam explosion converts biomass in a steam atmosphere at elevated temperatures ranging from 140–240 °C. The steam pressure is rapidly reduced to atmospheric pressure, whereby a mechanical disruption of the biomass occurs.

Acetic acid is formed during steam explosion via the cleavage of thermally labile acetyl groups in hemicelluloses [5]. The liberated acetic acid catalyzes the hydrolysis reactions of hemicelluloses. The solid residue after steam explosion pretreatment consists of cellulose, a chemically modified lignin and residual hemicelluloses. The sum of hemicelluloses in the residue and dissolved hemicellulose-derived sugars decreases with the pretreatment severity due to (1) condensation reactions leading to solid pseudolignin and (2) furfural formation by the dehydration of pentoses [10].

Steam explosion is a well-known effective pretreatment to increase the rate of enzymatic hydrolysis of lignocellulosic biomass [11–14]. For example, the enzymatic hydrolysis rate showed a 10-fold increase for steam-exploded hardwoods [15]. In addition, the glucose yields after enzymatic hydrolysis are higher, when a steam explosion pretreatment was applied. This increase in glucose yield is due to the removal of biomass components like hemicelluloses and lignin during the steam explosion pretreatment [16].

However, much less attention has been paid to the effect of steam explosion pretreatment on a hydrolysis step with diluted acids. Carrasco et al. [17] performed steam explosion pretreatments of different lignocellulosic biomasses followed by acid hydrolysis at 180 °C using diluted sulfuric acid. The steam explosion pretreatment lead to a decrease in the subsequent hydrolysis rate [17]. Schultz et al. [15,18,19] investigated the influence of steam explosion on concentrated acid hydrolysis. Among the lignocelluloses used, only steam-exploded rice husks showed a higher glucose yield after hydrolysis with concentrated sulfuric acid [15]. However, the concentration profiles of glucose were, in contrast to the present work, not recorded by Schultz et al. [15,18,19] and Carrasco et al. [17].

In this work, we evaluate the influence of a previous steam explosion on the acid hydrolysis of lignocellulosic biomass. Therefore, the steam explosion of beech wood as a representative of hardwood is performed in a batch system at different severities. The steam-exploded residues are characterized and, thereafter, subjected to acid hydrolysis. The acid hydrolysis step is done in a semi-continuous reactor at constant settings with diluted sulfuric acid. The concentration profiles of the main water-soluble hydrolysis products are recorded which also comprise glucose.

2. Materials and Methods

2.1. Pretreatment via Steam Explosion

As feedstock for pretreatment, pre-dried and bark-free beech wood is used. Beech trees are categorized as hardwoods and are one of the major forest trees in Europe. The dominant sugar in beech wood hemicelluloses is xylose [20]. The beech wood is used in chip size, has a moisture content of 8.0 wt.% and is obtained from Joh. Sinnerbrink GmbH & Co. (Verl, Germany). The chip dimensions are about 15 mm in width, 30 mm in length and 1–2 mm thick.

The steam explosion is performed in a self-constructed test rig, described in our previous work [21]. The stainless-steel reactor of the steam explosion has a volume of 1 L and is constantly agitated with a cross-arm stirrer at 8 min^{-1}. The reactor is filled completely with beech wood chips, referring to 82–94 g dry mass, before each experiment. Then the reactor is electrically preheated by a surrounding electrical heating jacket. Steam is introduced into the reactor with a flow of up to 5 g/min. The explosion step is performed by pneumatically opening a ball valve. Solid products and steam are discharged by the explosion step into a flash tank. The pretreated solid is manually collected from the flash tank and dried at 105 °C for 16 h.

The reaction temperature is measured by two thermocouples in the top and the bottom of the reactor. The severity parameter S_0 is calculated, combining temperature T and time t of the steam pretreatment in a single factor [10]. Equation (1) shows the time integral of S_0, which considers the non-isothermal character of the heating process [22].

$$S_0 = log \int_0^t exp\left(\frac{T\ (°C) - 100\ °C}{14.75}\right) dt \tag{1}$$

Pretreatment conditions with different reaction time (25–34 min), steam input mass (60–150 g) and maximum temperature (186–223 °C) are investigated. Figure 1 shows the temperature profile during heat up and Table 1 provides an overview of the experimental conditions.

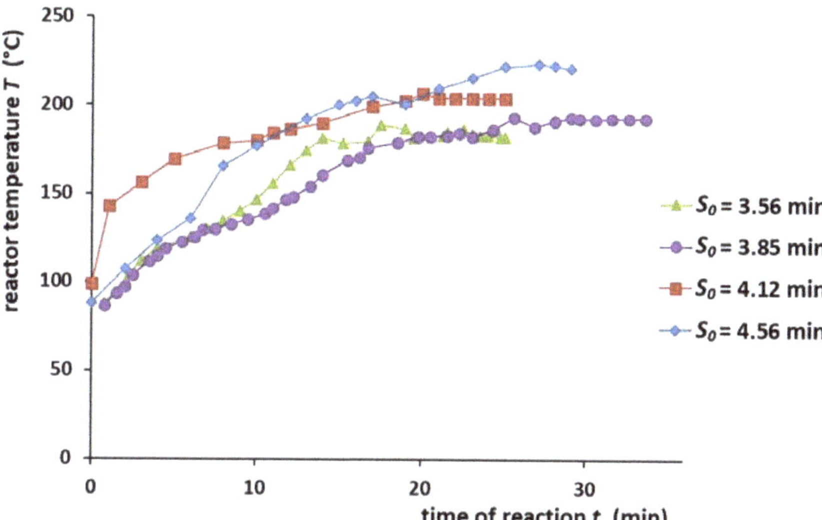

Figure 1. Temperature profile inside the steam explosion reactor, t = 0 min marks the beginning of steam input.

Table 1. Experimental setup for steam explosion of beech wood chips.

severity parameter S_0 (min)	3.56	3.85	4.12	4.56
beech wood mass (g_{dry})	91.6	81.8	83.4	93.8
steam input time (min)	25	33.5	25	29
steam input mass (g)	50	80	100	110
maximum reactor temperature (°C) [1]	186	193	206	223
maximum excess pressure (bar) [1]	11	13	19	26

[1] before explosion step.

2.2. Acid Hydrolysis of Steam-Exploded Residues

An acid-catalyzed hydrolysis of lignocellulose is performed to obtain a monosaccharide-containing product liquid, also called hydrolyzate. The solid residues after steam explosion as well as untreated beech wood chips are used as educts. Diluted sulfuric acid is used as a catalyst to hydrolyze hemicelluloses and cellulose polymers in the lignocellulose structure to water-soluble sugar monomers.

A semi-continuous test rig, described in our previous work [23], is used for acid hydrolysis, where the liquid phase is continuously exchanged. This has the advantage that liberated water-soluble molecules like sugar monomers are removed from the hot reactor and thus protected to a large extent from secondary reactions. The stainless-steel reactor has an internal volume of 100 mL and is loaded with 15.0 g of either untreated beech wood chips or steam-exploded residue as a fixed bed before the experiment. Demineralized water or dilute acid solution is continuously fed into the reactor at a volume flow of 15 mL/min. Demineralized water is pumped through the reactor during the heating phase. When the target temperature of 180 °C inside the reactor is reached, the feed stream is switched to a sulfuric acid solution of 0.05 mol/L. The hydrolyzate leaves the reactor continuously and is collected in interval samples of 5–8 min before storage at 4 °C.

2.3. Analytical Methods

After steam explosion, a wet solid residue is obtained, which is dried at 105 °C for 16 h to avoid microbiological degradation. After drying, the mass of the solid residue is measured. Then a sample of the solid residue is subjected to a two-hour Soxhlet extraction with water. In this way, a water-soluble and a water-insoluble fractions are obtained, which are examined for their composition. The mass of the water-insoluble fraction is determined gravimetrically, whereas the mass of the water-soluble fraction is calculated by difference.

The surfaces of dried beech wood and dried steam-exploded residues are examined via scanning electron microscopy (SEM) using a LEO 982 Gemini (Carl Zeiss, Jena, Germany) which is equipped with a Schottky-type thermal field emission cathode, a backscattered electron detector and two secondary electron detectors (inlens, Everhart-Thornley).

The Klason lignin content of steam-exploded residue and untreated beech wood is determined in triplicate according to the ASTM method D1106-96 (2013) [24]. In brief, two extractions of the biomass are performed with alcohol-benzene solution and with hot water. Then the residue is treated with 72 wt.% sulfuric acid which is afterwards diluted to 3 wt.% whereby the polysaccharides are completely hydrolyzed. The remaining solid gives, after correction with ash content, the Klason lignin content. The (polymeric) sugar content in water-insoluble steam-exploded residue and untreated beech wood is determined via a complete hydrolysis of polysaccharides to water-soluble monosaccharides according to Saeman [25].

Water-soluble sugar monomers are quantified via a gas chromatography method. As sugars cannot be evaporated, they are converted into stable, evaporable alditol acetate derivatives using a procedure described by Sawardeker et al. [26]. After extraction of the alditol acetates into chloroform, the sample is isothermally separated in a gas chromatograph type GC 5890A (Hewlett Packard, Palo Alto, CA, USA) at 240 °C and detection is performed via FID. The separation column RTX2330 (Restek, Bad Homburg, Germany) is used with a 30 m length and 25 µm diameter. The sugars, arabinose, xylose, rhamnose, mannose, galactose and glucose are calibrated. The accuracy of time-consuming derivatization and GC analysis is ensured by (1) performing a derivatization of a sugar standard solution in parallel for every analytical sequence and (2) the internal standard inositol, which is added in a known amount to any sample during derivatization.

The characterization of other water-soluble constituents in the hydrolyzate is conducted with two HPLC methods. To remove high-molecular-weight products, filtration is performed with syringe filters of type 0.45 µm GHP (Pall, New York, NY, USA). Furfural and hydroxymethylfurfural (HMF) are separated in a Lichrospher 100 RP-18 column (Merck, Darmstadt, Germany) at 20 °C and quantified by a UV detector at 290 nm. Therefore, an eluent of water-acetonitrile (9:1 *v/v*) is used at a flow rate of 1.4 mL/min. Formic acid, acetic acid, lactic acid and levulinic acid are separated with an Aminex HPX 87H column (Biorad, Hercules, CA, USA) at 25 °C. An eluent of 0.004 mol/L sulfuric acid is used at a flow rate of 0.65 mL/min and detection is performed by RI and DAD.

The cumulative yields of glucose and xylose after the hydrolysis step are calculated according to Equation (2). Thereby, the solid residue yield of the steam explosion step is considered. The volume flow of acid hydrolysis of 15 mL/min is multiplied with the measured concentration of glucose or xylose and integrated over reaction time.

$$cumulative\ yield\ =\ \frac{yield_{solid\ residue}}{m_{input\ hydrolysis}} \int^{t} \dot{V}{\cdot}c(t)\ dt \tag{2}$$

3. Results

3.1. Steam Explosion

Beech wood chips are pretreated via steam explosion. An increased severity parameter leads to a lower yield of solid residue (see Table 2). A Soxhlet extraction with water is performed whereby a water-soluble and a water-insoluble fraction are obtained. The yield of the water-insoluble fraction decreases with higher severity and fewer polymers of glucose are present (see Table 2). At the highest severity parameter of $S_0 = 4.56$ min, no polymeric bonded xylose can be detected in the solid residue. In contrast, the Klason lignin yield rises with severity. After all steam explosion runs, the Klason lignin yield is higher compared to untreated beech wood (0.229 g/$g_{beech\ wood}$).

Table 2. Solid residues after steam explosion of beech wood at different severities, fractionation of solid residues is performed via Soxhlet extraction with water, n.d.: not determined.

severity parameter S_0 (min)	3.56	3.85	4.12	4.56
solid residue mass (g)	85.5	74.8	61.8	64.0
solid residue yield (g/$g_{beech\ wood}$)	0.933	0.915	0.741	0.682
water-insoluble fraction (g/$g_{beech\ wood}$)	0.789	0.724	0.612	0.533
• polymeric bonded glucose (g/$g_{beech\ wood}$)	0.260	n.d.	0.210	0.192
• polymeric bonded xylose (g/$g_{beech\ wood}$)	0.103	n.d.	0.021	0.000
• Klason lignin (g/$g_{beech\ wood}$)	0.297	n.d.	0.353	0.346
water-soluble fraction (g/$g_{beech\ wood}$)	0.144	0.191	0.128	0.149
• glucose (g/$g_{beech\ wood}$)	0.002	0.001	0.002	0.000
• xylose (g/$g_{beech\ wood}$)	0.002	0.010	0.016	0.000

Figure 2 shows the topographical changes in the biomass macrostructure after steam explosion. Even in the photographs, a size reduction and defibration can be observed compared to the untreated wood chips. As the severity increases, the fibers become finer and the solid residue becomes darker. However, at the highest examined severity of $S_0 = 4.56$ min, no exposed fibers are visible and the biomass looks slightly carbonized. Additionally, the SEM images illustrate the defibration. Loosely present macrofibrils can be seen after intermediate severities (e.g., SEM at 100x at $S_0 = 3.85$ min). At the highest examined severity of $S_0 = 4.56$ min, the fiber structure can no longer be clearly recognized (see SEM image at 100× in Figure 2).

Figure 2. Beech wood and solid residues after steam explosion at different severity parameters S_0, photographs (**left**), SEM images at 100× (**center**) and 1000× (**right**).

3.2. Acid Hydrolyisis

To investigate the influence of the steam explosion pretreatment on acid hydrolysis, all influencing parameters during acid hydrolysis are kept constant and only the solid biomass input in the fixed bed reactor is varied. Acid hydrolysis is carried out with 0.05 mol/L sulfuric acid at a 180 °C reaction temperature. The volume flow of the diluted acid through the semi-continuous hydrolysis reactor is also constant. The biomass input originates from steam explosion runs with different severities. The solid residue after the highest examined severity of $S_0 = 4.56$ min is not subjected to acid hydrolysis, because the material appears to be already partly carbonized.

Figures 3 and 4 show the concentration profiles of the main water-soluble hydrolysis products during acid hydrolysis for different severities of steam explosion. The main hydrolysis products in our study using beech wood are the monosaccharides glucose, xylose and mannose, as well as furfural hydroxymethylfurfural and acetic acid. The principal trends in the concentration profiles of all main products are very similar regardless of the severity of pretreatment. However, there are big differences in the maximum concentration values. The maximum xylose concentration drops largely with increasing severity (from 4.5 g/L at $S_0 = 3.56$ min to 1.2 g/L at $S_0 = 4.12$ min). The maxima of acetic acid and furfural also decrease sharply with increasing severity parameters (see Figure 4). However, the influence of pretreatment severity on the glucose concentration is much less (see Figure 3). Figure 3 shows the concentration profiles of glucose and xylose during acid hydrolysis, where steam-exploded residue is compared with untreated beech wood.

Figure 5 shows the cumulative yields of glucose and xylose during acid hydrolysis for different steam explosion severities and a comparison to the untreated beech wood material is made. For the calculation of the cumulative yields according to Equation (2), the solid residue yield of stream explosion pretreatment from Table 2 is considered. For reaction times over 25 min, the yield of glucose is higher for untreated beech wood compared to all pretreated materials. Additionally, the yield of xylose is higher for untreated beech wood compared to stream-exploded residues.

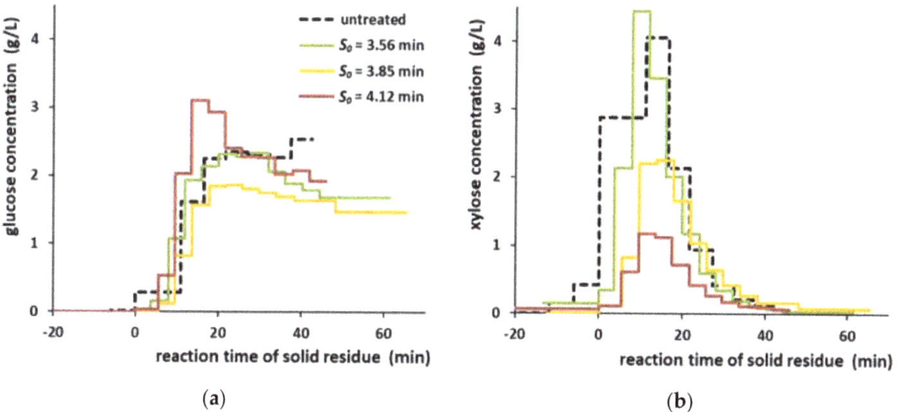

Figure 3. Formation of glucose (**a**) and xylose (**b**) during the hydrolysis of 15.0 g untreated beech wood or steam-exploded residue at different severity parameters S_0, hydrolysis with 0.05 mol/L sulfuric acid at 15 mL/min flow at 180 °C, $t = 0$ min marks the beginning of acid hydrolysis.

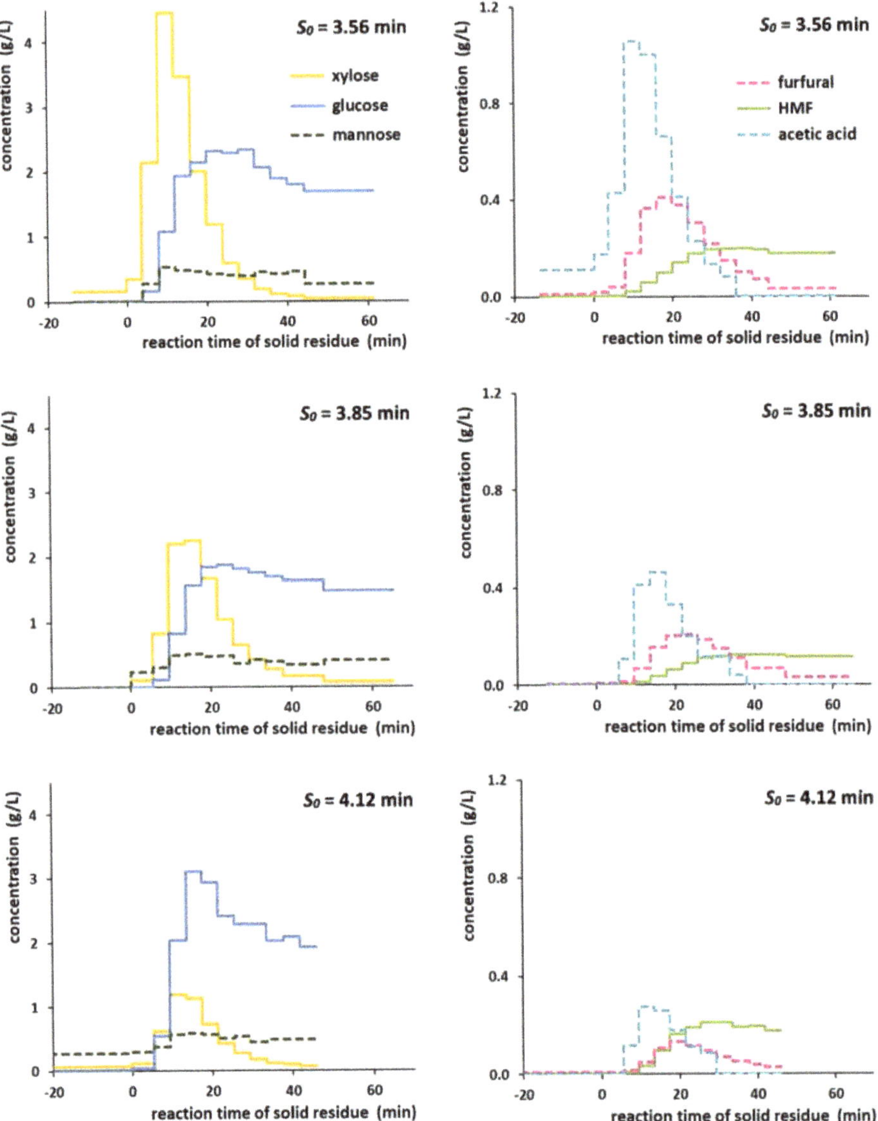

Figure 4. Formation of monosaccharides (**left**) as well as furfural, hydroxymethylfurfural and acetic acid (**right**) during the hydrolysis of 15.0 g steam-exploded residue at different severity parameters S_0, hydrolysis with 0.05 mol/L sulfuric acid at 15 mL/min flow at 180 °C, $t = 0$ min marks the beginning of acid hydrolysis.

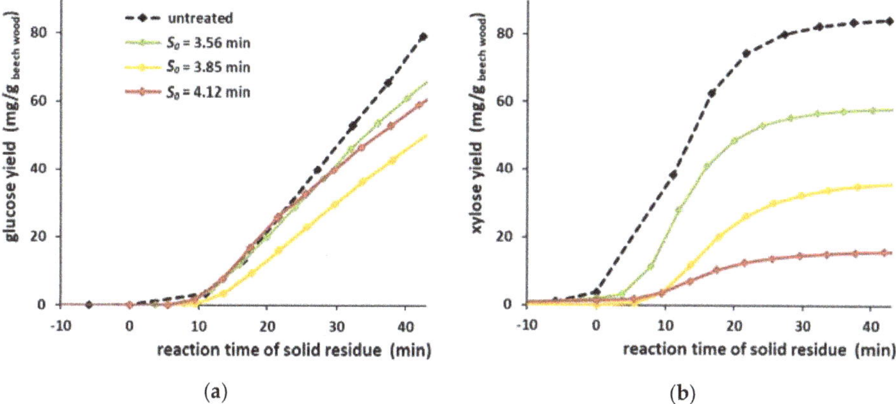

Figure 5. Cumulative yields of glucose (**a**) and xylose (**b**) during the hydrolysis of untreated beech wood or steam-exploded residue at different severity parameters S_0, hydrolysis with 0.05 mol/L sulfuric acid at 15 mL/min flow at 180 °C, $t = 0$ min marks the beginning of acid hydrolysis.

4. Discussion

The steam explosion increases the lignin mass according to the Klason method, which agrees with the results of other studies [18,27,28]. A lignin yield of 0.297–0.353 $g/g_{beech\ wood}$ is obtained after the steam explosion (see Table 2), while the Klason lignin content in beech wood is only 0.229 $g/g_{beech\ wood}$. Condensation reactions between hemicellulose constituents and lignin lead to the formation of an inert solid [29]. This solid is also measured in the gravimetric determination of Klason lignin and is therefore no longer distinguishable from lignin which was originally contained in beech wood. This is the reason why this newly formed solid is named pseudolignin. In principle, a reaction pathway via the repolymerization of water-soluble hemicellulose degradation products would also possibly form pseudolignin. This pathway could result in the formation of spherical structures on the solid product after steam explosion. This is known from the process of hydrothermal carbonization, where microspheres are formed by the polymerization of sugar degradation products [30,31]. However, in this study, such spherical structures were not found in the steam-exploded residues even at up to 50,000× magnification of the SEM.

The pretreatment via steam explosion was intended to increase the accessibility of the cellulose fibers so that the subsequent acid hydrolysis of the cellulose to glucose takes place in a shorter time. Consequently, a high glucose concentration should be achieved after a short time of acid hydrolysis, which then drops sharply after the cellulose has been completely converted. The intended effect of pretreatment on glucose formation did not occur. During acid hydrolysis, the steam-exploded residues begin to release glucose after roughly the same reaction time compared to untreated beech wood (see Figure 3). When steam-exploded residues are used, the maximum concentration of glucose is similar or even smaller (except for two measuring points at S_0 = 4.12 min). Additionally, no obvious drop in glucose concentration with longer reaction times can be observed. The comparison of glucose and xylose yields shows that a previous steam explosion pretreatment reduces the sugar yields, especially for xylose (see Figure 5).

Consequently, the defibration of the macrofibrils in the lignocellulose structure during steam explosion does not lead to an increased rate of cellulose hydrolysis. It can therefore be assumed that the rate-determining step in the hydrolysis is not the penetration of the acid to the individual macrofibrils. Rather, the penetration of the hydronium ions into the crystalline structure of cellulose can be assumed to determine the rate or hydrolysis might only occur at the exposed chain ends of the cellulose. This is in accordance with other studies [32]; the crystallinity of cellulose stabilized by hydrogen bonds is very strong and can only be broken by supercritical water, for example at 380 °C and 250 bar [33]. In the case of enzymatic hydrolysis reactions, the situation is different. An enzyme is a much larger molecule than a hydronium ion with solvation shell, therefore the increase in surface area by steam explosion has an impact.

Xylose, which is the main structural unit of the hemicelluloses in beech wood, is largely converted by steam explosion at a higher severity into other compounds (see Table 2). Thereby, xylose could be either decomposed to low-molecular-weight compounds or converted to pseudolignin. It is generally known that the total mass of hemicelluloses in the steam-exploded residues decreases with an increasing severity parameter [10]. Therefore, it is reasonable that in the product liquid of acid hydrolysis, less xylose can be detected with increasing severity of the pretreatment (see Figure 3), as less hemicelluloses are in the feedstock for hydrolysis, which could form xylose.

The concentrations of furfural and acetic acid during acid hydrolysis also decrease with the severity of the pretreatment (see Figure 4). The reduction of furfural concentration is a consequence of the lower xylose concentration, since furfural arises from the dehydration of pentoses [5]. The decline in acetic acid concentration with the increasing severity of the steam explosion can be explained as follows. The acetyl groups of the hemicelluloses are hydrolyzed to a greater extent during the steam explosion at high severity and are therefore removed from the biomass before the acid hydrolysis step begins.

5. Conclusions

In this study, beech wood lignocellulosic biomass was subjected to steam explosion pretreatment before an acid hydrolysis step. The pretreatment via steam explosion was intended to increase the accessibility of the cellulose fibers so that the subsequent acid hydrolysis of the cellulose to glucose would take place in a shorter time. We showed that the defibration of the macrofibrils in the lignocellulose structure during steam explosion does not lead to an increased rate of cellulose hydrolysis. Additionally, steam explosion causes a mass loss of solid material and large losses of the hemicellulose-derived sugar xylose, especially at a higher pretreatment severity. That is why steam-exploded material resulted in lower sugar yields based on lignocellulose input. So, steam explosion is not a suitable pretreatment for the acid hydrolysis of hardwood lignocellulosic biomass.

Author Contributions: D.S. and J.S. (Jonas Storz) designed and performed the experiments as well as analyzed the data. D.S., A.K. and J.S. (Jörg Sauer) wrote the article. All authors have read and agreed to the published version of the manuscript.

Funding: This work was financially supported by the German Federal Ministry of Food, Agriculture and Consumer Protection based on a decision of the German Bundestag (FNR project number 22027811). We acknowledge support by Deutsche Forschungsgemeinschaft and Open Access Publishing Fund of Karlsruhe Institute of Technology.

Acknowledgments: Thomas Tietz and Matthias Pagel built the steam explosion unit as well as the acid hydrolysis test rig. We thank Birgit Rolli for GC support, Sonja Habicht and Armin Lautenbach for HPLC support and Wilhelm Habicht for SEM measurements. We acknowledge Nicolaus Dahmen for project supervision.

Conflicts of Interest: The authors declare no conflict of interest. The funding sponsors had no role in the design of the study; in the collection, analyses, or interpretation of data; in the writing of the manuscript, and in the decision to publish the results.

References

1. Isikgor, F.H.; Becer, C.R. Lignocellulosic biomass: A sustainable platform for the production of bio-based chemicals and polymers. *Polym. Chem.* **2015**, *6*, 4497–4559. [CrossRef]
2. Serrano, D.; Coronado, J.; Melero, J. Conversion of cellulose and hemicellulose into platform molecules: Chemical routes. In *Biorefinery: From Biomass to Chemicals and Fuels*; Aresta, M., Dibenedetto, A., Dumeignil, F., Eds.; De Gruyter: Berlin, Germany, 2012.
3. Bobleter, O. Hydrothermal degradation of polymers derived from plants. *Prog. Polym. Sci.* **1994**, *19*, 797–841. [CrossRef]
4. Bozell, J.J.; Petersen, G.R. Technology development for the production of biobased products from biorefinery carbohydrates—The US Department of Energy's "Top 10" revisited. *Green Chem.* **2010**, *12*, 539–554. [CrossRef]

5. Steinbach, D.; Kruse, A.; Sauer, J. Pretreatment technologies of lignocellulosic biomass in water in view of furfural and 5-hydroxymethylfurfural production—A review. *Biomass Convers. Biorefinery* **2017**, *7*, 247–274. [CrossRef]

6. Fan, L.; Gharpuray, M.M.; Lee, Y.H. *Cellulose Hydrolysis*; Springer: Berlin/Heidelberg, Germany, 1987.

7. Saini, R.; Osorio-Gonzalez, C.S.; Hegde, K.; Brar, S.K.; Magdouli, S.; Vezina, P.; Avalos-Ramirez, A. Lignocellulosic Biomass-Based Biorefinery: An Insight into Commercialization and Economic Standout. *Curr. Sustain. Renew. Energy Rep.* **2020**. [CrossRef]

8. Schwald, W.; Breuil, C.; Brownell, H.H.; Chan, M.; Saddler, J.N. Assessment of pretreatment conditions to obtain fast complete hydrolysis on high substrate concentrations. *Appl. Biochem. Biotech.* **1989**, *20–21*, 29–44. [CrossRef]

9. Zhu, J.Y.; Pan, X.J. Woody biomass pretreatment for cellulosic ethanol production: Technology and energy consumption evaluation. *Bioresour. Technol.* **2010**, *101*, 4992–5002. [CrossRef]

10. Overend, R.P.; Chornet, E. Fractionation of lignocellulosics by steam-aqueous pretreatments. *Philos. Trans. R. Soc. A* **1987**, *321*, 523–536. [CrossRef]

11. Chiaramonti, D.; Prussi, M.; Ferrero, S.; Oriani, L.; Ottonello, P.; Torre, P.; Cherchi, F. Review of pretreatment processes for lignocellulosic ethanol production, and development of an innovative method. *Biomass Bioenergy* **2012**, *46*, 25–35. [CrossRef]

12. Jacquet, N.; Maniet, G.; Vanderghem, C.; Delvigne, F.; Richel, A. Application of Steam Explosion as Pretreatment on Lignocellulosic Material: A Review. *Ind. Eng. Chem. Res.* **2015**, *54*, 2593–2598. [CrossRef]

13. Alvira, P.; Tomas-Pejo, E.; Ballesteros, M.; Negro, M.J. Pretreatment technologies for an efficient bioethanol production process based on enzymatic hydrolysis: A review. *Bioresour. Technol.* **2010**, *101*, 4851–4861. [CrossRef] [PubMed]

14. Singh, J.; Suhag, M.; Dhaka, A. Augmented digestion of lignocellulose by steam explosion, acid and alkaline pretreatment methods: A review. *Carbohydr. Polym.* **2015**, *117*, 624–631. [CrossRef] [PubMed]

15. Schultz, T.P.; Templeton, M.C.; Biermann, C.J.; Mcginnis, G.D. Steam Explosion of Mixed Hardwood Chips, Rice Hulls, Corn Stalks, and Sugar-Cane Bagasse. *J. Agric. Food Chem.* **1984**, *32*, 1166–1172. [CrossRef]

16. Jacquet, N.; Vanderghem, C.; Danthine, S.; Quiévy, N.; Blecker, C.; Devaux, J.; Paquot, M. Influence of steam explosion on physicochemical properties and hydrolysis rate of pure cellulose fibers. *Bioresour. Technol.* **2012**, *121*, 221–227. [CrossRef]

17. Carrasco, J.E.; Sáiz, M.C.; Navarro, A.; Soriano, P.; Sáez, F.; Martinez, J.M. Effects of dilute acid and steam explosion pretreatments on the cellulose structure and kinetics of cellulosic fraction hydrolysis by dilute acids in lignocellulosic materials. *Appl. Biochem. Biotech.* **1994**, *45*, 23–34. [CrossRef]

18. Schultz, T.P.; Rughani, J.R.; Mcginnis, G.D. Comparison of the Pretreatment of Sweetgum and White Oak by the Steam Explosion and Rash Processes. *Appl. Biochem. Biotech.* **1989**, *20*, 9–27. [CrossRef]

19. Schultz, T.P.; Biermann, C.J.; Mcginnis, G.D. Steam Explosion of Mixed Hardwood Chips as a Biomass Pretreatment. *Ind. Eng. Chem. Prod. Res. Dev.* **1983**, *22*, 344–348. [CrossRef]

20. Willfor, S.; Sundberg, A.; Pranovich, A.; Holmbom, B. Polysaccharides in some industrially important hardwood species. *Wood Sci. Technol.* **2005**, *39*, 601–617. [CrossRef]

21. Steinbach, D.; Wüst, D.; Zielonka, S.; Krümpel, J.; Munder, S.; Pagel, M.; Kruse, A. Steam Explosion Conditions Highly Influence the Biogas Yield of Rice Straw. *Molecules* **2019**, *24*, 3492. [CrossRef]

22. Montané, D.; Overend, R.P.; Chornet, E. Kinetic models for non-homogeneous complex systems with a time-dependent rate constant. *Can. J. Chem. Eng.* **1998**, *76*, 58–68. [CrossRef]

23. Świątek, K.; Gaag, S.; Klier, A.; Kruse, A.; Sauer, J.; Steinbach, D. Acid Hydrolysis of Lignocellulosic Biomass: Sugars and Furfurals Formation. *Catalysts* **2020**, *10*, 437. [CrossRef]

24. ASTM Standard. *Standard Test Method for Acid-Insoluble Lignin in Wood*; ASTM Standard: West Conshohocken, PA, USA, 2013; Volume D1106–96.

25. Saeman, J.F.; Bubl, J.L.; Harris, E.E. Quantitative Saccharification of Wood and Cellulose. *Ind. Eng. Chem. Anal. Ed.* **1945**, *17*, 35–37. [CrossRef]

26. Sawardeker, J.S.; Sloneker, J.H.; Jeanes, A. Quantitative Determination of Monosaccharides as Their Alditol Acetates by Gas Liquid Chromatography. *Anal Chem* **1965**, *37*, 1602–1604. [CrossRef]

27. Negro, M.J.; Manzanares, P.; Oliva, J.M.; Ballesteros, I.; Ballesteros, M. Changes in various physical/chemical parameters of Pinus pinaster wood after steam explosion pretreatment. *Biomass Bioenergy* **2003**, *25*, 301–308. [CrossRef]

28. Springer, E.L.; Harris, J.F. Pre-Hydrolysis of Aspen Wood with Water and with Dilute Aqueous Sulfuric-Acid. *Sven. Papp.* **1982**, *85*, 152–154.
29. Ramos, L.P. The chemistry involved in the steam treatment of lignocellulosic materials. *Quim. Nova* **2003**, *26*, 863–871. [CrossRef]
30. Dinjus, E.; Kruse, A.; Tröger, N. Hydrothermal Carbonization–1. Influence of Lignin in Lignocelluloses. *Chem. Eng. Technol.* **2011**, *34*, 2037–2043. [CrossRef]
31. Titirici, M.M.; Antonietti, M.; Baccile, N. Hydrothermal carbon from biomass: A comparison of the local structure from poly- to monosaccharides and pentoses/hexoses. *Green Chem.* **2008**, *10*, 1204–1212. [CrossRef]
32. Tolonen, L.K.; Zuckerstätter, G.; Penttilä, P.A.; Milacher, W.; Habicht, W.; Serimaa, R.; Kruse, A.; Sixta, H. Structural Changes in Microcrystalline Cellulose in Subcritical Water Treatment. *Biomacromolecules* **2011**, *12*, 2544–2551. [CrossRef]
33. Tolonen, L.K.; Penttilä, P.A.; Serimaa, R.; Kruse, A.; Sixta, H. The swelling and dissolution of cellulose crystallites in subcritical and supercritical water. *Cellulose* **2013**, *20*, 2731–2744. [CrossRef]

Publisher's Note: MDPI stays neutral with regard to jurisdictional claims in published maps and institutional affiliations.

Article

Biomimetic Vanadate and Molybdate Systems for Oxidative Upgrading of Iono- and Organosolv Hard- and Softwood Lignins

Lucía Penín [1], Matteo Gigli [2,3,†], Federica Sabuzi [4,*], Valentín Santos [1], Pierluca Galloni [4], Valeria Conte [4], Juan Carlos Parajó [1], Heiko Lange [3,5,*,†] and Claudia Crestini [2,3,*,†]

[1] Chemical Engineering Department, Polytechnical Building, University of Vigo (Campus Ourense), As Lagoas, 32004 Ourense, Spain; lpenin@uvigo.es (L.P.); vsantos@uvigo.es (V.S.); jcparajo@uvigo.es (J.C.P.)

[2] Department of Molecular Science and Nanosystems, Ca' Foscari University of Venice, Via Torino 155, 30170 Venice Mestre, Italy; matteo.gigli@unive.it

[3] CSGI—Center for Colloid and Surface Science, Via della Lastruccia 3, 50019 Sesto Fiorentino, Italy

[4] Department of Chemical Science and Technologies, University of Rome 'Tor Vergata', Via della Ricerca Scientifica, 00133 Rome, Italy; galloni@scienze.uniroma2.it (P.G.); valeria.conte@uniroma2.it (V.C.)

[5] Department of Pharmacy, University of Naples 'Federico II', Via Domenico Montesano 49, 80131 Naples, Italy

* Correspondence: federica.sabuzi@uniroma2.it (F.S.); heiko.lange@unina.it (H.L.); claudia.crestini@unive.it (C.C.)

† Affiliated with (4) via NAST—Nanoscience & Nanotechnology & Innovative Instrumentation Center.

Received: 25 August 2020; Accepted: 13 September 2020; Published: 16 September 2020

Abstract: Recently reported acetosolv soft- and hardwood lignins as well as ionosolv soft- and hardwood lignins were transformed into monomeric aromatic compounds using either a vanadate or a molybdate-based catalyst system. Monomers were generated with remarkable, catalyst-dependent selectivity and high depolymerisation yields via oxidative exo- and endo-depolymerisation processes. Using the vanadate–hydrogen peroxide system on acetosolv pine lignin, vanillin and isovanillin were produced as main products with depolymerisation yields of 31%. Using the molybdate system on acetosolv and ionosolv lignin, vanillic acid was the practically exclusive product, with depolymerisation yields of up to 72%. Similar selectivities, albeit with lower depolymerisation yields of around 50% under standardised conditions, were obtained for eucalyptus acetosolv lignin, producing vanillin and syringaldehyde or vanillic acid as products, by using the vanadate- or the molybdate-based systems respectively.

Keywords: oxidative lignin upgrade; catalytic lignin oxidation; vanadate; molybdate; organosolv; ionosolv; lignin; biomimetic

---

1. Introduction

Lignocellulosic biomass received great interest as a sustainable and renewable source of fuel and platform chemicals in recent years [1,2]. Numerous studies target the conversion of cellulose and hemicelluloses into ethanol and other biofuels as well as platform chemicals for the chemical industries [3–5]. In sharp contrast, research on the conversion of lignin has often been limited to its removal from the other two principal biomass components either to enhance their chemical and/or enzymatic valorisation. Conversion of lignin—representing after all 30% of the weight and 40% of the energy content of lignocellulosic biomass and being isolated as a by-product in form of various technical lignins with different characteristics by cellulose-focused processes—is still a challenge [6,7]. Enzymatic and chemical reactions have been proposed for oxidative lignin valorisation. Several biocatalysts, mimics of biocatalysts, and inorganic catalysts have been studied regarding formation of aromatic monomers for the chemical industries. Nevertheless, both reductive and oxidative degradation methods have been studied and presented [8–12]. Mechanistic insights obtained using lignin model compounds to simulate the most abundant bonding motifs within the backbones of various technical lignins were only scarcely applicable to the complexity of a lignin oligomeric and/or polymeric material. In case of oxidative degradation, a series of oxidation products is obtained even in case of simple models, pointing at a lack of selectivity of the reactions triggered by the various catalytic sstems. As a noteworthy exception, only the methylrheniumoxide catalyst system has been reported [13,14].

Vanadium-based catalysts have been reported for lignin valorisation, interestingly both for oxidative and reductive depolymerisation approaches [8,9,11,15–20]. Molybdenum catalysts have been widely reported for the oxidative valorisation of technical lignins, especially in form of polyoxometalates (POMs) [21,22]. More recently, mixed, i.e., bifunctional catalyst systems such as copper-vanadium [23] and molybdenum-vanadium [24] systems have been reported for lignin valorisation.

In case of enzymatically mediated reactions, obtaining a large panel of different oxidation products—due to the natural evolution of phenoxy radical intermediates—is furthermore associated to a limited biocatalyst lifetime, which constitutes an additional problem for lignin valorisation. Potential ways to tackle both issues consist of supporting and/or encapsulating the enzymes or to mimic the catalytic centre of lignolytic enzymes using stable organometallic complexes that would eventually exhibit tuning possibilities towards a higher product selectivity [25–28].

In an effort to combine such an active centre-mimicking approach with our interest in evaluating the use of non-lignolytic enzymes in lignin valorisation [29], we tested a vanadate (**V**)-based and a molybdate (**Mo**)-based catalytic system [**V**] and [**Mo**], respectively, as mimics of the reactive vanadate and molybdate centres in bromide peroxidases [EC 1.11.1.18] and xanthine oxidase [EC 1.17.3.2], respectively. The vanadate and molybdate catalyst systems can, based on their reactivity, be eventually considered suitable non-lignolytic biocatalyst mimics for lignin degradation. The reactivity of the catalyst systems holds the promise that they are capable of oxidising lignin in a more selective way than the typical lignolytic enzymes, i.e., laccases or manganese peroxidases due to the different range of electric potentials [30]. The present study deals with the investigation of the oxidation potential of vanadate and molybdate catalysts toward lignin oxidation.

Two hardwood and two softwood lignins, isolated from *Eucalyptus nitens* and *Pinus pinaster*, in form of both organosolv and ionosolv lignins have been selected. These lignins have been reported and characterised before [31–33] and were chosen as starting materials for this study also because the structural characterization had revealed noteworthy structural differences between the acetosolv and ionosolv lignin of each biomass. These lignins were used in this study without any further refinements or fractional purifications. We refrained in this study from following the wide-spread approach in which the catalyst activity is demonstrated using monomeric, dimeric, or sometimes trimeric lignin models. While this approach allows relatively facile mechanistic understanding of product formation, applicability of the results to whole lignin degradation is often scarce, due to the high density of reactive sites along the lignin backbone prone to undergo inter- and intramolecular reactions not delineable using lignin models.

2. Materials and Methods

General information: Reagents and solvents were purchased from Sigma-Aldrich/Merck KGaA, Darmstadt, Germany and Carlo Erba, Milano, Italy, and used without further purification, if not stated otherwise. Acetosolv pine lignin (**AP**) and acetosolv eucalyptus lignin (**AE**), as well as ionosolv pine lignin (**IP$_B$**) and ionosolv eucalyptus lignin **IE$_B$** were produced and characterised as described elsewhere [31–33]. Ammonium vanadate (**V**), NH_4VO_3, and ammonium molybdate tetrahydrate (**Mo**), $(NH_4)_6Mo_7O_{24} \cdot 4H_2O$ were purchased from Sigma Aldrich/Merck KGaA and used without further purification.

Oxidation of lignins: 50 mg of lignin were weighted in a screw cap vial and suspended in 1 mL of a pre-made stock solution containing the catalyst. Stock solutions comprised: (i) 30 µL perchloric acid; (ii) 450 µL hydrogen peroxide 30% (*w/v*); (iii) 2520 µL distilled water; (iv) the vanadate or molybdate catalyst at concentrations of 5%, 7.5%, or 10% (*w/w*) of liquid phase. Reactions were heated to 80 °C for 9 h while being continuously stirred and allowed to cool down to room temperature overnight.

Degradation products were isolated in an extraction process. The reaction mixture was centrifuged to separate any solids, and the liquid phase was taken and mixed with an equal volume of ethyl acetate. As internal standard, 50 µL of a solution of 4-ethoxy-3-methoxy benzaldehyde in ethyl acetate at a concentration of typically 50 µM were mixed into the solution. The biphasic system was vigorously shaken for 30 s. The organic phase was separated, dried over $MgSO_4$, and centrifuged. An aliquot was sampled for analysis by gas chromatography coupled to mass spectrometry (GC-MS) as described below.

Solid lignin residues were isolated by centrifugation and washed four times using a 1 M aqueous solution of sulphuric acid. Lignin residues were dried to constant weight for ^{31}P NMR analysis as described below. Experiments were run in duplicate by default, and selected experiments were additionally repeated for verification of results. An error margin for solid mass returns/depolymerisation yields of max ±6% was encountered.

Quantitative ^{31}P NMR spectroscopy: In general, a procedure similar to the one originally published and previously applied was used [34]: approx. 30 mg of the lignin were accurately weighed for analysis after phosphitylation using an excess of 2-chloro-4,4,5,5-tetramethyl-1,3,2-dioxa-phospholane (**Cl-TMDP**). ^{31}P NMR spectra were recorded on a Bruker, (Billerica, MA, USA, 300 MHz or Bruker 700 MHz NMR spectrometer controlled by TopSpin software, using an inverse gated decoupling technique with the probe temperature set to 20 °C. The maximum standard deviation of the reported data is 0.02 mmol g^{-1}, while the maximum standard error is 0.01 mmol g^{-1} [35,36]. NMR data were processed with MestreNova (Version 8.1.1, Mestrelab Research, Santiago di Compostella, Spain). Technical loadings are determined by comparing the abundancies of total aromatic hydroxyl groups of the product lignin with the starting lignin.

Gel permeation chromatography (GPC): For GPC measurements, approx. 2–3 mg of solid were dissolved in HPLC-grade dimethylsulphoxide (DMSO) (Chromasolv®, Sigma-Aldrich/Merck KGaA) containing 0.1% (*m/v*) lithium chloride (LiCl). A Shimadzu, Kyoto, Japan, instrument was used consisting of a controller unit (CBM-20A), a pumping unit (LC 20AT), a degasser (DGU-20A3), a column oven (CTO-20AC), a diode array detector (SPD-M20A), and a refractive index detector (RID-10A) and was controlled by Shimadzu LabSolutions (Version 5.42 SP3). For separation, a PLgel 5 µm MiniMIX-C column (Agilent, Santa Clara, CA, USA, 250 × 4.6 mm) was eluted at 70 °C at 0.25 mL min^{-1} flow rate with DMSO containing 0.1% lithium chloride for 20 min. Standard calibration is performed with polystyrene sulfonate standards (Sigma Aldrich/Merck KGaA, MW range 0.43–2.60 × 10^6 g mol^{-1}) in acid form; lower calibration limits are verified by the use of monomeric and dimeric lignin models. Final analyses of each sample were performed using the intensities of the UV signal at λ = 280 nm employing the Shimadzu analysis software.

Gas chromatography coupled to mass spectrometry (GC-MS): A prepared sample of extractives of each reaction was analysed by gas chromatography coupled with mass spectrometry, using a Shimadzu, Kyoto, Japan, GCMS QP2010 Ultra equipped with an AOi20 autosampler unit. A SLB®-5ms Capillary GC Column (L × I.D. 30 m × 0.32 mm, df 0.50 µm) was used as the stationary phase, ultrapure helium as the mobile phase. The Shimadzu LabSolutions GCMS Solution software (Version 2.61) was used. The various components were identified by comparison against the NIST11 library. For control of sensitivity of analysis, selected samples were analysed after silylating the OH-groups present in the analytes: after this first analysis, 100 µL dry pyridine and 100 µL of N,O-bis(trimethylsilyl)-trifluoroacetamide were added to the aliquot, in order to repeat the analysis, after gently stirring the mixture for 30 min at room temperature, using identical GC-MS conditions. Reproducibility of single sample measurements was encountered with an error of maximum ±0.5%. Reproducibility across duplicated experiments was approx. ±5%.

3. Results and Discussion

3.1. Catalytic Systems

Taking inspiration from natural enzymes, several **V**- and **Mo**-based systems have been proposed in the literature, to perform sustainable oxidations on various substrates [37–39]. Basically, the mechanism of action of the [**V**]- and [**Mo**]-catalysts here adopted tracks the one of vanadium-dependent haloperoxidase enzymes [38,40]. Dissolving a [**V**]-catalyst precursor, such as NH_4VO_3, in acidic aqueous solution (pH = 1), in the presence of H_2O_2, a vanadium-monoperoxido complex forms, which is the effective catalytic species in solution. Conversely, at higher pH values, the formation of the diperoxido vanadium species occurs, which is less effective than the mono-peroxido in oxidation and oxybromination reactions. Therefore, in this study, pH = 1 was chosen to perform the oxidative lignin degradation.

Similarly, dissolution of ammonium molybdate at pH = 1 in water, with H_2O_2, leads to the formation of the diperoxido–molybdenum complex, which is much more stable than the molybdenum monoperoxo derivative [38].

Importantly, the reactivity of vanadium-peroxo and molybdenum-diperoxo complexes is definitely superior than that of H_2O_2 in oxidation reactions, therefore the oxidation of different alkanes, alkenes, alcohols, aromatic substrates, sulphides, as well as oxybromination reactions were successfully achieved with such catalytic systems [41–44]. Literature reports suggest that vanadate/peroxide oxidation system acts following substrate specifics in radical or ionic modes [38,41,45,46].

3.2. Lignin Starting Materials

The lignins chosen for this study are representative samples of a series of lignins isolated in course of a process optimisation and have recently been presented and characterised [31–33]. More specifically, they comprised two softwood lignins isolated from *Pinus pinaster* (acetosolv, **AP**, and ionosolv isolated using [bmim] HSO_4, **IP$_B$**), and two hardwood lignins isolated from *Eucalyptus nitens* (acetosolv, **AE**, and ionosolv isolated using [bmim] HSO_4, **IE$_B$**).

With respect to the acetosolv lignins that exhibited structural features typical for this type of lignin and close to those attributed to pristine lignin, ionosolv species showed clear signs of structural degradation when compared to the milder isolation process, i.e., the organosolv treatments. **IP$_B$** and **IE$_B$** presented lower molecular weight distributions and comparatively lower contents of aliphatic OH-groups (Table 1), a fact ascribed to a partial degradation of the backbone [31]. The use of these lignin samples was envisaged to allow for delineating the effect of a 'pre-degradation' to the oxidative valorisation process.

The catalytic activity of the chosen oxidative systems was expected to take effect mainly via the OH-groups present in the lignin structures; changes in relative abundances of hydroxyl groups based on consumption upon degradation and/or polymerisation is indicative of catalyst activity and allows mechanistic insight. Table 1 details again the OH-group contents as determined by ^{31}P NMR spectroscopy.

3.3. Catalytic Degradation of Softwood Lignins

Degradation studies started using the softwood samples reported in Table 1, i.e., **AP** and **IP$_B$**, and the Vanadate system discussed above. In order to test for a potential direct downstream application of the degrading catalyst in an industrial set-up, lignins were used as obtained in the initial isolation process. Effectiveness of treatments was tested via the analysis of newly generated low molecular weight components extractable in ethyl acetate. Results are summarised in Table 2.

Using the softwood acetosolv lignin **AP**, an initial screening of conditions revealed that optimum conversions are obtained employing catalyst loadings of 7.5% or 10% (*w/w*) at elevated temperatures of 80 °C for reaction times of 9 h. Lower catalyst loadings of 5% (*w/w*) resulted in lower conversions under otherwise unchanged conditions.

Table 1. Distribution of OH-groups and molecular weight key data of acetosolv and ionosolv lignins tested for oxidative valorisation.

Lignins [a]	AP	IP$_B$	AE	IE$_B$	IP$_B$[V]10	IP$_B$[Mo]10
OH-group	Abundance (mmol/g)					
aliphatic OH	0.23	1.36	0.34	1.28	1.31	0.80 [b]
Condensed	0.96	0.95	2.90	3.28	0.89	0.06 [b]
o-disub. phenols (S units)	—	—	1.25	1.23	—	—
4-*O*-5′ + 5-5′	0.96	0.95	1.65	2.05	0.89	0.06 [b]
o-monosub. phenols (G units)	1.14	1.56	0.60	0.73	1.22	0.07 [b]
p-OH phenols	0.21	0.24	0.15	0.14	0.36	0.04 [b]
total phenolic OH	2.32	2.75	3.65	4.16	2.47	0.17 [b]
carboxylic OH	0.24	0.22	0.31	0.05	0.33	0.75 [b]
total phenolic OH/aliphatic OH	9.9	2.0	10.8	3.2	1.9	0.2 [b]
total phenolic OH/ condensed phenolic OH	1.9	2.2	1.2	1.2	2.8	2.8 [b]
Mn [kDa]	1.75	1.40	1.20	1.20	n.d. [d]	n.d. [d]
(PDI) [c]	(7.3)	(5.8)	(2.4)	(2.5)		

[a]: **AP**: acetosolv pine lignin; **IP$_B$**: ionosolv pine lignin isolated using [bmim] HSO$_4$; **AE**: acetosolv eucalyptus lignin; **IE$_B$**: ionosolv eucalyptus lignin isolated using [bmim]HSO$_4$; **IP$_B$**[V]10: ionosolv pine lignin using [bmim] HSO$_4$ treated with 10% **V**; **IP$_B$**[V]10: ionosolv pine lignin using [bmim] HSO$_4$ treated with 10% **Mo**. [b]: Compound not fully soluble. [c]: PDI: polydispersity index. [d]: Not determined, sample not soluble.

Under the chosen conditions, the vanadate catalyst system **[V]** generated a series of oxidation products on the basis of **AP** that are comparable to products observed in the archival literature (Table 2). The molybdate catalyst system **[Mo]**, on the other hand, led to detectable product formation on the basis of **AP** when applied in loadings of 5–10% (*w/w*). In any case, the molybdate system revealed a higher selectivity as fewer products were formed in comparatively higher amounts (Table 2). In both cases, catalytic activity of the present metal species is evident from the significantly increased depolymerisation yields compared to the blank sample.

In case of both **[V]**- and **[Mo]**-catalysed degradation, lignin polymerisation takes place as a background reaction, as indicated by the increase in both the number average molecular weight (Mn) and the polydispersity as compared to the starting material. This polymerisation can be explained by the activation of the lignin by the catalyst system by formation of radicals (vide infra, mechanistic discussion). Importantly, also in case a metal species is absent, polymerisation is observed (Table 2, blank sample), pointing towards an expected activation of the lignin by the perchloric acid alone. Control of this intrinsic activation is obtained, however, only when a metal catalyst is present as well. Polydispersities in case of **[V]**- and **[Mo]**-mediated reactions are significantly lower than in the case of the blank sample. Overall, both depolymerisation yields as well as the characteristics of the reisolated lignins thus indicate a beneficial role of the metal catalysts in the investigated valorisation system.

When subjecting ionosolv pine lignin **IP$_B$** to different loadings of vanadate and molybdate systems under otherwise unchanged conditions, product formation is actually overall enhanced (Table 2). Both catalyst systems show an overall comparable activity, under various loadings, but a rather remarkable selectivity in product formation is observed. The **[V]**-system, at a loading of 10% (*w/w*), delivers as dominant product of more than 90% relative abundance vanillin. The **[Mo]**-system, on the other hand, delivers isovanillic acid as the most abundant species under otherwise unchanged conditions.

When estimating the absolute amount of these products against an internal standard during work-up and GC-analysis, a depolymerisation yield can be calculated based on the amount of total extractives generated upon the reaction. This depolymerisation yield was found to be 34% in case of the vanadate system and, remarkably, 62% in case of molybdate-based catalyst applied each at a loading of 10% (*w/w*). Also, in case of **IP$_B$**, the presence of the metal catalysts led to a significant increase in depolymerisation yields compared to the blank.

Table 2. Results obtained in the oxidative degradation of aceto- and ionosolv pine lignins, **AP** and **IP**, respectively, using various loadings of vanadate- and molybdate catalyst systems.

Lignin [a]	AP							IP_B						
Catalyst System [b]	[V]				[Mo]			[V]				[Mo]		
Loading [% (*w/w*)]	Blank	5	7.5	10	5	7.5	10	Blank	5	7.5	10	5	7.5	10
compound [c]	abundance [%] [d]							abundance [%] [d]						
vanillin (**2a**)	—[e]	—[e]	—[e]	50	—[e]	—[e]	—[e]	—[e]	67	90	100	7.4	10	11
vanillic acid (**3a**)	—[e]	33	33	—[e]	100	100	100	—[e]	24	—[e]	—[e]	93	90	—[e]
isovanillin (**4**)	—[e]	50	17	17	—[e]	—[e]	—[e]	—[e]	—[e]	—[e]	—[e]	—[e]	—[e]	—[e]
isovanillic acid (**5**)	—[e]	—[e]	—[e]	—[e]	—[e]	—[e]	—[e]	—[e]	—[e]	—[e]	—[e]	—[e]	—[e]	89
1-(4-hydroxy-3-methoxy-phenyl)-2-methylprop-2-en-1-one (**6**)	—[e]	—[e]	17	33	—[e]	—[e]	—[e]	—[e]	9	10	—[e]	—[e]	—[e]	—[e]
4-acetoxy-3-methoxyacetophenone (**7**)	—[e]	17	33	—[e]	—[e]	—[e]	—[e]	—[e]	—[e]	—[e]	—[e]	—[e]	—[e]	—[e]
depolymerisation yield [%] [f]	6.0	23	26	31	26	69	72	9.8	31	33	34	58	57	62
Mn [kDa] (PDI) [g]	2.80 (6.0)	n.d. [h]	n.d. [h]	1.95 (2.6) [i]	n.d. [h]	n.d. [h]	3.20 (4.8) [i]	1.80 (8.3)	n.d. [h]	1.95 (3.2) [i]	n.d. [h]	n.d. [h]	2.50 (3.7) [i]	n.d. [h]

[a]: **AP**: acetosolv pine lignin and **IP_B**: ionosolv pine lignin isolated using [bmim]HSO4. [b]: **[V]**: vanadate catalyst system and **[Mo]**: molybdate catalyst system. [c]: Determined after extraction using GC-MS analyses, only lignin-stemming products are listed. [d]: Product abundancies normalised considering all lignin-derived compounds present in abundances higher than 0.05% in the chromatogram. [e]: Not present or present at quantities lower than 0.05%. [f]: Calculated on the basis of reisolated lignins. [g]: PDI: polydispersity index. [h]: Not determined. [i]: Sample not fully soluble under analysis conditions.

Residues of the **IP_B** lignin successfully depolymerised using 10% of vanadate-based catalyst or molybdate-based catalyst were subsequently isolated and analysed for structural changes using quantitative ^{31}P NMR. Results are given in Table 1; Figure 1 shows a comparison of the ^{31}P NMR spectra of starting **IP_B** and residues isolated after treatment with the **[V]**- and **[Mo]**-based catalysts systems.

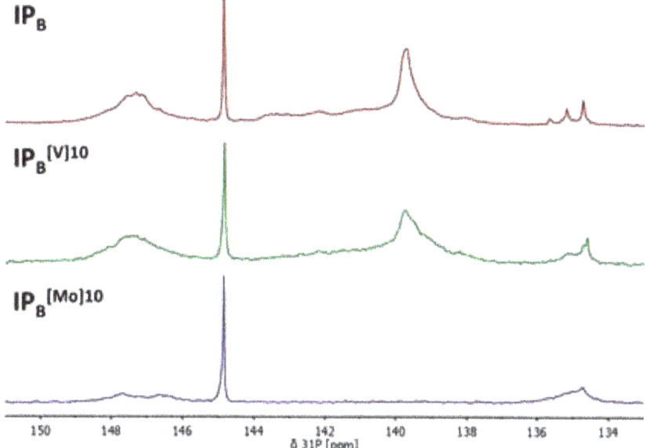

Figure 1. Comparison of the ^{31}P NMR spectra of starting **IP_B** and residues isolated after treatment with the **[V]**-based catalyst system and the **[Mo]**-based catalyst system at 10% (*w/w*).

Results show a drastic decease of OH-groups, independent of the catalyst type. This suggests significant oxidation along the lignin backbone. An increase in carboxylic acid group content as seen in the ^{31}P NMR of **IP_B** subjected to 10% (*w/w*) supports this analysis (Figure 1). Analysis of the number and the weight average molecular weights of reisolated lignins sustain the picture that emerged during the analysis of the **AP** system: the lignins are mainly activated towards polymerisation by perchloric acid, while in presence of the metal species, depolymerisation is favoured and re-polymerisation is present as a background reaction (compare mechanistic discussion below).

3.4. Catalytic Degradation of Hardwood Lignins

Given the promising results obtained with the softwood lignin, attention was turned towards *Eucalyptus nitens* lignins, both in acetosolv and ionosolv form. Maximum catalyst loadings were studied using identical conditions for the oxidative degradation and valorisation of **AE** and **IE_B**; the results are summarised in Table 3.

The vanadate-based system led, in case of acetosolv hardwood lignin **AE**, to the formation of a seemingly homogenised product portfolio. Depolymerisation yields were found to be ranging from 38 to 53% (Table 3). Highest selectivity was found at a catalyst loading of 7.5% (*w/w*) with essentially only syringaldehyde being formed. Lower depolymerisation yields of around 38% are observed independent of catalyst loading in the case of **IE_B** lignin (Table 3). When using 5% (*w/w*) vanadate-based catalyst, vanillin, and syringaldehyde were formed as single products; at higher catalyst loadings, depolymerisation yields remained constant, while product diversity increased. Together with 1-(4-hydroxy-3-methoxy-phenyl)-2-methylprop-2-en-1-one (**6**), 2,6-dimethoxybenzoquinone (**8**) was found. Interestingly, in none of the experiments the corresponding acids were found; the aqueous phase was controlled during extraction for sufficient acidity. The findings thus correspond to the activity of the vanadate system towards the pine lignin samples discussed above.

Given the overall better results obtained when applying lower concentrations of the **[Mo]** catalyst system in case of the softwood lignin, the eucalyptus acetosolv lignin **AE** and the ionosolv lignin **IE_B** were, respectively, treated with only 7.5% and 5% (*w/w*) of molybdate catalyst, as these concentrations led to more selective product formation with the other catalyst. In both cases, vanillic acid was detected as the only significantly abundant depolymerisation product, achieving depolymerisation yields of 84 and 78%, respectively (Table 3).

Table 3. Results obtained in the oxidative degradation of aceto- and ionosolv eucalyptus lignins, **AE** and **IE$_B$**, respectively, using various loadings of vanadate and molybdate catalyst systems.

Lignin [a]	AE					IE$_B$			
Catalyst System [b]	blank	[V]			[Mo]	blank	[V]		[Mo]
Loading [% (*w/w*)]	—	5	7.5	10	7.5	—	5	7.5	10	5
compound [c]	abundance [%] [d]					abundance [%] [d]			
vanillin (**2a**)	— [e]	— [e]	— [e]	— [e]	— [e]	— [e]	23	24	49	— [e]
syringaldehyde (**2b**)	— [e]	61	100	69	— [e]	— [e]	77	65	42	— [e]
vanillic acid (**3a**)	— [e]	— [e]	— [e]	— [e]	100	— [e]	— [e]	— [e]	— [e]	100
syringic acid (**3b**)	— [e]	— [e]	— [e]	— [e]	— [e]	— [e]	— [e]	— [e]	— [e]	— [e]
isovanillin (**4**)	— [e]	12	— [e]	— [e]	— [e]	— [e]	— [e]	— [e]	— [e]	— [e]
1-(4-hydroxy-3-methoxy-phenyl)-2-methylprop-2-en-1-one (**6**)	— [e]	27	— [e]	31	— [e]	— [e]	— [e]	6	9	— [e]
2,6-dimethoxy-benzoquinone (**8**)	— [e]	— [e]	— [e]	— [e]	— [e]	— [e]	— [e]	5	— [e]	— [e]
depolymerisation yield [%] [f]	22	46	38	53	84	25	38	37	38	78
Mn [kDa] (PDI) [g]	2.00 (3.8)	n.d. [h]	n.d. [h]	3.35 (9.3)	5.40 (3.8)	2.10 (3.5)	n.d. [h]	n.d. [h]	3.00 (14)	3.60 (9.9)

[a]: **AE**: acetosolv eucalyptus lignin and **IE$_B$**: acetosolv eucalyptus lignin isolated using [bmim] HSO$_4$. [b]: [V]: vanadate catalyst system and [Mo]: molybdate catalyst system. [c]: Determined after extraction using GC-MS analyses, only lignin-stemming products are listed. [d]: Product abundancies normalised considering all lignin-derived compounds present in abundances higher than 0.05% in the chromatogram. [e]: Not present or present at quantities lower than 0.05%. [f]: Calculated on the basis of reisolated lignins. [g]: PDI: polydispersity index. [h]: Not determined.

The overall more difficult situation encountered in case of the investigated oxidative hardwood valorisation is reflected as well in the molecular weight analyses. As in the case of the softwood lignins, the presence of the metal catalysts helps to increase depolymerisation, but in contrast to the situation found in the softwood cases, re-polymerisation is less effectively suppressed. Rather high values for the polydispersities indicate that re-polymerisation poses a significant problem during the oxidative degradation, which is likely to contribute to the overall lower depolymerisation yields seen for the hardwood lignins.

3.5. Mechanistic Considerations

Given the basic reactivities of the catalysts and the polyphenolic substrates, different reaction pathways can be assumed. One route comprises an initial attack of the catalytic system on the phenolic OH-group [44,46,47]. The phenoxy radical (**I**) formed initially can be stabilised by the aromatic system before eventually reaching the ipso-position where it can trigger an exo-depolymerisation pathway of the lignin chains (Scheme 1). In the case of the [**V**]-mediated oxidation of **IE$_B$** isolated monomeric quinone **8** supports this thesis.

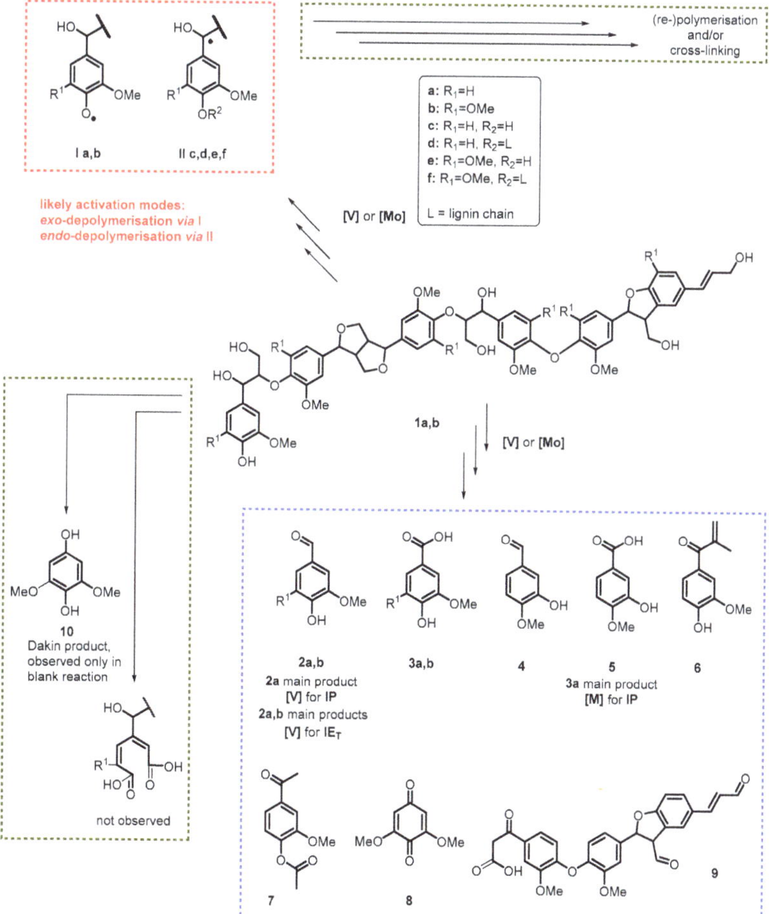

Scheme 1. Proposed activation modes based on observed products in [**V**]- and [**Mo**]-catalysed oxidation of soft and hardwood aceto- and ionosolv lignins [48].

Formation of vanillin (**2a**) and syringaldehyde (**2b**) in the case of [**V**]-mediated pine and eucalyptus ionosolv lignin degradation can be seen as immediate products originating from an exo-depolymerisation process triggered by oxidative activation of the benzylic position (**II**) in terminal β-*O*-4′ motifs; previous in silico studies had revealed that the benzylic position is similarly susceptible to an activation via radicals [48]. Formation of alkenone **7** can be explained by this route as well. Isolation of isovanillic acid (**5**), as in case of [**Mo**]-mediated valorisation of **IP$_B$**, and formation of vanillic acid, in case of [**Mo**]-treated **IE$_B$**, suggests a similar mechanism under eventually less controlled conditions favouring rearrangement reactions. Functionalisation of the benzylic position is an important aspect for rendering the transformation of the lignin aromatic monomers effective by giving rise to an effective endo-polymerisation happening in parallel.

Importantly, hydroquinones as products deriving from a Dakin reaction [49] have not been observed.

^{31}P NMR data (Table 1) indicate the presence of significant amounts of condensed units in the starting lignins, whereas material isolated after oxidative valorisation using the [**V**] and [**Mo**] catalysts systems contains less of these groups. In addition, the loss of these motifs can be interpreted in favour of an endo-depolymerisation. Repolymerisation reactions via the phenoxy radical could lead to an increase i 4-*O*-5′ groups, however this is not observed in this study.

Reduced amounts of aliphatic OH-groups (Table 1, Figure 1) show the activity of the catalyst systems on the aliphatic OH-groups along the lignin chain (Scheme 1); this activity is more pronounced in case of the [**Mo**]-based catalyst system, especially in case of softwood ionosolv lignin in which essentially all OH-groups must have been oxidised in remaining oligomeric structures (Scheme 1); structure **9** in Scheme 1 illustrates such a putatively 'per-oxidised' lignin fragment.

An interesting insight into the tighter interplay of the involved redox potentials can be delineated from the presence of vanillic acid (**3a**) in case of the [**Mo**]-mediated oxidation of pine and eucalyptus lignin, while the [**V**]-mediated valorisation of both soft- and hardwoods is possible yielding preferentially vanillin (**2a**) and syringaldehyde (**2b**), respectively. The data further suggest that the oxidation potential of the phenolic OH-group in initially formed vanillin still fits the dynamic range of the [**Mo**]-catalyst, while it is outside of the oxidation range of the [**V**]-catalyst.

Observed selectivities can be reasoned based on structural features of the lignins involved. However, the current study does not allow making clear connections between the catalyst type, the biomass source, and/or the isolation process and the observed activities and selectivities. Additional studies targeting this aspect are currently being pursued.

4. Conclusions

Acetosolv soft- and hardwood lignins as well as ionosolv soft- and hardwood lignins were transformed into monomeric aromatic compounds using either a vanadate or molybdate-based catalyst system. Monomers were generated with remarkable, catalyst-dependent selectivity and high depolymerisation yields via oxidative exo- and endo-depolymerisation processes. Using the vanadate–hydrogen peroxide system on acetosolv pine lignin, vanillin and isovanillin were produced as main products with depolymerisation yields of 31%. Using the molybdate system on the same lignin, vanillic acid was the practically exclusive product, with a depolymerisation yield of 72%. Ionosolv pine could be valorised into vanillin and vanillic acid as practically exclusive products in depolymerisation yields of 34 and 57%, respectively, using either the vanadate or the molybdate catalyst system. Similar selectivities, albeit with lower depolymerisation yields of around 50% under standardised conditions, could be obtained for eucalyptus acetosolv lignin, producing vanillin and syringaldehyde or vanillic acid as products, using either the vanadate or the molybdate system. Omitting the metal species in the reactions mixture, acetosolv eucalyptus was converted into benzoquinones as effectively only isolable aromatic monomer. Interestingly, the ionosolv hardwood lignins did not perform as well as the ionosolv softwood under the chosen conditions. A partial backbone degradation, as induced during the isolation of ionosolv lignins employing [bmim]HSO$_4$, does not fundamentally enhance oxidative depolymerisation. In all cases, (re-)polymerisation reactions in form of elevated number average molecular weights in combination with significantly increased polydispersities of re-isolated materials have been observed. In case of softwood lignins, the presence of the metal catalysts led to a partial control/suppression of such (re-)polymerisation; in case of the hardwood samples, however, metal catalysts appeared less effective in this respect.

Author Contributions: Conceptualization, F.S., V.C., H.L. and C.C.; data curation, L.P., M.G. and H.L.; investigation, L.P., M.G. and F.S.; methodology, F.S., P.G., V.C., H.L. and C.C.; project administration, V.C., J.C.P. and C.C.; resources, P.G., V.C., V.S., J.C.P. and C.C.; supervision, V.S., J.C.P., F.S., P.G., H.L. and C.C.; validation, M.G., H.L. and C.C.; writing—original draft, M.G., F.S. and H.L.; writing—review & editing, J.C.P., V.C., H.L. and C.C. All authors have read and agreed to the published version of the manuscript.

Funding: This research was funded by the "Ministry of Economy and Competitiveness" of Spain (research project "Modified aqueous media for wood biorefineries", reference CTQ2017-82962-R).

Acknowledgments: H.L. acknowledges the MIUR Grant 'Dipartimento di Eccellenza 2018–2022' to the Department of Pharmacy of the University of Naples 'Federico II'. C.C. acknowledges the Ca'Foscari FPI 2019 funding.

Conflicts of Interest: The authors declare no conflict of interest.

References

1. Aresta, M.; Dibenedetto, A.; Dumeignil, F. *Biorefineries, An Introduction*; De Gruyter: Berlin, Germany; Boston, MA, USA, 2015; ISBN 978-3-11-033158-5.
2. Fernando, S.; Adhikari, S.; Chandrapal, C.; Murali, N. Biorefineries: Current Status, Challenges, and Future Direction. *Energy Fuels* **2006**, *20*, 1727–1737. [CrossRef]
3. Takkellapati, S.; Li, T.; Gonzalez, M.A. An overview of biorefinery-derived platform chemicals from a cellulose and hemicellulose biorefinery. *Clean Technol. Environ. Policy* **2018**, *20*, 1615–1630. [CrossRef] [PubMed]
4. Yabushita, M.; Kobayashi, H.; Fukuoka, A. Catalytic transformation of cellulose into platform chemicals. *Appl. Catal. B Environ.* **2014**, *145*, 1–9. [CrossRef]
5. Delidovich, I.; Leonhard, K.; Palkovits, R. Cellulose and hemicellulose valorisation: An integrated challenge of catalysis and reaction engineering. *Energy Environ. Sci.* **2014**, *7*, 2803–2830. [CrossRef]
6. Argyropoulos, D.S.; Crestini, C. A Perspective on Lignin Refining, Functionalization, and Utilization. *ACS Sustain. Chem. Eng.* **2016**, *4*, 5089. [CrossRef]
7. Fiorani, G.; Crestini, C.; Selva, M.; Perosa, A. Advancements and Complexities in the Conversion of Lignocellulose Into Chemicals and Materials. *Front. Chem.* **2020**, *8*. [CrossRef]
8. Mobley, J.K. Conversion of Lignin to Value-added Chemicals via Oxidative Depolymerization. In *Chemical Catalysts for Biomass Upgrading*; Crocker, M., Santillan-Jimenez, E., Eds.; Wiley: Hoboken, NJ, USA, 2020; pp. 357–393. ISBN 978-3-527-34466-6.
9. Lange, H.; Decina, S.; Crestini, C. Oxidative upgrade of lignin—Recent routes reviewed. *Eur. Polym. J.* **2013**, *49*, 1151–1173. [CrossRef]
10. Crestini, C.; Crucianelli, M.; Orlandi, M.; Saladino, R. Oxidative strategies in lignin chemistry: A new environmental friendly approach for the functionalisation of lignin and lignocellulosic fibers. *Catal. Today* **2010**, *156*, 8–22. [CrossRef]
11. Zakzeski, J.; Bruijnincx, P.C.A.; Jongerius, A.L.; Weckhuysen, B.M. The Catalytic Valorization of Lignin for the Production of Renewable Chemicals. *Chem. Rev.* **2010**, *110*, 3552–3599. [CrossRef]
12. Argyropoulos, D.S. *Oxidative Delignification Chemistry: Fundamentals and Catalysis*; ACS Symposium Series; American Chemical Society: Washington, DC, USA, 2001; ISBN 978-0-8412-3738-4.
13. Crestini, C.; Pro, P.; Neri, V.; Saladino, R. Methyltrioxorhenium: A new catalyst for the activation of hydrogen peroxide to the oxidation of lignin and lignin model compounds. *Bioorg. Med. Chem.* **2005**, *13*, 2569–2578. [CrossRef]
14. Crestini, C.; Caponi, M.C.; Argyropoulos, D.S.; Saladino, R. Immobilized methyltrioxo rhenium (MTO)/H2O2 systems for the oxidation of lignin and lignin model compounds. *Bioorg. Med. Chem.* **2006**, *14*, 5292–5302. [CrossRef]
15. Guadix-Montero, S.; Sankar, M. Review on Catalytic Cleavage of C–C Inter-unit Linkages in Lignin Model Compounds: Towards Lignin Depolymerisation. *Top. Catal.* **2018**, *61*, 183–198. [CrossRef]

16. Chan, J.M.W.; Bauer, S.; Sorek, H.; Sreekumar, S.; Wang, K.; Toste, F.D. Studies on the Vanadium-Catalyzed Nonoxidative Depolymerization of Miscanthus giganteus-Derived Lignin. *ACS Catal.* **2013**, *3*, 1369–1377. [CrossRef]

17. Son, S.; Toste, F.D. Non-Oxidative Vanadium-Catalyzed C-O Bond Cleavage: Application to Degradation of Lignin Model Compounds. *Angew. Chem. Int. Ed.* **2010**, *49*, 3791–3794. [CrossRef] [PubMed]

18. Amadio, E.; Di Lorenzo, R.; Zonta, C.; Licini, G. Vanadium catalyzed aerobic carbon–carbon cleavage. *Coord. Chem. Rev.* **2015**, *301–302*, 147–162. [CrossRef]

19. Bozell, J.J.; Hoberg, J.O.; Dimmel, D.R. Heteropolyacid Catalyzed Oxidation of Lignin and Lignin Models to Benzoquinones. *J. Wood Chem. Technol.* **2000**, *20*, 19–41. [CrossRef]

20. Mobley, J.K.; Jennings, J.A.; Morgan, T.; Kiefer, A.; Crocker, M. Oxidation of Benzylic Alcohols and Lignin Model Compounds with Layered Double Hydroxide Catalysts. *Inorganics* **2018**, *6*, 75. [CrossRef]

21. Voitl, T.; Rudolf von Rohr, P. Oxidation of Lignin Using Aqueous Polyoxometalates in the Presence of Alcohols. *ChemSusChem* **2008**, *1*, 763–769. [CrossRef]

22. Weinstock, I.A.; Atalla, R.H.; Reiner, R.S.; Moen, M.A.; Hammel, K.E.; Houtman, C.J.; Hill, C.L.; Harrup, M.K. A new environmentally benign technology for transforming wood pulp into paper. Engineering polyoxometalates as catalysts for multiple processes. *J. Mol. Catal. A Chem.* **1997**, *116*, 59–84. [CrossRef]

23. Mottweiler, J.; Puche, M.; Räuber, C.; Schmidt, T.; Concepción, P.; Corma, A.; Bolm, C. Copper- and Vanadium-Catalyzed Oxidative Cleavage of Lignin using Dioxygen. *ChemSusChem* **2015**, *8*, 2106–2113. [CrossRef]

24. Hao, K.; Zhang, L.-L.; Song, L.; Guan, H.-Y.; Li, C.-M.; Liu, T.; Yu, Q.; Zeng, J.-M.; Wang, Z.-W. Highly active Mo-V-based bifunctional catalysts for catalytic conversion of lignin dimer model compounds at room temperature. *Inorg. Chem. Commun.* **2020**, *116*, 107910. [CrossRef]

25. Drago, G.A.; Gibson, T.D. Enzyme Stability and Stabilisation: Applications and Case Studies. In *Engineering and Manufacturing for Biotechnology*; Hofman, M., Thonart, P., Eds.; Focus on Biotechnology; Springer: Dordrecht, The Netherlands, 2002; pp. 361–376. ISBN 978-0-306-46889-6.

26. Crestini, C.; Pastorini, A.; Tagliatesta, P. Metalloporphyrins immobilized on motmorillonite as biomimetic catalysts in the oxidation of lignin model compounds. *J. Mol. Catal. A Chem.* **2004**, *208*, 195–202. [CrossRef]

27. Crestini, C.; Tagliatesta, P. Metalloporphyrins in the Biomimetic Oxidation of Lignin and Lignin Model Compounds: Development of Alternative Delignification Strategies. *Porphyr. Handb. Bioinorg. Bioorg. Chem.* **2003**, *11*, 161.

28. Crestini, C.; Tagliatesta, P.; Saladino, R. A biomimetic approach to lignin degradation, metalloporphyrins catalysed oxidation of lignin and lignin model compounds. *Oxid. Delignification Chem. Fundam. Catal.* **2001**, 213–225.

29. Giannì, P.; Lange, H.; Crestini, C. Lipoxygenase: Unprecedented Carbon-Centered Lignin Activation. *ACS Sustain. Chem. Eng.* **2018**, *6*, 5085–5096. [CrossRef]

30. Wong, D.W.S. Structure and Action Mechanism of Ligninolytic Enzymes. *Appl. Biochem. Biotechnol.* **2009**, *157*, 174–209. [CrossRef]

31. Penín, L.; Lange, H.; Santos, V.; Crestini, C.; Parajó, J.C. Characterization of Eucalyptus nitens Lignins Obtained by Biorefinery Methods Based on Ionic Liquids. *Molecules* **2020**, *25*, 425. [CrossRef]

32. Penín, L.; Peleteiro, S.; Santos, V.; Alonso, J.L.; Parajó, J.C. Selective fractionation and enzymatic hydrolysis of Eucalyptus nitens wood. *Cellulose* **2019**, *26*, 1125–1139. [CrossRef]

33. Penín, L.; Santos, V.; del Río, J.C.; Parajó, J.C. Assesment on the chemical fractionation of Eucalyptus nitens wood: Characterization of the products derived from the structural components. *Bioresour. Technol.* **2019**, *281*, 269–276. [CrossRef]

34. Meng, X.; Crestini, C.; Ben, H.; Hao, N.; Pu, Y.; Ragauskas, A.J.; Argyropoulos, D.S. Determination of hydroxyl groups in biorefinery resources via quantitative 31 P NMR spectroscopy. *Nat. Protoc.* **2019**, *14*, 2627–2647. [CrossRef]

35. Granata, A.; Argyropoulos, D.S. 2-Chloro-4,4,5,5-tetramethyl-1,3,2-dioxaphospholane, a Reagent for the Accurate Determination of the Uncondensed and Condensed Phenolic Moieties in Lignins. *J. Agric. Food Chem.* **1995**, *43*, 1538–1544. [CrossRef]

36. Argyropoulos, D.S. 31P NMR in Wood Chemistry: A Review of Recent Progress. *Res. Chem. Intermed.* **1995**, *21*, 373–395. [CrossRef]

37. Amini, M.; Haghdoost, M.M.; Bagherzadeh, M. Oxido-peroxido molybdenum(VI) complexes in catalytic and stoichiometric oxidations. *Coord. Chem. Rev.* **2013**, *257*, 1093–1121. [CrossRef]

38. Conte, V.; Floris, B. Vanadium and molybdenum peroxides: Synthesis and catalytic activity in oxidation reactions. *Dalton Trans.* **2011**, *40*, 1419–1436. [CrossRef] [PubMed]

39. Schwendt, P.; Tatiersky, J.; Krivosudský, L.; Šimuneková, M. Peroxido complexes of vanadium. *Coord. Chem. Rev.* **2016**, *318*, 135–157. [CrossRef]

40. Conte, V.; Floris, B. Vanadium catalyzed oxidation with hydrogen peroxide. *Inorg. Chim. Acta* **2010**, *363*, 1935–1946. [CrossRef]

41. Conte, V.; Coletti, A.; Floris, B.; Licini, G.; Zonta, C. Mechanistic aspects of vanadium catalysed oxidations with peroxides. *Coord. Chem. Rev.* **2011**, *255*, 2165–2177. [CrossRef]

42. Galloni, P.; Mancini, M.; Floris, B.; Conte, V. A sustainable two-phase procedure for V-catalyzed toluene oxidative bromination with H2O2–KBr. *Dalton Trans.* **2013**, *42*, 11963–11970. [CrossRef]

43. Langeslay, R.R.; Kaphan, D.M.; Marshall, C.L.; Stair, P.C.; Sattelberger, A.P.; Delferro, M. Catalytic Applications of Vanadium: A Mechanistic Perspective. *Chem. Rev.* **2019**, *119*, 2128–2191. [CrossRef]

44. Floris, B.; Sabuzi, F.; Coletti, A.; Conte, V. Sustainable vanadium-catalyzed oxidation of organic substrates with H2O2. *Catal. Today* **2017**, *285*, 49–56. [CrossRef]

45. Kirillova, M.V.; Kuznetsov, M.L.; Romakh, V.B.; Shul'pina, L.S.; Fraústo da Silva, J.J.R.; Pombeiro, A.J.L.; Shul'pin, G.B. Mechanism of oxidations with H2O2 catalyzed by vanadate anion or oxovanadium(V) triethanolaminate (vanadatrane) in combination with pyrazine-2-carboxylic acid (PCA): Kinetic and DFT studies. *J. Catal.* **2009**, *267*, 140–157. [CrossRef]

46. Garau, G.; Palma, A.; Lauro, G.P.; Mele, E.; Senette, C.; Manunza, B.; Deiana, S. Detoxification Processes from Vanadate at the Root Apoplasm Activated by Caffeic and Polygalacturonic Acids. *PLoS ONE* **2015**, *10*, e0141041. [CrossRef] [PubMed]

47. Linskens, H.F.; Jackson, J.F. (Eds.) *Plant Cell Wall Analysis*; Modern Methods of Plant Analysis; Springer: Berlin/Heidelberg, Germany, 1996; Volume 17, ISBN 978-3-642-64644-7.

48. Crestini, C.; Jurasek, L.; Argyropoulos, D.S. On the Mechanism of the Laccase-Mediator System in the Oxidation of Lignin. *Chem. Eur. J.* **2003**, *9*, 5371–5378. [CrossRef] [PubMed]

49. Dakin, H.D. The oxidation of hydroxy derivatives of benzaldehyde, acetophenone and related substances. *Am. Chem. J.* **1909**, *42*, 477–498.

 *processes*

Article

Melt Stable Functionalized Organosolv and Kraft Lignin Thermoplastic

Shubhankar Bhattacharyya [1], Leonidas Matsakas [2], Ulrika Rova [2] and Paul Christakopoulos [2,*]

[1] Chemistry of Interfaces, Division of Chemical Engineering, Department of Civil, Environmental and Natural Resources Engineering, Luleå University of Technology, 97187 Luleå, Sweden; shubhankar2chem@gmail.com

[2] Biochemical Process Engineering, Division of Chemical Engineering, Department of Civil, Environmental and Natural Resources Engineering, Luleå University of Technology, 97187 Luleå, Sweden; leonidas.matsakas@ltu.se (L.M.); ulrika.rova@ltu.se (U.R.)

* Correspondence: paul.christakopoulos@ltu.se; Tel.: +46-(0)920-492510

Received: 23 July 2020; Accepted: 3 September 2020; Published: 5 September 2020

Abstract: A shift towards an economically viable biomass biorefinery concept requires the use of all biomass fractions (cellulose, hemicellulose, and lignin) for the production of high added-value products. As lignin is often underutilized, the establishment of lignin valorization routes is highly important. *In-house* produced organosolv as well as commercial Kraft lignin were used in this study. The aim of the current work was to make a comparative study of thermoplastic biomaterials from two different types of lignins. Native lignins were alkylate with two different alkyl iodides to produce ether-functionalized lignins. Successful etherification was verified by FT-IR spectroscopy, changes in the molecular weight of lignin, as well as ^{13}C and ^{1}H Nuclear Magnetic Resonance (NMR). The thermal stability of etherified lignin samples was considerably improved with the $T_{2\%}$ of organosolv to increase from 143 °C to up to 213 °C and of Kraft lignin from 133 °C to up to 168 °C, and glass transition temperature was observed. The present study shows that etherification of both organosolv and Kraft lignin with alkyl halides can produce lignin thermoplastic biomaterials with low glass transition temperature. The length of the alkyl chain affects thermal stability as well as other thermal properties.

Keywords: lignin; organosolv; Kraft lignin; etherification; lignin functionalization; thermoplastics

---

1. Introduction

Lignin is an aromatic heteropolymer, the second most abundant biopolymer in the world after cellulose. It constitutes 15–35% *w/w* of a plant's cell wall and plants are estimated to generate 0.5–3.6 billion tons of lignin annually [1,2]. The global annual production of lignin is estimated at approximately 100 million tons, of which only 2% is used commercially (primarily in dispersants, adhesives, and surfactants); whereas the rest is burned as low-value fuel [3–5]. The main sources of lignin are technical lignin from the pulp industry, namely Kraft and ligno-sulphonate. Organosolv lignin (accounting for approximately 2% of the total lignin) has started to gain attention owing to advances in biomass biorefinery technology, its high purity, and a chemical structure close to that of natural lignin [3,6]. The ongoing worldwide construction of second-generation cellulosic ethanol plants is expected to further increase the availability of lignin [7]. Kraft lignin is produced through the sulphate cooking process, in which fibers are treated at temperatures of 165–175 °C for 1–2 h in the presence of sodium hydroxide and sodium sulphite [5]. Ligno-sulphonates are a by-product of sulphite cooking in which the fibers are treated by HSO_3^- and SO_3^{2-} ions and the digestion is typically operated at 120–180 °C for 1–5 h [1,5].

On the contrary, organosolv lignin is produced through organosolv pulping, in which the biomass is treated in aqueous-organic solvent mixtures with or without the addition of an inorganic catalyst [8,9].

Ligno-sulfonates and Kraft lignin are sulfur-containing lignin, with the sulfur to vary in the range 1–3% for Kraft and 3.5–8% for lignosulfonates and they consist of a by-product of chemical pulping process from the paper industry [3]. On the contrary, organosolv lignin is practically sulfur-free with a very low ash content [10]. Another major difference between the first two lignins and organosolv is that organosolv retains a considerable fraction of the β-O-4 linkages, whereas in lignin from chemical pulping, the β-O-4 linkages represent fewer than 10% of the connections and promote the formation of carbon–carbon bonds [5,11]. Due to the high lignin content of lignocellulosic biomass, its valorization is crucial for the financial viability of biorefineries. However, existing technologies for lignin valorization are still not as advanced as those applied to carbohydrate fractions [1]. New technologies for the conversion of lignin into high added-value products represent an important cornerstone towards establishing economically viable biomass biorefinery processes that can reduce our dependency on fossil feedstock.

Thermoplastic-based polymers are widely used in various applications due to their flexible nature. The development of polymer technology has seen a continuous demand for eco-friendly, biodegradable thermoplastic materials. Thermoplasticity is defined as the use of heating to weaken intermolecular forces that connect polymer chains, without altering the chemical structure of the polymer. Thus, thermoplastic materials can be melted and shaped multiple times into desired structures without eliciting any chemical changes. In this regard, lignin is a very promising raw material as it is readily available, sustainable, and can be functionalized to achieve the desired physical properties.

Lignin, however, presents also some challenges as it is brittle, exhibits poor mechanical properties, has a relative high glass transition temperature (for Kraft lignin 124–174 °C, organosolv lignin 91–97 °C [12]), and its thermal processing at elevated temperatures is often impaired by radical-induced self-condensation, which limits its application to thermosets [13,14]. Self-polymerization of lignin during heating is often manifested by an increase in molecular weight, changes in chemical behavior (e.g., loss of solubility in common organic solvents such as tetrahydrofuran—THF) or 'loss' of Tg after repeated heating and cooling cycles [15]. The elevated hydroxyl content of lignin could induce swelling due to water adsorption, causing expansions in the composite [16] and, in turn, poor dimensional stability. However, the high reactivity of hydroxyl groups allows also lignin functionalization, which can significantly improve Tg, thermal stability, and compatibility with polymers [13,14].

Various efforts have been made to achieve functionalized lignins with low Tg and improved thermal properties for thermoplastic application [13,17]. Lignin esterification is one of the oldest methods to decrease Tg, whereas acetylation augments the solubility of lignin for molecular weight and structural analysis. Lewis and Brauns were the first to describe the solubility and thermal properties of lignin-based esters [18]. Later, Glasser and Jain reported that Tg of lignin esters decreased linearly with increasing length of the acyl group [19]. In most cases, lignin esterification was achieved using carboxylic acids and particularly acid chlorides or acid anhydrides. Bi-functional reagents, such as dicarboxylic aliphatic (sebacoyl) or aromatic (terephthaloyl) acid chlorides, are also promising in the production of lignin-based polyester network materials [20,21]. Another such reagent is dimer acid [22], which can be produced during the processing of soybean, tall, and cottonseed oil by a Diels–Alder cycloaddition reaction. Dimer acid-based lignin polyesters are fully renewable materials for polymer applications. Recently, polybutadiene-based lignin polyester thermoplastic with network structure and very low Tg was reported [2].

Graft modification is another efficient way to produce integrated lignin-polymer grafted materials with improved thermal processing and stability. Both "grafting onto" and "grafting from" modification methods have been reported. Korich et al. were the first to introduce boronic-acids based reagents for "grafting onto" modification of lignin, resulting in promising thermal properties and Tg in the range of −30 °C to −50 °C [23]. A similar lignin-based copolymer with low Tg was obtained by using azide–alkyne Huisgen cycloaddition "click chemistry" tools [24]. In comparison, the "grafting from" modification method allowed de Oliveira and Glasser to chemically modify hydroxypropylated lignin with ring-opening polymerization of ε-caprolactone [25] and subsequently lower Tg. Another widely used "grafting from" method is atom transfer radical polymerization (ATRP) [26]. ATRP was used by Hilburg et al. to prepare lignin-based thermoplastic with improved mechanical properties, and by Kim and Kadla to obtain lignin-based thermoresponsive thermoplastic [27,28].

Two other prominent etherification methods for obtaining lignin-based thermoplastic include methylation and hydroxyalkylation [29,30]. However, no systematic studies of etherification reactions with higher alkyl chain congeners have been reported [13,31]. Furthermore, comparative studies regarding hydrophobic lignin-based thermoplastic from different types of lignins are also scarcely documented. Thus, in this work we have chosen our in house organosolv lignin as well as commercial Kraft lignin for comparative studies. Both lignins were alkylated by etherification reaction with two different alkyl iodides to produce hydrophobic thermoplastic. All lignin based hydrophobic thermoplastics were characterized and comparative studies has been discussed in this work.

2. Materials and Methods

2.1. Materials and Chemicals

Kraft lignin (product no. 370959; Sigma-Aldrich, St. Louis, MO, USA) was used without any further treatment. Organosolv lignin was produced from a hybrid organosolv—steam explosion reactor as described previously [32]. Specifically, birch chips from mills in northern Sweden were milled to <1 mm particle size and used as raw material for organosolv treatment at 200 °C for 30 min with 30% (*v/v*) ethanol. The resulting pretreated solids were separated from the liquor by vacuum filtration and the liquor was collected for lignin isolation. The ethanol was removed under vacuum in a rotary evaporator and, finally, lignin was separated from the liquor by centrifugation at 29,416× *g* for 1 min at 4 °C. The lignin was air-dried until further use. Approximately 12 g of lignin were recovered per 100 g treated birch chips (dry basis). Contaminants in lignin were analyzed as previously described [32]; they amounted to 2.73% *w/w* hemicellulose and 0.12% *w/w* ash, but no cellulose was detected.

Anhydrous potassium carbonate was purchased from Sigma-Aldrich and dried overnight in the oven at 130 °C before every reaction. Dimethylformamide (DMF) was dried through a 4-Å molecular sieve. 1-iododecane and 1-iodohexadecane were purchased from Sigma-Aldrich and used without further purification.

2.2. Etherification Reaction

Prior to etherification, ~200 mg of lignin was weighted and heated in the oven (up to 60 °C) for 1 h, followed by drying under vacuum in a two-neck round-bottom flask for 2 h. Then, 5 mL of dry DMF was added under nitrogen atmosphere and the mixture was stirred at room temperature until a homogeneous solution was formed. Subsequently, 4 mol equivalents of anhydrous K_2CO_3 were added and the mixture was heated up to 65 °C under nitrogen atmosphere. Next, 1.2 mol equivalents of alkyl iodide were added dropwise with continuous stirring and heating, after which the mixture was heated at 60–65 °C for 16 h under an inert atmosphere. Upon completion, the reaction was quenched by the addition of water (50 mL). The reaction mass was stirred and a solid precipitate was formed. The solid mass was washed several times with water, dried under vacuum, and mixed with diethyl ether to dissolve the alkylated hydrophobic lignin. The ether-soluble part was separated and, upon evaporation, a solid mass of 250–260 mg was recovered. The sample was kept under vacuum for 48 h before further characterization.

2.3. Nuclear Magnetic Resonance (NMR) Characterization

A Bruker Ascend Aeon WB 400 (Bruker BioSpin AG, Fällanden, Switzerland) NMR spectrometer was used with a working frequency of 400.22 MHz for ^{1}H, 100.65 MHz for ^{13}C, and 162.02 MHz for ^{31}P with a 10-mm Z-gradient Broadband Observe (BBO) probe. Bruker Topspin 3.5pl2 software was used for data processing. All spectra were recorded at 25 °C. Chemical shifts were expressed in ppm (δ) downfield from tetramethylsilane (TMS), using the solvent as internal standard (CDCl₃,

$\delta = 7.26$). ^{31}P NMR spectra were acquired at an inverse gated pulse sequence with 90° pulse and Waltz 16 decoupling technique. A total of 72 scans were acquired with a 10-s recycle delay. 1H NMR spectra were analyzed with the Bruker standard pulse program based on a total of 32 scans, while ^{13}C NMR spectra included a total of 1024 scans.

2.3.1. Quantitative ^{31}P NMR

^{31}P NMR analysis was performed according to a previously published method [33], modified so as to be analyzed in 10 mm NMR probe. Approximately 120 mg of lignin was dissolved in 1.6 mL anhydrous $CDCl_3$/pyridine (1:1.6 *v/v*) solution. A 0.1 M standard solution containing cholesterol as internal standard was prepared in anhydrous $CDCl_3$/pyridine solution and 5 mg/mL Cr(III) acetylacetonate was added as relaxation reagent. Then, 400 µL of this solution was added to the above prepared lignin solution. The mixture was stirred vigorously and 400 µL of phosphitylating reagent II (2-chloro-4,4,5,5-tetramethyl-1,2,3-dioxaphospholane) was added. The reaction mixture was stirred for 2 h at room temperature and then transferred to a 10-mm NMR tube for ^{31}P NMR analysis. ^{31}P NMR measurements were performed to determine hydroxyl and carboxyl group content in organosolv as well as in commercial Kraft lignin. ^{31}P NMR results were expressed in mmol g^{-1} and were used to calculate the reactants' molar ratio for the etherification reaction.

2.3.2. ^{1}H and ^{13}C NMR

Approximately 200 mg of ether-functionalized lignin was dissolved in 3 mL $CDCl_3$ and then transferred to a 10-mm NMR tube. NMR spectra were acquired at 25 °C with sample spinning frequency of 20 Hz. All ether-functionalized lignin samples were referenced with solvent signal at 7.26 ppm ($CDCl_3$).

2.4. Fourier Transform Infrared (FT-IR) Measurements

FT-IR spectra were recorded on a Bruker IFS 80v vacuum FT-IR spectrometer equipped with a deuterated triglycine sulphate detector. Samples were prepared with KBr as disks. All spectra were recorded at room temperature (~22 °C) using the double-side forward-backward acquisition mode under vacuum. A total of 128 scans were co-added and signal-averaged at an optical resolution of 4 cm^{-1}.

2.5. Thermogravimetric Analysis (TGA)

TGA of lignin samples was performed on a PerkinElmer 8000 TGA instrument (Waltham, MA, USA) at 30–800 °C with a heating ramp of 10 °C/min under nitrogen atmosphere. Approximately 1–2 mg of lignin sample was used for each experiment and placed on PerkinElmer ceramic pans. All lignin samples were dried for 48 h under vacuum before TGA analysis to remove moisture from the samples.

2.6. Differential Scanning Calorimetry (DSC) Analysis

DSC of lignin samples was determined on a PerkinElmer DSC 6000 single-furnace instrument (Waltham, MA, USA) between −80 °C and 100 °C using an intracooler with a heating and cooling ramp of 5 °C/min under nitrogen atmosphere. Approximately 1–2 mg of lignin sample was placed in aluminum pans and sealed manually. At the beginning of the analysis, the sample was heated to 100 °C to form a thin layer inside the aluminum pan and then cooled gradually to −80 °C.

2.7. Gel Permeation Chromatography (GPC)

To determine the molecular weight of lignin samples, GPC was performed on a PerkinElmer Flexar HPLC apparatus equipped with a UV detector (PerkinElmer, Waltham, MA, USA) and a Waters (Milford, MA, USA) Styragel HR 4E column at 40 °C. THF was used as mobile phase at a flow rate of 0.6 mL/min. The detector was set at 280 nm. Prior to analysis, lignin was acetobrominated as previously described [34]. The calibration curve was acquired with polystyrene standards (Sigma-Aldrich, St. Louis, MO, USA).

3. Results and Discussion

3.1. Selection of Appropriate Functionalization Conditions

Etherification is one of the most well-known functionalization methods for generating lignin-based thermoplastic. Lignin ether linkages are more thermally stable and chemically resistant than lignin ester linkages, making them more suitable for further thermoplastic applications. Our initial objective was to prepare hydrophobic lignin thermoplastic by functionalization of organosolv lignin under mild reaction conditions. The most commonly used etherification procedure involves the reaction of an alkyl halide with lignin in the presence of a base. Lignin contains both aliphatic and phenolic hydroxyl groups that are susceptible to etherification with the alkyl halide. However, due to the high pKa value of the aliphatic hydroxyl group, etherification occurs exclusively at the phenolic hydroxyl position under normal reaction conditions. We hypothesized that the use of strong base (e.g., NaOH or KOH) at an elevated temperature could result in partial decomposition of native lignin. In addition to this, a strong base can promote further competitive side hydrolysis reaction of alkyl halide with strongly nucleophilic hydroxide ions (-OH). In this regard, K_2CO_3 is the best choice of base when aiming at milder reaction condition and it has been widely used for the deprotonation of phenolic compounds. However, as K_2CO_3 is highly soluble in water, it may also generate hydroxide ions under these conditions and hydrolyze alkyl halides. To avoid this side reaction, etherification was carried out under a nitrogen atmosphere with the use of anhydrous K_2CO_3 and DMF. DMF is the most compatible solvent for the partial dissolution of K_2CO_3 particles, and allows for good solubility of lignin and most alkyl halides.

Previously Ren et al. carried out the alkylation of alkaline lignin with *n*-dodecyl bromide for the preparation of dodecylated lignin under reflux conditions in water for 2.5 days [35], which further suggest the slower reactivity of alkyl bromide. Thus, the selection of alkyl halides was narrowed to alkyl iodides, as these are more reactive compared to other alkyl halide congeners. Most alkyl iodides undergo reaction with phenolic compounds at room temperature, but long-chain alkyl iodides require elevated temperatures (60–65 °C) perhaps due to size flexibility or steric hindrance. We selected two long-chain alkyl iodides, 1-iodohexadecane and 1-iododecane, to evaluate the effect of alkyl chain length on the properties of functionalized lignin. Finally, to examine how the source of lignin affected the etherification process, we compared in-house organosolv birch lignin with commercially available Kraft lignin.

^{31}P NMR analysis provided an estimate of hydroxyl as well as carboxyl group content. As summarized in Table 1, total phenol and carboxyl content amounted to 2.50 and 0.19 mmol g^{-1} in organosolv lignin and 3.92 and 0.35 mmol g^{-1} in Kraft lignin. In contrast, the aliphatic hydroxyl group content was higher in organosolv lignin (3.00 mmol g^{-1}) than in Kraft lignin (2.18 mmol g^{-1}). Etherification of lignin was carried out on the basis of total phenol as well as carboxyl group content.

Table 1. ^{31}P NMR (Nuclear Magnetic Resonance) data.

Lignin	Total Phenol (mmol g^{-1})	Carboxyl (mmol g^{-1})	Aliphatic (mmol g^{-1})
Organosolv	2.50	0.19	3.00
Kraft	3.92	0.35	2.18

3.2. Characterization of Lignins

Lignin is one of the most complex biopolymers as it contains various kinds of bonds and functional groups. Its complex nature often results in overlapped infrared spectra, which make it difficult to assign the correct identities. In spite of band overlapping in the fingerprint region, FT-IR spectra of native organosolv lignin and its ether functionalized forms (Figure 1), as well as the corresponding Kraft lignins (Figure 2), allowed bands to be assigned according to previously published data (Table 2) [36–44].

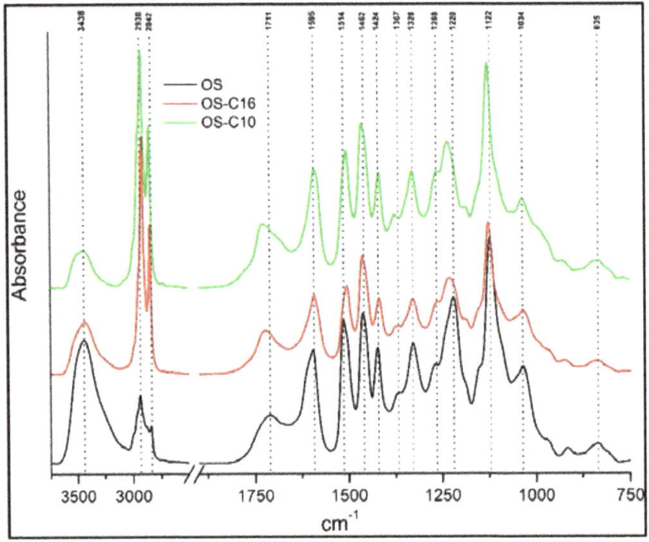

Figure 1. FT-IR spectra of nonfunctionalized organosolv (OS) lignin etherified with C10 (OS-C10) and C16 (OS-C16) alkyl chains.

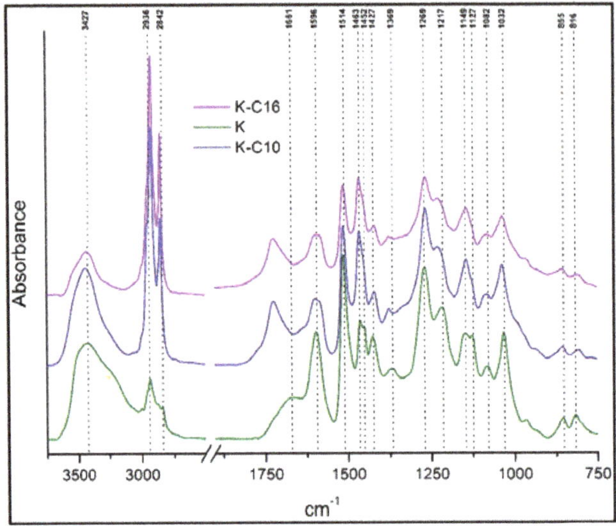

Figure 2. FT-IR spectra of nonfunctionalized Kraft (K) lignin etherified with C10 (K-C10) and C16 alkyl chains.

Table 2. IR band assignments [36–44].

Wave Number (cm^{-1})	IR Band Assignments
3427–3442	O-H stretching in aliphatic and phenolic -OH
2924–2938	C-H stretching in methyl groups
2842–2854	C-H stretch in methylene groups
1711–1730	C=O stretching in unconjugated ketones and carboxyl groups; saturated esters
1661	Stretching of C=O conjugated to aromatic rings (conjugated carbonyl)
1590–1598	Aromatic skeletal ring vibration (S > G) + C=O stretch
1506–1514	Aromatic skeletal ring vibrations (G > S)
1452–1466	C-H asymmetric deformations in methyl and methylene groups
1419–1427	Aromatic skeletal ring vibrations
1367–1378	Aliphatic C-H symmetric deformation in methyl (not methoxy) + O-H deformation in phenols
1328–1330	S ring breathing vibration + G ring substituted in position 5
1264–1269	G ring breathing vibration and C-O stretching
1217–1235	C-O stretching in phenols and ethers
1122–1149	C-H stretching in G ring and Aromatic C–H in plane deformation (S)
1082	C-O stretch of secondary alcohols and aliphatic ethers
1034	Aromatic C-H in-plane deformations in G units + C-O deformations in primary alcohols
835/855	Aromatic C-H out-of-plane deformation (only in GS and H lignin types)

The band around 3438 cm^{-1} in organosolv (OS) lignin and 3427 cm^{-1} in Kraft (K) lignin corresponded to the O-H stretching frequency in aliphatic or phenolic hydroxyl groups. This band decreased significantly after functionalization in both OS and K lignin due to formation of an ether linkage with the aromatic hydroxyl group. The bands in the 2924 cm^{-1} to 2938 cm^{-1} and 2842 cm^{-1} to 2854 cm^{-1} regions corresponded to C-H stretching in methyl and methylene groups, respectively. A sharp increase in the C-H stretching frequency after functionalization confirmed the presence of an alkyl chain, demonstrating the successful etherification of lignin. The carbonyl/carboxyl region between 1711 cm^{-1} and 1730 cm^{-1} exhibited a different pattern in OS and K lignin before and after etherification. In the case of OS lignin, the band around 1711 cm^{-1} demonstrated the presence of aryl saturated carboxylic acids, accompanied by a shift to ~1725 cm^{-1} that suggested esterification at the carboxylic acid groups. In the case of K lignin, a broad aromatic conjugated carboxylate band around 1661 cm^{-1} was observed. After etherification, the carboxylate band shifted to 1725 cm^{-1}, perhaps due to side esterification of the carboxylate group with alkyl iodide.

Aromatic skeletal vibration at 1590 cm^{-1} to 1598 cm^{-1}, as well as aromatic ring vibration at 1506 cm^{-1} to 1514 cm^{-1} and 1419 cm^{-1} to 1427 cm^{-1} were common for all types of lignin. Band intensity varied between samples depending on texture. Asymmetric deformations of C-H bonds in methoxy and methylene groups appeared at 1452 cm^{-1} to 1466 cm^{-1}, but displayed no change in intensity and position for both lignin types before and after etherification. Common weak bands at around 1367 cm^{-1} to 1378 cm^{-1} were found in all lignin samples; they originated from aliphatic C-H symmetric deformation in methyl groups and O-H deformation in phenolic groups. Interestingly, the band around 1328 cm^{-1} corresponding to the syringyl ring breathing vibration was present only in organosolv samples (Figure 1). This is probably due to the fact that Kraft lignin originated from spruce and the syringyl unit is absent from this source. The bands around 1264 cm^{-1} to 1269 cm^{-1} were assigned to the guaicyl ring breathing vibration and were present in all lignin samples; however, their intensity was higher in K lignin, confirming its provenance from spruce, which contains mainly guaicyl units.

Aromatic C-H in-plane deformation was observed at 1122 cm^{-1} to 1127 cm^{-1} in all lignin samples with different intensities. On the one hand, the small hump of C-O stretching from secondary alcohols and aliphatic ethers at 1082 cm^{-1} was found in all Kraft lignin samples. On the other hand, aromatic C-H in-plane deformations in G units along with C-O deformations in primary alcohols at 1034 cm^{-1} were found in all lignin samples. The same was true for C-H out-of-plane deformations corresponding to a weak band around 835 cm^{-1} to 855 cm^{-1}. Overall, no significant changes were observed in the region from 835 cm^{-1} to 1750 cm^{-1}, demonstrating that the core structure of lignin was maintained after mild etherification reaction.

NMR analysis based on ^{1}H and ^{13}C NMR spectra is a powerful tool for characterizing lignin functionalization. After etherification, lignin samples become highly hydrophobic in nature and thus highly soluble in low-polarity solvents, such as chloroform, diethylether. In this case CDCl$_3$ was used as NMR solvent due to high solubility of etherified lignin samples. In ^{1}H NMR spectra, the broad triplet signal around 0.87 ppm was assigned to methyl protons of C10–C16 alkyl chain units (Figure 3). The strong broad peaks around 1.2 ppm were assigned to all methylene protons (except α, β methylene protons) of C10–C16 alkyl chain units. The broad signal around 1.5–1.8 ppm was attributed to β methylene protons of C10–C16, whereas that of α methylene protons appeared around 3.8 ppm as a broad signal together with lignin methoxy groups. Aromatic and vinylic protons of lignin appeared around 6.1–7.0 ppm, although signal intensity was very low compared to aliphatic signals of C10–C16 chain units.

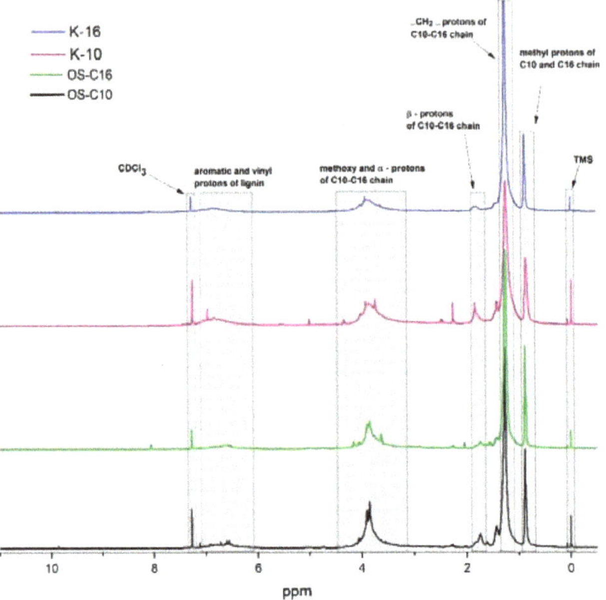

Figure 3. ^{1}H NMR spectra of functionalized organosolv (OS) and Kraft (K) lignins.

In ^{13}C NMR spectra, the methyl carbon of C10–C16 chain units appeared around 14.4 ppm (Figure 4). Signals from methylene carbons (except α, β methylene carbons) of C10–C16 alkyl chains appeared between 22.9–30 ppm. The signals between 32.1–32.2 ppm were assigned to the β carbon of C10–C16 aliphatic chain units, whereas that of the α carbon was found around 69 ppm. The methoxy carbons of lignin appeared around 56.6 ppm. Aromatic carbons of lignin showed a very low signal between 102–156 ppm compared to C10–C16 aliphatic carbons. NMR signals of etherified lignin samples reflected previously reported O-alkyl phenolic compounds and thus further confirmed the successful etherification of lignin [45,46], as suggested also by high solubility in low-polarity solvents, such as chloroform and diethylether.

Molecular weights of lignin samples before and after etherification were compared and are reported in Table 3. Kraft lignin displayed generally higher molecular weight compared to organosolv lignin and was accompanied by a high polydispersity index. In all cases, etherification increased the molecular weight as a result of the newly attached alkyl chains, further attesting to successful etherification of both lignin types. The increase in lignin molecular weight has been previously used to verify the success of esterification reactions [14,19,47]. Finally, etherification was seen to lower the polydispersity index, as observed for the esterification of lignin [19,47].

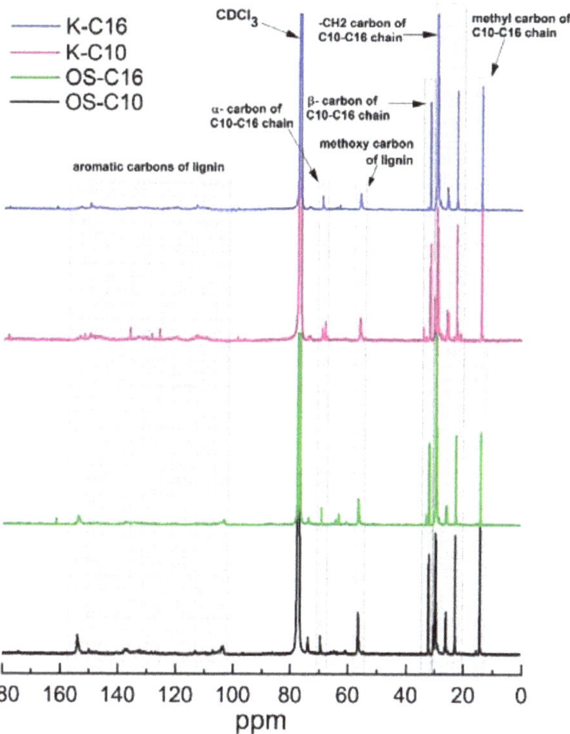

Figure 4. ^{13}C NMR spectra of functionalized organosolv (OS) and Kraft (K) lignins.

Table 3. Molecular weight values for non-functionalized organosolv (OS) and Kraft (K) lignin, as well as for their derivatives after etherification.

Lignin	Number Average Molecular Weight, Mn (Da)	Weight Average Molecular Weight, Mw (Da)	Polydispersity Index
OS	1757	4603	2.62
OS-C10	1859	3719	2.00
OS-C16	2661	6219	2.34
K	1924	7759	4.03
K-C10	3452	11,664	3.38
K-C16	3493	10,925	3.13

3.3. Thermal Characterization of Functionalized Lignins

Thermal stability is one of the main parameters for thermoplastic applications, which can be assessed by measuring TGA analysis. There are very few reports available on systematic thermal analysis data of alkylated lignins with higher alkyl chains. Previously Chen et al. reported on C12 alkylated lignin for PP blends and observed 10% weight loss at 200 °C [48]. Ramp TGA is a widespread method for the initial thermal stability screening of any compound. Short-term thermal stability based on $T_{wt\%}$ from ramped TGA data represents the fastest way to measure thermal stability of any materials [49]. Here, short-term thermal stability of $T_{0.5\%}$, $T_{1\%}$, and $T_{2\%}$ corresponded to the temperature at which a weight loss of 0.5%, 1%, and 2% was observed in lignin samples (Table 4).

Thermal stability of all lignin samples was analyzed by TGA with a heating ramp of 10 °C/min up to 800 °C under nitrogen atmosphere. In this study, two sets of lignin samples were analyzed by thermogravimetric method.

Table 4. $T_{wt\%}$ values of ether-functionalized organosolv (OS) and Kraft (K) lignin samples from the TGA plot.

Sample	$T_{0.5\%}$ (°C)	$T_{1\%}$ (°C)	$T_{2\%}$ (°C)
OS	50	65	143
OS-C10	174	193	213
OS-C16	135	140	148
K	54	66	133
K-C16	148	156	168
K-C10	75	88	104

In the first case, organosolv lignin and its alkylated thermoplastic lignins were analyzed. In our case degradation of native organosolv lignin was observed relatively easily, even at <80 °C, possibly due to internal radical coupling reactions [50]. Initial 2% weight loss of native organosolv lignin were observed around 143 °C. From the TGA plot (Figure 5), it is evident that thermal stability increased significantly after ether functionalization and was highest in case of organosolv lignin etherified with C10. $T_{0.5\%}$ of OS-C10 was observed around 174 °C; whereas OS-C16 reached 135 °C. The highest $T_{1\%}$ was observed at 193 °C for OS-C10; whereas $T_{1\%}$ for OS-C16 was 140 °C. Similarly, $T_{2\%}$ of OS-C10 was around 213 °C, which is substantially more than the values for OS-C16, which was 148 °C.

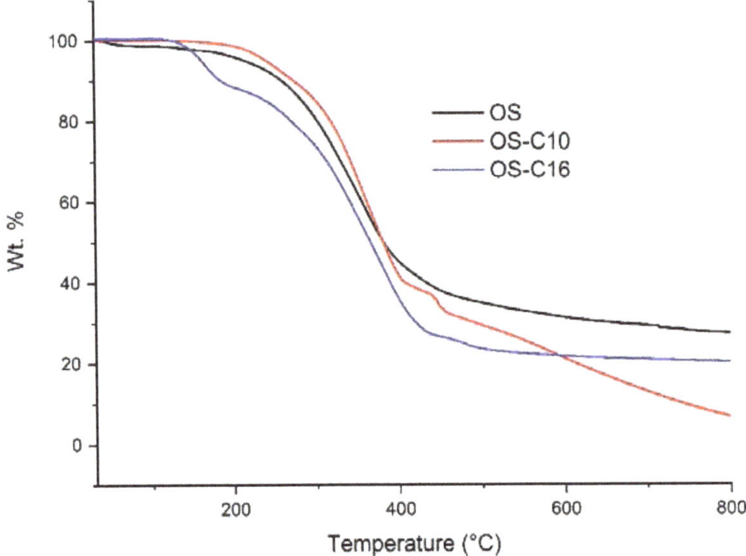

Figure 5. Thermal decomposition graphs obtained under nitrogen atmosphere of non-functionalized and etherified organosolv lignin.

On the other hand, the degradation of native Kraft lignin was also relatively facile like organosolv lignin and started even <80 °C. However, a 2% weight loss in native Kraft lignin was observed around 133 °C. Interestingly, etherified Kraft lignin also displayed good thermal stability (Figure 6), with the K-C16 sample exhibiting 0.5% decomposition at 148 °C, whereas K-C10 showed very poor thermal stability $T_{0.5\%}$ at 75 °C. Furthermore, $T_{1\%}$ of K-C16 and K-C10 were found to be 156 °C and 88 °C.

Although 2% degradation of K-C16 was comparatively high 168 °C but K-C10 showed lower thermal stability 104 °C than its native Kraft lignin 133 °C. Finally, non-functionalized lignins (especially the Kraft one) produced a higher amount of char at 800 °C compared to etherified lignins. This is probably due to formation of highly fused polycyclic aromatic hydrocarbon compounds via coupling of free phenolic hydroxyl groups.

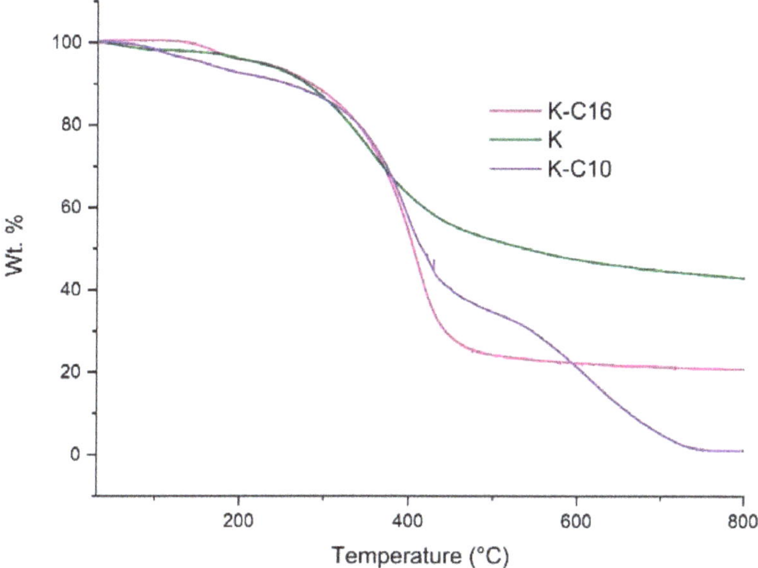

Figure 6. Thermal decomposition graphs obtained under nitrogen atmosphere of nonfunctionalized and etherified Kraft lignin.

Tg and melting temperature (Tm) are crucial for choosing the most appropriate candidate thermoplastic application and can be measured by DSC analysis. Tg of non-functionalized organosolv lignin was found to be very broad and centered at 117.1 °C and for commercial Kraft lignin is reported at 153 °C [51], although both lignin starts to decompose also at lower temperature. This finding indicated that non-functionalized lignin did not exhibit true Tg and could not be used in thermoplastic applications, as it would decompose during the thermoforming process. On the contrary, ether functionalization of organosolv lignin led to a sharp Tg peak and the detection of Tm (Figure 7). Importantly, controlling the alkyl chain length during etherification allowed the control of Tg, opening the opportunity for fine-tuning the thermal properties of the generated thermoplastics in accordance with the required application. During DSC analysis, samples were initially heated to 100 °C to form a thin film and then gradually cooled to −80 °C prior to starting up the heating/cooling cycles. During the cooling part, a small dip around 40–43 °C was observed for almost in all samples and it was ascribed to the endothermic peak of crystallization (Tc). As all lignin samples were highly amorphous, the endothermic peak of Tc was small and was observed only during the cooling cycle. This small endothermic peak of crystallization occurred possibly due to the introduction of alkyl chain into the lignin structure, which induces alkyl stacking in lignin thermoplastic [52–55]. Another endothermic peak, corresponding to Tg, was observed when lignins were cooled further to a lower temperature. OS-C10 presented the lowest Tg (−45.6 °C), while OS-C16 showed a sharp endothermic peak around 9 °C, further confirming how variations in alkyl chain length could serve to fine-tune the thermoplastic properties. Finally, the Kraft lignin K-C16 sample exhibited glass transition around −7.5 °C (Figure 7), whereas K-C10 failed to show any Tg throughout the heating or cooling cycle. This was probably due to its broad molecular weight distribution with high polydispersity index (3.38) compared to other lignin samples.

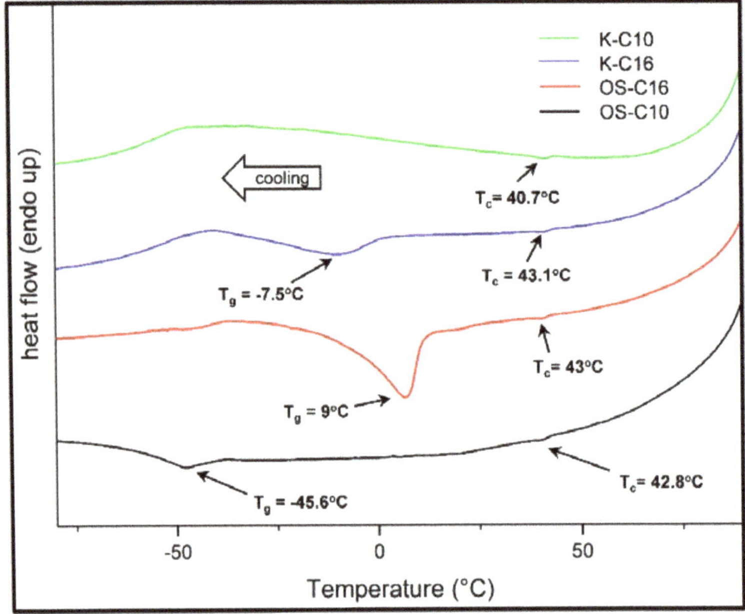

Figure 7. DSC analysis of functionalized lignins indicating glass transition (Tg) and crystallization (Tc) temperatures.

Based on the above, it was shown that lignin can serve as raw material for the production of thermoplastic. Utilization of all the biomass fractions towards the production of bio-based products is important for an economically viable biomass biorefinery. In this context, organosolv fractionation serves as an excellent option to deliver fractions of cellulose, hemicellulose and lignin that can be used at different applications. Recently, in a techno-economic analysis of a biomass biorefinery based on organosolv fractionation of birch and spruce biomass it was shown that the process can be economically viable for the production of ethanol from cellulose, when hemicellulose and lignin are also products [56]. As it was discussed previously, lignin is often underutilized and there is a need to further develop a process for its conversion to high added-value products that aim to provide additional profit to the process. In this context, the production of thermoplastic materials from lignin can serve as a promising alternative towards establishing economically viable biorefinery processes.

4. Conclusions

Valorization of lignin towards high added-value products is a requirement for establishing an economically viable biorefinery paradigm. Most traditional processes for the valorization of lignocellulosic biomass result in the production of lignin contaminated with sugars, ash or microbial cells. Organosolv fractionation can produce elevated yields of highly pure and practically ash-free lignin with minimal sugar contaminations. Here, we demonstrated the comparative functionalization characteristics of organosolv-isolated lignin as well as Kraft lignin. Organosolv lignin can also serve as a promising material towards production of green thermoplastics with tunable characteristics compared to commercial lignin.

Author Contributions: Conceptualization, S.B., L.M., U.R. and P.C.; methodology, S.B. and L.M.; investigation, S.B. and L.M.; resources, U.R. and P.C.; writing—original draft preparation, S.B. and L.M.; writing—review and editing, U.R. and P.C. All authors have read and agreed to the published version of the manuscript.

Funding: This research received no external funding.

Acknowledgments: The authors would like to thank Sveaskog, Sweden, for providing birch chips. L.M., U.R., and P.C. would like to thank Bio4Energy, a strategic research environment appointed by the Swedish government, for supporting this work.

Conflicts of Interest: The authors declare no conflict of interest.

References

1. Azadi, P.; Inderwildi, O.R.; Farnood, R.R.; King, D.A. Liquid fuels, hydrogen and chemicals from lignin: A critical review. *Renew. Sustain. Energy Rev.* **2013**, *21*, 506–523. [CrossRef]
2. Saito, T.; Perkins, J.H.; Jackson, D.; Trammel, N.E.; Hunt, M.A.; Naskar, A.K. Development of lignin-based polyurethane thermoplastics. *RSC Adv.* **2013**, *3*, 21832. [CrossRef]
3. Bajwa, D.; Pourhashem, G.; Ullah, A.; Bajwa, S. A concise review of current lignin production, applications, products and their environmental impact. *Ind. Crop. Prod.* **2019**, *139*, 111526. [CrossRef]
4. Hrůzová, K.; Matsakas, L.; Sand, A.; Rova, U.; Christakopoulos, P. Organosolv lignin hydrophobic micro- and nanoparticles as a low-carbon footprint biodegradable flotation collector in mineral flotation. *Bioresour. Technol.* **2020**, *306*, 123235. [CrossRef]
5. Gillet, S.; Aguedo, M.; Petitjean, L.; Morais, A.R.C.; Lopes, A.M.D.C.; Lukasik, R.M.; Anastas, P.T. Lignin transformations for high value applications: Towards targeted modifications using green chemistry. *Green Chem.* **2017**, *19*, 4200–4233. [CrossRef]
6. Mu, L.; Wu, J.; Matsakas, L.; Chen, M.; Rova, U.; Christakopoulos, P.; Zhu, J.; Shi, Y. Two important factors of selecting lignin as efficient lubricating additives in poly (ethylene glycol): Hydrogen bond and molecular weight. *Int. J. Boil. Macromol.* **2019**, *129*, 564–570. [CrossRef]
7. Beckham, G.T.; Johnson, C.W.; Karp, E.M.; Salvachúa, D.; Vardon, D.R. Opportunities and challenges in biological lignin valorization. *Curr. Opin. Biotechnol.* **2016**, *42*, 40–53. [CrossRef] [PubMed]
8. Matsakas, L.; Karnaouri, A.; Cwirzen, A.; Rova, U.; Christakopoulos, P. Formation of Lignin Nanoparticles by Combining Organosolv Pretreatment of Birch Biomass and Homogenization Processes. *Molecules* **2018**, *23*, 1822. [CrossRef] [PubMed]
9. Thoresen, P.P.; Matsakas, L.; Rova, U.; Christakopoulos, P. Recent advances in organosolv fractionation: Towards biomass fractionation technology of the future. *Bioresour. Technol.* **2020**, *306*, 123189. [CrossRef] [PubMed]
10. Kalogiannis, K.G.; Matsakas, L.; Aspden, J.; Lappas, A.; Rova, U.; Christakopoulos, P. Acid Assisted Organosolv Delignification of Beechwood and Pulp Conversion towards High Concentrated Cellulosic Ethanol via High Gravity Enzymatic Hydrolysis and Fermentation. *Molecules* **2018**, *23*, 1647. [CrossRef] [PubMed]
11. Rinaldi, R.; Jastrzebski, R.; Clough, M.T.; Ralph, J.; Kennema, M.; Bruijnincx, P.C.; Weckhuysen, B.M. Paving the Way for Lignin Valorisation: Recent Advances in Bioengineering, Biorefining and Catalysis. *Angew. Chem. Int. Ed.* **2016**, *55*, 8164–8215. [CrossRef] [PubMed]
12. Glasser, W.G. *Classification of Lignin According to Chemical and Molecular Structure*; American Chemical Society (ACS): Washington, DC, USA, 1999; Volume 742, pp. 216–238.
13. Wang, C.; Kelley, S.S.; Venditti, R. Lignin-Based Thermoplastic Materials. *ChemSusChem* **2016**, *9*, 770–783. [CrossRef] [PubMed]
14. Gordobil, O.; Robles, E.; Egüés, I.; Labidi, J. Lignin-ester derivatives as novel thermoplastic materials. *RSC Adv.* **2016**, *6*, 86909–86917. [CrossRef]
15. Sen, S.; Patil, S.; Argyropoulos, D.S. Thermal properties of lignin in copolymers, blends, and composites: A review. *Green Chem.* **2015**, *17*, 4862–4887. [CrossRef]
16. Ashori, A.; Sheshmani, S. Hybrid composites made from recycled materials: Moisture absorption and thickness swelling behavior. *Bioresour. Technol.* **2010**, *101*, 4717–4720. [CrossRef]
17. Laurichesse, S.; Averous, L. Chemical modification of lignins: Towards biobased polymers. *Prog. Polym. Sci.* **2014**, *39*, 1266–1290. [CrossRef]

18. Lewis, H.F.; Brauns, F.E. Esters of Lignin Material. U.S. Patent Application No. 2,429,102, 14 October 1944.

19. Glasser, W.G.; Jain, R.K. Lignin derivatives. I. Alkanoates. *Holzforschung* **1993**, *47*, 225–233. [CrossRef]

20. Guo, Z.-X.; Gandini, A.; Pla, F. Polyesters from lignin. 1. The reaction of Kraft lignin with dicarboxylic acid chlorides. *Polym. Int.* **1992**, *27*, 17–22. [CrossRef]

21. Guo, Z.-X.; Gandini, A. Polyesters from lignin—2. The copolyesterification of Kraft lignin and polyethylene glycols with dicarboxylic acid chlorides. *Eur. Polym. J.* **1991**, *27*, 1177–1180. [CrossRef]

22. Fang, R.; Cheng, X.-S.; Lin, W.-S. Preparation and application of dimer acid/lignin graft copolymer. *BioResources* **2011**, *6*, 2874–2884.

23. Korich, A.L.; Fleming, A.B.; Walker, A.R.; Wang, J.; Tang, C.; Iovine, P.M. Chemical modification of organosolv lignin using boronic acid-containing reagents. *Polymer* **2012**, *53*, 87–93. [CrossRef]

24. Han, Y.; Yuan, L.; Li, G.; Huang, L.; Qin, T.; Chu, F.; Tang, C. Renewable polymers from lignin via copper-free thermal click chemistry. *Polymer* **2016**, *83*, 92–100. [CrossRef]

25. De Oliveira, W.; Glasser, W.G. Multiphase materials with lignin. 11. Starlike copolymers with caprolactone. *Macromolecules* **1994**, *27*, 5–11. [CrossRef]

26. Chung, H.; Al-Khouja, A.; Washburn, N.R. Lignin-Based Graft CopolymersviaATRP and Click Chemistry. In *Green Polymer Chemistry: Biocatalysis and Materials II*; American Chemical Society (ACS): Washington, DC, USA, 2013; Volume 1144, pp. 373–391.

27. Kim, Y.S.; Kadla, J.F. Preparation of a Thermoresponsive Lignin-Based Biomaterial through Atom Transfer Radical Polymerization. *Biomacromolecules* **2010**, *11*, 981–988. [CrossRef]

28. Hilburg, S.L.; Elder, A.N.; Chung, H.; Ferebee, R.; Bockstaller, M.; Washburn, N.R. A universal route towards thermoplastic lignin composites with improved mechanical properties. *Polymer* **2014**, *55*, 995–1003. [CrossRef]

29. Sadeghifar, H.; Cui, C.; Argyropoulos, D.S. Toward Thermoplastic Lignin Polymers. Part 1. Selective Masking of Phenolic Hydroxyl Groups in Kraft Lignins via Methylation and Oxypropylation Chemistries. *Ind. Eng. Chem. Res.* **2012**, *51*, 16713–16720. [CrossRef]

30. Glasser, W.G.; Barnett, C.A.; Rials, T.G.; Saraf, V.P. Engineering plastics from lignin II. Characterization of hydroxyalkyl lignin derivatives. *J. Appl. Polym. Sci.* **1984**, *29*, 1815–1830. [CrossRef]

31. Kazzaz, A.E.; Feizi, Z.H.; Fatehi, P. Grafting strategies for hydroxy groups of lignin for producing materials. *Green Chem.* **2019**, *21*, 5714–5752. [CrossRef]

32. Matsakas, L.; Nitsos, C.; Raghavendran, V.; Yakimenko, O.; Persson, G.; Olsson, E.; Rova, U.; Olsson, L.; Christakopoulos, P. A novel hybrid organosolv: Steam explosion method for the efficient fractionation and pretreatment of birch biomass. *Biotechnol. Biofuels* **2018**, *11*, 160. [CrossRef]

33. Meng, X.; Crestini, C.; Ben, H.; Hao, N.; Pu, Y.; Ragauskas, A.J.; Argyropoulos, D.S. Determination of hydroxyl groups in biorefinery resources via quantitative 31P NMR spectroscopy. *Nat. Protoc.* **2019**, *14*, 2627–2647. [CrossRef] [PubMed]

34. Lange, H.; Rulli, F.; Crestini, C. Gel Permeation Chromatography in Determining Molecular Weights of Lignins: Critical Aspects Revisited for Improved Utility in the Development of Novel Materials. *ACS Sustain. Chem. Eng.* **2016**, *4*, 5167–5180. [CrossRef]

35. Ren, W.; Pan, X.; Wang, G.; Cheng, W.; Liu, Y. Dodecylated lignin-g-PLA for effective toughening of PLA. *Green Chem.* **2016**, *18*, 5008–5014. [CrossRef]

36. Hergert, H.L. Infrared Spectra of Lignin and Related Compounds. II. Conifer Lignin and Model Compounds 1,2. *J. Org. Chem.* **1960**, *25*, 405–413. [CrossRef]

37. Faix, O. Classification of Lignins from Different Botanical Origins by FT-IR Spectroscopy. *Holzforschung* **1991**, *45*, 21–28. [CrossRef]

38. Boeriu, C.G.; Bravo, D.; Gosselink, R.J.; Van Dam, J.E. Characterisation of structure-dependent functional properties of lignin with infrared spectroscopy. *Ind. Crop. Prod.* **2004**, *20*, 205–218. [CrossRef]

39. Derkacheva, O.Y.; Sukhov, D. Investigation of Lignins by FTIR Spectroscopy. *Macromol. Symp.* **2008**, *265*, 61–68. [CrossRef]

40. Gabov, K.; Gosselink, R.J.A.; Smeds, A.I.; Fardim, P. Characterization of Lignin Extracted from Birch Wood by a Modified Hydrotropic Process. *J. Agric. Food Chem.* **2014**, *62*, 10759–10767. [CrossRef]

41. Zhang, L.; Yan, L.; Wang, Z.; Laskar, D.D.; Swita, M.S.; Cort, J.R.; Yang, B. Characterization of lignin derived from water-only and dilute acid flowthrough pretreatment of poplar wood at elevated temperatures. *Biotechnol. Biofuels* **2015**, *8*, 203. [CrossRef]

42. Todorciuc, T.; Căpraru, A.-M.; Kratochvílová, I.; Popa, V.I. Characterization of non-wood lignin and its hydoxymethylated derivatives by spectroscopy and self-assembling investigations. *Cellul. Chem. Technol.* **2009**, *43*, 399–408.

43. Sammons, R.J.; Harper, D.P.; Labbé, N.; Bozell, J.J.; Elder, T.; Rials, T.G. Characterization of Organosolv Lignins using Thermal and FT-IR Spectroscopic Analysis. *BioResources* **2013**, *8*, 2752–2767. [CrossRef]

44. Hergert, H.L. Infrared spectra. In *Lignins: Occurrence, Formation, Structures and Reactions*; Sarkanen, K.V., Hergert, H.L., Eds.; John Wiley & Sons, Inc.: New York, NY, USA, 1971; pp. 267–297.

45. Swapna, K.; Murthy, S.N.; Jyothi, M.T.; Nageswar, Y.V.D. Recyclable heterogeneous copper oxide on alumina catalyzed coupling of phenols and alcohols with aryl halides under ligand-free conditions. *Org. Biomol. Chem.* **2011**, *9*, 5978–5988. [CrossRef] [PubMed]

46. Jammi, S.; Sakthivel, S.; Rout, L.; Mukherjee, T.; Mandal, S.; Mitra, R.; Saha, P.; Punniyamurthy, T. ChemInform Abstract: CuO Nanoparticles Catalyzed C-N, C-O, and C-S Cross-Coupling Reactions: Scope and Mechanism. *ChemInform* **2009**, *40*, 1971–1976. [CrossRef]

47. Hult, E.-L.; Koivu, K.; Asikkala, J.; Ropponen, J.; Wrigstedt, P.; Sipilä, J.; Poppius-Levlin, K. Esterified lignin coating as water vapor and oxygen barrier for fiber-based packaging. *Holzforschung* **2013**, *67*, 899–905. [CrossRef]

48. Chen, F.; Dai, H.; Dong, X.; Yang, J.; Zhong, M. Physical properties of lignin-based polypropylene blends. *Polym. Compos.* **2011**, *32*, 1019–1025. [CrossRef]

49. Bhattacharyya, S.; Shah, F.U. Thermal stability of choline based amino acid ionic liquids. *J. Mol. Liq.* **2018**, *266*, 597–602. [CrossRef]

50. Durbeej, B.; Eriksson, L. A Density Functional Theory Study of Coniferyl Alcohol Intermonomeric Cross Linkages in Lignin—Three-Dimensional Structures, Stabilities and the Thermodynamic Control Hypothesis. *Holzforschung* **2003**, *57*, 150–164. [CrossRef]

51. An, Y.-X.; Li, N.; Wu, H.; Lou, W.-Y.; Zong, M.-H. Changes in the Structure and the Thermal Properties of Kraft Lignin during Its Dissolution in Cholinium Ionic Liquids. *ACS Sustain. Chem. Eng.* **2015**, *3*, 2951–2958. [CrossRef]

52. Cao, Z.; Galuska, L.; Qian, Z.; Zhang, S.; Huang, L.; Prine, N.; Li, T.; He, Y.; Hong, K.; Gu, X. The effect of side-chain branch position on the thermal properties of poly(3-alkylthiophenes). *Polym. Chem.* **2020**, *11*, 517–526. [CrossRef]

53. Prasad, S.; Jiang, Z.; Sinha, S.K.; Dhinojwala, A. Partial Crystallinity in Alkyl Side Chain Polymers Dictates Surface Freezing. *Phys. Rev. Lett.* **2008**, *101*, 065505. [CrossRef]

54. Jones, A.T. Crystallinity in isotactic polyolefins with unbranched side chains. *Makromol. Chem.* **1964**, *71*, 1–32. [CrossRef]

55. Cowie, J.M.G.; Reid, V.M.C.; McEwen, I.J. Effect of side chain length on the glass transition of copolymers from styrene with n -alkyl citraconimides and with n -alkyl itaconimides. *Br. Polym. J.* **1990**, *23*, 353–357. [CrossRef]

56. Mesfun, S.; Matsakas, L.; Rova, U.; Christakopoulos, P. Technoeconomic Assessment of Hybrid Organosolv– Steam Explosion Pretreatment of Woody Biomass. *Energies* **2019**, *12*, 4206. [CrossRef]

Article

Structural and Thermal Characterization of Novel Organosolv Lignins from Wood and Herbaceous Sources

Anna Trubetskaya [1,*,†], Heiko Lange [2,*,†], Bernd Wittgens [3], Anders Brunsvik [3], Claudia Crestini [4], Ulrika Rova [5], Paul Christakopoulos [5], J. J. Leahy [1] and Leonidas Matsakas [5,*,†]

[1] Bernal Center, University of Limerick, V94 T9PX Castletroy, Ireland; J.J.Leahy@ul.ie
[2] Department of Pharmacy, University of Naples 'Federico II', 80131 Naples, Italy
[3] SINTEF Industry, Richard Birkelands vei 2B, 7034 Trondheim, Norway; Bernd.Wittgens@sintef.no (B.W.); Anders.Brunsvik@sintef.no (A.B.)
[4] Department of Molecular Science and Nanosystems, University of Venice Ca' Foscari, Via Torino 155, 30170 Venice Mestre, Italy; claudia.crestini@unive.it
[5] Biochemical Process Engineering, Division of Chemical Engineering, Department of Civil, Environmental and Natural Resources Engineering, Luleå University of Technology, 97187 Luleå, Sweden; Ulrika.Rova@ltu.se (U.R.); paul.christakopoulos@ltu.se (P.C.)
* Correspondence: atrubetskaya@gmx.de (A.T.); heiko.lange@unina.it (H.L.); leonidas.matsakas@ltu.se (L.M.)
† These authors contributed equally to this work.

Received: 26 June 2020; Accepted: 13 July 2020; Published: 17 July 2020

Abstract: This study demonstrates the effects of structural variations of lignins isolated via an organosolv process from different woody and herbaceous feedstocks on their thermal stability profiles. The organosolv lignins were first analysed for impurities, and structural features were determined using the default set of gel permeation chromatography, FT-IR spectroscopy, quantitative ^{31}P NMR spectroscopy and semi-quantitative ^{1}H-^{13}C HSQC analysis. Pyrolysis-, O_2- and CO_2-reactivity of the organosolv lignins were investigated by thermogravimetric analysis (TGA), and volatile formation in various heating cycles was mapped by head-space GC-MS analysis. Revealed reactivities were correlated to the presence of identified impurities and structural features typical for the organosolv lignins. Data suggest that thermogravimetric analysis can eventually be used to delineate a lignin character when basic information regarding its isolation method is available.

Keywords: lignin; organosolv fractionation; TGA; ^{31}P NMR; HSQC; heat treatment

---

1. Introduction

Lignocellulosic feedstocks hold a potential for large-scale production of second-generation biofuels leading to a decarbonization of the energy sector [1]. The fractionation of biomass into cellulose, hemicellulose, lignin and extractives provides valuable feedstocks for the energy sector and the chemical industries [2]. While cellulose and hemicellulose represent structurally regular polymers, extractives and lignin represent a more difficult starting material for valorisation. In case of lignin, not only the (presumably) uncontrolled biosynthesis leading to random polymer linkages, but also the biomass-dependent distribution of monomeric building blocks mark challenging aspects. The relative abundance of these monomers within a lignin polymer leads to the common differentiation between lignins isolated from softwood, hardwood or herbaceous biomass [3]. The polymeric lignin structure present in the plants is more resistant to most forms of biological attacks, sunlight and temperature changes compared to polysaccharides, contributing thus to the industrial challenges faced in lignin valorisation [4]. Once separated from the other biopolymers, lignin has nevertheless been utilized as a primary feedstock for composites, PU-based foams, films, paints, and plastics[5]. From a different technical point of view, lignin rich feedstocks present challenges for gas cleaning units of gasification reactors as high concentration of lignin may increase soot yields in addition to increasing formation of PAH precursors[6]. Thus, removal of lignin from biomass is important to prevent outlet blockages and to ensure steady syngas production increasing overall efficiency of the gasification process.

When it comes to the production of biofuels from low cost lignocellulosic biomass, such as agricultural or forestry residues, a typical process consists of biomass pretreatment, enzymatic saccharification of the carbohydrates and microbial conversion of the sugars to biofuels[7]. The main aim of pretreatment is to efficiently remove hemicellulose and lignin from cellulose and increase the susceptibility of cellulose to enzymatic hydrolysis[8]. Removal of lignin from cellulose is particularly important as lignin has a negative influence on enzymatic saccharification due to the irreversible adsorption of cellulolytic enzymes onto lignin and their inhibition from soluble lignin-derived molecules[9]. Hydrothermal pretreatment is one of the most common methods to pretreat lignocellulosic biomass, which effectively degrades hemicelluloses, even without chemicals, increases the biomass porosity, thus enhancing the enzymatic hydrolysis of the pre-treated biomass[10]. The common challenge of hydrothermal pretreatment is that lignin cannot be directly removed and hence partly rearranges on the surface of the lignocellulosic biomass, causing an inhibitory effect on the saccharification process[11,12]. Organosolv pretreatment is known as an effective method to fractionate biomass into cellulose, hemicellulose and lignin streams by using aqueous-organic solvent mixtures, with high solvent concentration (30–70%) at temperatures of 100–220 °C, with or without the addition of catalysts[13,14]. One of the main benefits of organosolv pretreatment is the isolation of high-quality lignin and cellulose fractions[15,16]. Another two advantages of the organosolv process are related to the relative easy recovery and re-use of the commonly used organic solvents (such as ethanol or acetone) and improved mass transfer and dissolution of lignin in the presence of an organic solvent[17–19]. Previous research showed that organosolv pulps have bleachability and viscosity retention which are comparable to those of cellulose soda and kraft pulps[20]. Most studies[21–23] investigated the effect of feedstock on the chemical properties of lignocellulosic fractions from the organosolv wood pre-treatment, whereas the chemical properties and reactivity of lignocellulosic fractions after organosolv treatment of herbaceous biomass are rarely studied in the literature.

In a previous work[7], the efficiency of organosolv pretreatment on wood at different operating conditions such as type of solvent, lack or presence of homogeneous catalyst, and type of homogeneous acid catalyst was investigated. The aim of the present study was to evaluate the properties of lignins isolated from various feedstocks via the previously described novel organosolv processes[7,9]. Structural and physicochemical characteristics of five different lignins were elucidated using established analysis methods: Gel permeation chromatography for molecular mass features, quantitative ^{31}P NMR and comparative two-dimensional ^{1}H-^{13}C HSQC analyses for more detailed structural aspects. Another aim of this work was to investigate to which extent the structural differences would be reflected in gas chromatographic analysis for determining initially present volatiles and inducible volatile contents, as well as in thermogravimetric analysis for delineating polymer characteristics and lignin reactivity in different atmospheres[12]. Structural information obtained by the various methods was correlated to findings in head-space GC-MS and thermal analyses. The presented analyses offer a solid base on which a specific lignin could be chosen for a value-added application.

2. Materials and Methods

2.1. Raw Materials

In the present study, both wood (pine sawdust and spruce bark) and herbaceous biomass (cotton stalks and sweet sorghum bagasse) were used as raw materials. Pine sawdust and spruce bark were obtained from mills from Northern Sweden, cotton stalks were obtained from fields in Thessaly (Greece), and sweet sorghum (Keller cultivar) was obtained from fields in the Kopaida plain (Central Greece). The materials, apart from sweet sorghum, were immediately air-dried upon receive, milled to particle size $<$ 1 mm and stored at room temperature. For the preparation of sweet sorghum bagasse, the sugars were extracted from the sweet sorghum stalks as previously described [24] and bagasse was dried, milled and stored at room temperature. Finally, spruce bark was treated in hot-water extraction (75 °C for 2 h) with the addition of sodium bisulfite (2 % w/w$_{biomass}$) and sodium carbonate (0.5 % w/w$_{biomass}$) to remove the tannin and other water soluble extractives [25]. After hot-water extraction, the tannin-extracted bark solids were removed from the process by vacuum filtration, washed with water and air dried.

2.2. Organosolv Fractionation

Pine sawdust and cotton stalks were treated in a hybrid organosolv: steam explosion reactor which was previously described [9], whereas sweet sorghum bagasse and tannin-extracted spruce bark were treated in an autoclave organosolv reactor, as reported previously [7]. The organosolv treatment conditions for the different feedstock were as follows: pine sawdust was treated at 190 °C for 60 min in a 60 % v/v ethanol solution with the addition of 1 % w/w$_{biomass}$ sulfuric acid; cotton stalks were treated at 200° for 45 min in a 50 % v/v ethanol solution with the addition of 1 % w/w$_{biomass}$ sulfuric acid; sweet sorghum bagasse was treated at 180 °C for 30 min in a 60 % v/v ethanol solution; tannin-extracted bark was treated at 180 °C for 1 h with 60 % v/v ethanol content, with and without the use of 1 % w/w$_{biomass}$ sulfuric acid. At the end of the organosolv treatment, the pretreated solids were separated from the liquor by vacuum filtration. Lignin was recovered from the liquor by centrifugation after ethanol removal in a rotary evaporator. Finally, lignin was air-dried and stored at room temperature. In the manuscript, the abbreviations of lignin fractionated from pine sawdust (PL), cotton stalks (CL), sweet sorghum bagasse (SSL), spruce bark with (SBAL) and without acid addition (SBNL) were used.

2.3. Lignin Characterization

2.3.1. Elemental Analysis

The elemental analysis was performed on two instruments of the same model (Eurovector, model EA3000). Acetanilide was used as a reference standard. The ash content was determined using a standard ash test at 550 °C, according to the procedure described in DIN EN 14775.

2.3.2. Headspace Gas Chromatography-Mass Spectroscopy

Approximately 20 mg of a lignin sample was accurately weighed and directly sealed into a 20 mL headspace vial. Headspace gas chromatography-mass spectroscopy (HSGC-MS) analysis was performed using an Agilent 7694E Headspace sampler (Agilent Technologies, Santa Clara, CA, US), connected to an Agilent 7890B series gas chromatograph coupled with an Agilent 5977A series mass spectrometer and equipped with a HP-5MS Agilent column (0.25 mm $\times$ 30 m $\times$ 0.25 μm). The headspace operating conditions were as follows: The equilibration time was 20 min; the headspace oven, loop, and transfer line temperatures were 110, 180 and 270 °C; 130, 200 and 280 °C; and 130, 200 and 280 °C respectively; the shaking time was 2 min at low intensity; the injecting time was 2 min. GC operating conditions were as follows: the carrier gas (helium) was set at a flow rate of 1.0 mL min^{-1} with the split ratio was 5:1; the column temperature program was initially set at 50 °C for 1 min, and was gradually increased to 100 °C at 3 °C min^{-1}, then kept for 3 min before being gradually increased to 110, 180, or 270 °C at 10 °C min^{-1}; for MS detection, an electron ionization (EI) system was used with the ionization energy at 70 eV; the temperature of the ion source and the quadrupole temperature was 230 and 150 °C, respectively; the mass range was 50–550 amu in the full-scan acquisition mode with 3 min of solvent delay. The HSGC-MS collected data were processed by MassHunter Qualitative Analysis B.06.00 for the peak deconvolution. The mass spectra with the well-resolved overlapping peaks were imported into the mass spectra library software NIST MS Search 2.3 [26].

2.3.3. Molecular Weight Determination of Lignin

Lignin molecular weight and molecular weight distribution were determined by gel permeation chromatography (GPC) as described previously [27] after being acetobrominated according to the protocol proposed by Asikkala et al. [28]. More specifically, approximately 5 mg of lignin was initially reacted in a 1 mL solution of 9:1 v/v of glacial acetic acid/acetyl bromide (Sigma Aldrich, St. Louis, MO, USA) for 2 h, followed by removal of the solvent in a rotary evaporator. The sample was then dissolved in THF (tetrahydrofuran; VWR Chemicals, Radnor, PA, USA), followed by solvent removal. This process was done twice, to ensure that glacial acetic acid and acetyl bromide were properly removed. Finally, the sample was dissolved in 1 mL of THF and used for the determination of the molecular weight. GPC analysis was done in an HPLC apparatus equipped with a UV detector (PerkinElmer Flexar, Watham, MA, USA) and a Styragel HR 4E (Waters; Miford, MA, USA) chromatographic column. THF was used as mobile phase at 0.6 mL min^{-1} flow rate, and the column was kept at 40 °C. The UV detector was set at 280 nm. The calibration curves were prepared by using polystyrene standards (Sigma Aldrich). Molecular masses calculated based on the calibration were rounded to the full hundreds.

2.3.4. FTIR Spectroscopy

The lignin samples were analyzed by a Cary 630 FTIR spectrometer (Agilent Technologies, Santa Clara, CA, USA). All absorption spectra were obtained in the 4000–600 cm^{-1} range by 100 scans at 4 cm^{-1} resolution. For background, 200 scans were acquired. Good contact between sample and ATR-crystal surface was ensured before all measurements. All samples were measured in triplicate.

2.3.5. Quantitative ^{31}P NMR Analysis

In general, a procedure similar to the one originally published and previously applied was used [29]. Approximately 30 mg of the lignin were accurately weighed for analysis in a volumetric flask and suspended in 400 μL of a solvent mixture of pyridine and deuterated chloroform (dCDCl$_3$) (1.6:1, v/v) the above prepared solvent solution. One hundred microliters of the internal standard solution, i.e., cholesterol at a concentration of 0.1 M in the aforementioned NMR solvent mixture, were added. 50 mg of Cr(III) acetyl acetonate were added as relaxation agent to this solution, followed by 100 μL of 2-chloro-4,4,5,5-tetramethyl-1,3,2-dioxa-phospholane (Cl-TMDP). After stirring for 120 min at ambient temperature, ^{31}P NMR spectra are recorded on a Bruker 400 MHz NMR spectrometer controlled by TopSpin software, using an inverse gated decoupling technique with the probe temperature set to 20 °C. The maximum standard deviation of the reported data is 0.02 mmoL g^{-1}, while the maximum standard error is 0.01 mmoL g^{-1}. NMR data were processed with MestreNova (Version 8.1.1, Mestrelab Research, Santiego de Compostela, Spain). Technical loadings are determined by comparing the abundancies of total aromatic hydroxyl groups of the product lignin with the starting lignin.

2.3.6. Qualitative ^{1}H-^{13}C HSQC Analysis

Samples of around 90 mg were dissolved in 600 μL DMSO-d6; chromium acetyl acetonate was added as spin-relaxing agent at a final concentration of ca. 1.5–1.75 mg mL^{-1}. HSQC spectra were recorded at 27 °C on a Bruker 400 MHz instrument equipped with TopSpin 2.1 software. ^{1}H-^{13}C HSQC spectra were obtained applying the following parameters for acquisition: TD = 2048 (F2), 512 (F1). The Bruker hsqcetgp pulse program in DQD acquisition mode was used, with NS = 32; TD = 2048 (F2), 512 (F1); SW = 15.0191 ppm (F2), 149.9819 ppm (F1); O2 (F2) = 2000.65 Hz, O1 (F1) = 7545.96 Hz; D1 = 2 s; CNST2 (1J(C-H) = 145; acquisition time F2 channel = 85.1968 ms, F1 channel = 8.4818 ms; pulse length of the 90° high power pulse P1 was optimised for each sample. NMR data were processed with MestreNova; spectra were referenced to the residual signals of DMSO-d6 (2.49 ppm for ^{1}H and 39.5 ppm for ^{13}C domain, respectively).

2.3.7. Thermogravimetric Analysis

The lignin samples were firstly crushed to a fine powder in a mortar with a ceramic pestle. The thermal decomposition of lignin samples was determined using an atmospheric thermogravimetric instrument (Mettler Toledo, Columbus, OH, USA). The pyrolysis of lignin samples was investigated in 100 % volume fraction N_2 (100 cm^3 min^{-1} of N_2 measured at 20 °C and 101.3 kPa). The reactivity of lignin samples in 20 % volume fraction CO_2 (20 cm^3 min^{-1} of CO_2 and 80 cm^3 min^{-1} of N_2 measured at 20 °C and 101.3 kPa) and 5 % volume fraction O_2 (5 cm^3 min^{-1} of O_2 and 95 cm^3 min^{-1} of N_2 measured at 20 °C and 101.3 kPa) was determined by loading 5 mg of sample in an Al_2O_3 crucible. The lignin samples were firstly heated up to 110 °C and kept for 30 min isothermally for drying. The dried samples were subsequently heated to 1100 °C at a constant heating rate of 10 °C min^{-1}. All measurements were conducted in duplicate to verify sufficient reproducibility.

3. Results and Discussion

3.1. Ultimate and Proximate Analysis

The ultimate and proximate analysis of fractionated lignin was carried out and the results are shown in Table 1. Organosolv lignin fractions were free from sulphur, confirming previous results [30]. Table 1 shows carbon (C), oxygen (O), hydrogen (H), nitrogen (N) and sulphur (S) contents of all lignins according to CHNS analyzer. The carbon content of lignin from sweet sorghum bagasse was slightly lower than that of lignin samples from other feedstocks. Nitrogen content found in lignin reflects contamination by residues of proteins or industrial fertilisers, especially in the case of the annual plant-derived sorghum and cotton lignins. In most lignins except pine sawdust lignin, nitrogen amount is in the range of 0.4–1.1 wt. % [31,32].

Table 1. Proximate and ultimate analysis of lignin that was fractionated from cotton, sweet sorghum, pine sawdust, spruce bark (with and without the acid catalyst). The abbreviations of lignin fractionated from pine sawdust (PL), cotton stalks (CL), sweet sorghum bagasse (SSL), spruce bark with (SBAL) and without acid addition (SBNL) were used. The standard error for all measurements was <10 % of the value.

Properties	PL	CL	SSL	SBAL	SBNL
Proximate and ultimate analysis (% on dry basis)					
Moisture [a]	2.1	0.1	0.4	0.1	0.2
C	65.2	64.9	61.9	66.5	64.6
H	6.3	7.0	5.9	6.7	6.4
O	28.0	26.3	31.0	25.8	27.8
N	0.4	1.7	1.1	0.9	1.1
S	0.1	0.1	0.1	0.1	0.1

[a] wt. % (as received).

Table 2 shows the results of analyses of impurities, cellulose and hemicellulose sugars and ash in the various lignins, determined according to the NREL technical report [33]. The results indicate that lignin samples are of high purity, exhibiting only low ash and carbohydrate contents. Specifically, the highest carbohydrate impurities were observed in SSL and reached 7 % *w/w*, whereas for the rest of lignins it was less than 2.3 % *w/w*. The ash content in all lignin samples was less than 0.6 % *w/w*, with most of the samples to be below 0.3 % *w/w* (Table 2).

Table 2. Composition in impurities ($\%$, w/w) and GPC analysis of the lignin samples. The abbreviations of lignin fractionated from pine sawdust (PL), cotton stalks (CL), sweet sorghum bagasse (SSL), spruce bark with (SBAL) and without acid addition (SBNL) were used. The standard error for all measurements was $<10\%$ of the value.

Feedstock	Cellulose	Hemicellulose	Ash	M_n	M_w	PDI
	%, *w/w*			Da		
PL	0.2	0.9	0.2	1900	7700	4.4
CL	0.3	0.6	0.2	3400	16,800	4.9
SSL	4.1	2.9	0.3	1600	6600	4.1
SBAL	1.5	0.7	0.6	1600	10,600	6.6
SBNL	1.5	0.6	0.2	1600	8900	5.6

3.2. GPC Analysis

The molecular weight of lignin samples is given in Table 2. The lignin number average molecular weight (M_n) varied from 1600 to 3400 Da with the highest weight for the cotton stalk lignin. The size exclusion analysis showed a broad variation of the weight average molecular weight (M_w) from 6600 to 16,800 Da, emphasizing eventually the effect of feedstock and the structural characteristics between the lignins of the various feedstocks in combination with the fractionation protocol applied. In general, the molecular weight found for the lignin samples can be attributed to the milder reaction conditions of the organosolv treatment used in the present study. Overall, lower number average molecular weight samples exhibited also a lower polydispersity, corresponding to previous results [34–37].

3.3. FTIR Analysis

FTIR analysis was conducted to investigate the differences in main functional groups and monomer composition in the tested lignin samples, as shown in Figure 1; band assignments are summarized in the Appendix A (Table A1).

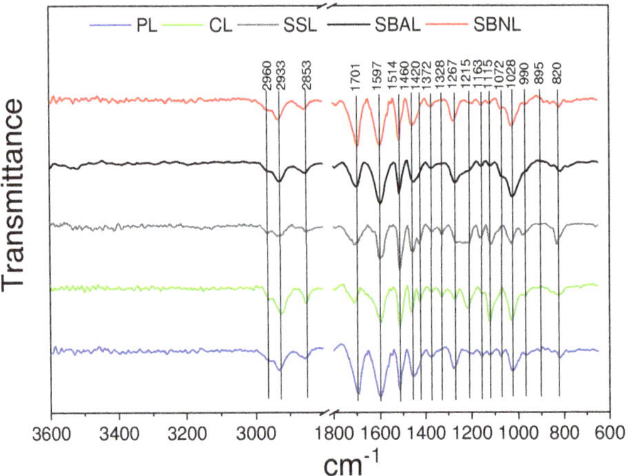

Figure 1. Experimental IR spectra of lignin samples. The abbreviations of lignin fractionated from pine sawdust (PL), cotton stalks (CL), sweet sorghum bagasse (SSL), spruce bark with (SBAL) and without acid addition (SBNL) were used.

The bands at 1514 and 1597 cm^{-1} were significantly pronounced in all lignin samples and represent the aromatic skeletal vibrations [38,39]. All organosolv lignins from the different feedstocks presented the vibration of C-H stretching in $-CH_2-$ and $-CH_3$ group at 2960–2933 and 2853 cm^{-1}, respectively, as well as the C=O group stretching of carbonyl groups of α-oxidized structural motifs with bands at 1694–1701 cm^{-1}. Aromatic skeletal vibrations at 1514 cm^{-1}, C-H deformations in CH_2 and CH_3 group (1450–1460 cm^{-1}), aliphatic C-H stretch in CH_3, not in OCH_3 (1372 cm^{-1}), G ring breathing (1267–1272 cm^{-1}), and aromatic C-H in plane deformation (G > S) (1028 cm^{-1}) can be observed in all lignin samples with different intensities. The intensity of the guaiacyl unit in CL and SSL was represented by the peak at C-C and C-O stretch. FTIR absorptions indicative of syringyl units were found to be stronger in CL and SSL, as one would expect based on the type of feedstock.

3.4. ^{31}P NMR and ^{1}H-^{13}C HSQC Analysis

Quantitative ^{31}P NMR analyses of the lignins under study reveal structural differences and dominating types of aromatic structures given the starting biomasses. Results are summarized in Table 3.

Table 3. ^{31}P NMR analysis of lignin samples. The abbreviations of lignin fractionated from pine sawdust (PL), cotton stalks (CL), sweet sorghum bagasse (SSL), spruce bark with (SBAL) and without acid addition (SBNL) were used.

Lignin	Aliph OH	Aromatic OH				Acidic	Total	Arom/Aliph
		Cond	G	p-OH	Total	OH	OH	OH
	$mmoL\ g^{-1}$	$mmoL\ g^{-1}$				$mmoL\ g^{-1}$	$mmoL\ g^{-1}$	
PL	1.41	0.32	0.58	0.06	0.96	0.16	2.37	0.68
CL	1.06	0.46	0.41	0.14	1.00	0.26	2.06	0.95
SSL	1.64	0.55	0.39	0.45	1.39	0.27	3.03	0.85
SBAL	0.83	0.39	0.45	0.36	1.21	0.32	2.04	1.46
SBNL	0.87	0.25	0.43	0.32	1.01	0.31	1.88	1.16

Noteworthy, only the two bark lignin samples exhibit more phenolic OH than aliphatic OH, with a significant difference in the SBAL sample. The positive effect of the presence of the acid catalyst with respect to generating free phenolics is obvious in this case. The higher amounts of S-type-phenolics found for CL and SSL reflect the presence of the syringyl monomers in the structure, while overall notable presence of condensed units suggests that the treatment might cause, to a low extend, intramolecular condensations.

HSQC spectra of lignins were acquired using fixed, standardised conditions in terms of concentrations of samples and acquisition parameters to allow a comparative analysis without quantification. Non-acetylated samples were dissolved in DMSO for analysis. Results are summarized in Table 4. HSQC-analyses essentially confirm structural differences of the lignins as it can be expected based on the starting biomasses. In terms of monomer composition, PL which has been chosen as a base for its essentially 'pure G' character compared to the other samples, shows the typical G-units as essentially only monomer type present, and the presence of typical interunit binding motifs. Cross peaks indicative for α-oxidized β-O-4' motifs are present. Cinnamyl alcohol and aldehyde are detectable end motifs in PL. The overall intensities of cross-peaks in the lower aliphatic region, eventually attributable to the presence of further extractable aliphatic impurities and aliphatic end groups of various nature are comparable to those found for CL and SSL, but inferior to those of the bark samples. Traces of hemicellulose residues, especially xylan residues, in the lignin are detectable, signs indicative of lignin-carbohydrate complexes (LCCs) in PL are de facto absent. CL shows the typical distribution of monomer types for herbaceous lignins, and standard interunit motifs are present as well as typical termination motifs. Cross peaks indicative for α-oxidized β-O-4′ motifs are less intense than for PL. Traces of hemicellulose residues in the lignin are detectable, signs indicative of lignin-carbohydrate complexes (LCCs) are absent here as well. The sample contains a significant amount of para-coumarates as one could expect compared to PL.

A more densely populated aliphatic region indicates the presence of larger amounts of extractives in the sample compared to PL. Moreover, SSL contains still some extractives/aliphatics but in overall lower concentrations when normalizing abundancies to PL. Apart from standard monomer units for this lignin, the sweet sorghum bagasse sample contains the standard bonding motifs, seemingly in less abundance than PL and CL. However, given the fact that the sample due to the presence of the S-units, contains significantly more methoxy groups per C9-unit, differences as highlighted in Table 4 must not be overinterpreted. Nevertheless, coumarate residues can be seen as significantly enhanced also with respect to CL. The sample contains significantly higher concentrations of hemicellulose residues than PL and CL.

Both bark extracts, i.e., SBAL and SBNL, give rather different HSQC spectra compared to the other lignin samples as is expectable. The two samples are overall very similar, showing mainly β-O-4′ and β-O-5′ interunit bondings. The clear presence of LCC-indicating cross-peaks in SBNL vs. traces of these peaks in SBAL is in agreement with the ^{31}P NMR findings discussed above, hinting at the effectiveness of the acid treatment for eventually cleaving LCC bonding motifs and facilitating thus removal of carbohydrates. Both bark samples are characterized by high intensities of cross-peaks in the lower aliphatic region, eventually attributable to the presence of higher amounts of polar extractable impurities.

Table 4. Summary of identified structural motifs in the lignins under study based on HSQC-analyses. PL has been set as base for comparison across all samples; SBAL and SBNL compared directly. The abbreviations of lignin fractionated from pine sawdust (PL), cotton stalks (CL), sweet sorghum bagasse (SSL), spruce bark with (SBAL) and without acid addition (SBNL) were used.

Bonding Motif	$\delta\,^1$H	$\delta\,^{13}$C	Comparison to PL, %					Comparison SBAL, SBNL, %	
	ppm		PL	CL	SSL	SBAL	SBNL	SBAL	SBNL
H2,6-H	7.23	128.23	0.86	0.85	1.13	1.52	1.27	2.17	−0.10
G2-H	7.00	110.64	13.34	−0.47	−0.74	−0.29	−0.45	9.47	−0.23
S2,6-H	6.72	103.65	1.71	5.10	3.29	0.28	0.02	2.19	−0.20
Hγ in cinn-OH	4.04	59.85	1.34	1.82	0.75	4.72	4.37	7.67	−0.06
Hα in β-O-4'	4.87	71.35	3.33	0.26	−0.47	−0.46	−0.47	1.79	−0.02
Hβ in β-O-4'	4.34	83.12	3.57	−0.41	−0.69	−0.55	−0.55	1.62	−0.01
Hβ in β-O-4' α-C=O G	7.48	110.83	1.14	−0.15	−0.54	−0.18	−0.43	0.94	−0.31
Hα in β-5'	5.47	86.74	2.35	−0.51	−0.84	−0.50	−0.55	1.18	−0.10
Hα in β-β'	4.65	85.11	0.68	0.34	−0.71	−0.40	−0.22	0.41	0.29
Hβ in β-β'	3.06	53.66	1.21	0.31	−0.67	0.03	−0.04	1.25	−0.07
Hβ in epi-β-β'	2.86	53.61	0.55	−0.04	−0.73	−0.09	−0.29	0.50	−0.22
benzaldehyde	6.84	126.49	0.57	−0.84	−0.81	4.98	4.89	3.41	−0.01
G-hydroxyethylketone	1.24	21.93	7.49	0.16	−0.74	0.95	0.58	14.57	−0.19
Aryl ethyl ketone	2.21	33.2	2.37	3.27	0.02	7.57	5.80	20.31	−0.21
Cinnamyl aldehyde	6.96	123.8	0.61	−0.79	−0.87	4.41	4.15	3.30	−0.05
FA-H6	7.16	123.97	0.92	−0.60	−0.61	3.14	3.01	3.81	−0.03
PCE-H2,6	7.47	129.91	0.23	0.65	21.17	1.35	2.04	0.54	0.30
xylan signals	3.05	72.47							
	3.29	73.64	0.36	0.31	2.57	0.99	1.11	0.73	0.10
	3.52	75.26							

3.5. Headspace Gas Chromatography-Mass Spectrometry

The formation of main compounds during HSGC-MS treatment of lignin was investigated at treatment temperatures of 110, 180 and 270 °C. Only compounds with a spectral match quality greater than 85 % and an abundance of greater than 0.5 % are listed in the Table 5.

The HSGC-MS analysis at 110 °C indicated only a few compounds in the vapor of lignin samples: Isovanilline in CL, PL, SSL, vanillin in SBNL and SBAL, eugenol in PL, coumaran, α-curcumene, o-guaiacol and succinic acid in CL, cadelene in SBAL, CL and PL. The presence of coumaran in CL agrees with the literature as lignin in grasses generally contains significant greater amount of coumaryl (H) (5–35 %) than in softwoods [40]. The detected vanillin-based compounds represent the G-group unit [41]. The increase of vanillin concentration during the heating is the result of the oxidative degradation of guaiacyl structures [42]. The other released aromatic compounds, i.e., isovanilline, etc., have other aromatic origins or can be seen as products of more complex degradation/migration and rearrangements occuring upon heating under air. The relative amounts of vanillin were less abundant in CL and SSL for the reason of smaller proportion of guaiacyl units in herbaceous biomass than in softwood. Both bark lignin samples retain greater concentrations of extractives bonded to their structure than other samples, as the results listed in Table 5 show. Even using the additional ethanol-water or water-diethyl ether washing of lignin indicated that extractives can remain partially linked to lignin fibers due to their affinity [43]. In addition, previous results showed that α-curcumene and cadelene could be formed during low temperature heating of resins [44]. The most abundant fragment released during HSGC-MS analysis was calamenene in SBNL and SBAL and PL samples, which were previously found in the released compound vapor of coniferous wood [45]. This result agrees with the thermogravimetric analysis of lignin samples, whereas lignins fractionated from softwood showed a pronounced DTG peak at low temperatures, as discussed in Section 3.6. Overall, cotton stalks lignin showed the broadest distribution of released compounds during headspace analysis at 110 °C.

An increase in heat treatment temperature from 110 °C to 180 °C and then 270 °C during HSGC-MS analysis led to an increase in aromatic compounds and esters, and decrease in aldehydes which mainly originated from remaining impurities of hemicellulose and cellulose, as previously discussed [46]. The ethyl levulinate found in SSL, SBNL, SBAL and CL and dehydroabietal detected in SSL were released at 180 °C, emphasizing the presence of cellulose-related compounds, which were, however, not observed any more at 270 °C. Moreover, HSGC-MS treatment at 180 °C led to the formation of 5-formylfurfural in SBNL, SBAL, and SSL, confirming the presence of hemicellulose-related species, in accordance with HSQC data. This result agrees with the thermogravimetric analysis (see Section 3.6), whereas both bark lignins showed a pronounced DTG peak at low temperatures that indicates the presence of remaining extractives and carbohydrates. Results further indicate that compounds in the vapor phase at 270 °C analysis were mainly oxygenated aromatic chemicals. The main products in the HSGC-MS analysis of SSL, PL, SBAL, SBNL and CL at 270 °C were vanillin, acetovanillone and o-guaiacol, stemming from G-based structural units. Isoeugenol, 4-ethylguaiacol, vanillin and methyl vanillate were present in all lignin samples, with noteworthy exception of SBAL, whereas 4-vinylguaiacol was detected in all lignin samples with the exception of both spruce bark lignins. These results are in agreement with data of thermogravimetric analyses indicating an increased release of guaiacol and its derivatives with higher heating temperatures. At 180 and 270 °C, all lignin samples, except PL at both temperatures and CL at 270 °C, formed succinic acid, stemming from oxidative cleavage under the severe conditions.

Table 5. List of identified compounds from HSGC-MS analysis at headspace oven temperature of 110, 180, 270 °C of lignin fractionated from pine sawdust (PL), cotton stalks (CL), sweet sorghum (SSL), spruce bark with an acid catalyst (SBAL) and spruce bark without the acid catalyst (SBNL).

No	Compound	Presence with More than 0.5 % of Total Amount in Lignin Samples at T Equal		
		110 °C	180 °C	270 °C
1	Acetosyringone			PL
2	Acetovanillone			SSL, PL, CL, SBNL, SBAL
3	5-Acetoxymethyl-2-furaldehyde			PL
4	4-Acetylsyringol		SSL, CL, PL, BNL	SSL, CL
5	1,2-Benzenediol		CL, SBNL, SBAL	
6	Benzoic acid			PL
7	Cadelene	SBAL, CL, PL	SBNL, SBAL	SBNL
8	α-Calacorene		SBAL	
9	Calamenene	SBAL, SBNL, PL	SBAL, SBNL, PL	SBAL, SBNL, PL
10	γ-Caprolactone			CL
11	γ-Carboethoxy-γ-butyrolactone		SBAL, CL	
12	Cembrene		SBNL	
13	Coumaran	CL, SSL	CL, SSL	
14	α-Curcumene	CL		
15	Decanal			SBAL, SBNL
16	2-Decanone			CL
17	Dehydroabietal		SBAL, SBNL	
18	Dehydroabietan		SBAL, SSL	
19	Dihydroeugenol			PL
20	Ethyl coumarate			SSL, SBNL
21	Ethyl DL-malate		SSL, SBAL, SBNL	
22	Ethyl elaidate		SSL, CL	
23	Ethyl heptadecanoate		SBAL, SBNL	

Table 5. *Cont.*

No	Compound	Presence with More than 0.5 % of Total Amount in Lignin Samples at T Equal		
		110 °C	180 °C	270 °C
24	Ethyl homovanillate			SSL
25	Ethyl levulinate		CL, SBAL, SBNL, SSL	
26	Ethyl linolenate		SBAL, CL	CL
27	Ethyl oleate		SBNL, CL	
28	Ethyl pentadecanoate		SSL	
29	Ethyl pyroglumate		SSL	
30	4-Ethylguaiacol		SSL, SBNL, SBAL	SSL, SBNL, CL, PL
31	Ethylhexyl benzoate			PL
32	Eugenol	PL		
33	4-Ethylphenol		SSL, SBNL, SBAL, CL	SBAL, CL, SSL
34	Ferulic acid ethyl ester		SSL	SSL, CL
35	5-Formylfurfural		SBAL, SBNL, SSL	
36	4-Formylphenol		PL	SSL, PL
37	2-Furoic acid		SBAL	SBNL
38	Guaiacylacetone		SSL, CL	SSL, SBNL, CL
39	5-Hydroxymethylfurfural		SSL, PL, SBAL	SSL, PL
40	Isoeugenol	CL	CL, SBNL, SBAL, PL, SSL	SSL, SBNL, CL, PL
41	8-Isopropyl-1,3-dimethylphenanthrene			PL
42	Isovanilline	CL, PL, SSL		PL
43	Levoglucosan			SBNL
44	Methoxyeugenol			SSL, CL, SBNL
45	(1-Methoxy-pentyl)-cyclopropane	SBAL, SBNL	SBNL	
46	Methyl dehydroabiatate		PL	SBAL, SBNL
47	Methyl pimarate			SBNL, SSL
48	Methyl vanillate			SSL, PL, CL, SBNL
49	Mono(2-ethylhexyl)phthalate		SSL, PL, SBAL, CL	SSL, PL, SBAL, CL

Table 5. *Cont.*

No	Compound	Presence with More than 0.5 % of Total Amount in Lignin Samples at T Equal		
		110 °C	180 °C	270 °C
50	Naphthalene			SBAL
51	Nonanal		SBAL	SBAL, PL
52	Nonanoic acid		CL	CL
53	9,12-Octadecadienoic acid		SSL	SSL
54	Octadecenoic acid			SSL, SBNL
55	*o*-Guaiacol	CL	CL, PL, SSL, SBNL, SBAL	CL, PL, SSL, SBNL, SBAL
56	*o*-Pyrocatechualdehyde		SBAL, SBNL	
57	4-Oxononanal			CL
58	Palmitic acid			SBAL, CL
59	Palmitic acid ethyl ester	CL	SBAL, SSL, SBNL, CL	SBAL, SSL, SBNL, CL
60	*p*-cresol			PL, SSL
61	*p*-Formylphenol			CL
62	Propenal			SSL, PL
63	3-Pyridinol		SSL	SBNL, CL
64	Retene		SSL, SBAL, SBNL, CL, PL	SSL, SBAL, SBNL, CL, PL
65	Succinic acid	CL	CL, SBNL, SBAL, SSL	SBAL, SBNL, SSL
66	Syringa aldehyde		SSL, CL	SSL
67	Syringaldehyde		SSL, CL, PL	
68	Syringol		CL	CL
69	Tetradecanal			CL
70	2-Tetradecanone			CL
71	2-Undecanone			CL
72	Vanillin	SBNL, SBAL	SSL, CL, PL	SBNL, SSL, SBAL, CL, PL
73	4-Vinylguaiacol	CL	SSL, CL, PL, SBNL, SBAL	SSL, CL, PL

The major difference of CL and SSL to other samples lies in the presence of acetylsyringol and ferulic acid ethyl ester in their structure confirming the typical presence of dimethoxyphenols for these lignins. Ferulate and coumarate esters present the major part of LCC linkages in herbaceous biomass which can produce ester linkage with polysaccharides and proteins due to the presence of carboxylic acid groups at the end of propenyl groups [47]. In comparison to previous results, esterified fatty alcohols were detected in significant smaller numbers in this study, hinting at the effectiveness of the treatment in terms of removing these impurities [48,49].

No *p*-cresol was detected in cotton and spruce bark lignins. Free fatty acids were present in the range from palmitic to 13-octadecadienoic acid. The esterified fatty acids were detected in all lignin samples as impurities. In addition, the HSGC-MS analysis of all lignin samples at 180 and 270 °C showed the presence of retene in the vapor phase, indicating the presence of such lipid components as impurities in all lignin samples, including the lignin from herbaceous cotton stalks (CL). The presence of dihydroabietal at 180 °C in both spruce bark lignin samples and absence at 270 °C emphasizes the impact of temperature on the released products during HSGC-MS analysis as well as their potential stability limits.

3.6. Thermogravimetric Analysis

The thermogravimetric analysis showed that the conversion of all lignin samples was similar. The main difference in O_2 and CO_2 reactivity of lignin samples was observed in the higher maximum temperature of PL compared to other lignin samples due to the shift of the DTG peak to the higher temperatures. The DTG curves show a double broad peak that indicates a heterogeneous lignocellulosic mixture with respect to O_2 and CO_2 reactivity [50,51].

The DTG peaks of lignin degradation can be interpreted on the basis of the above discussed structural differences between various lignin samples, taking into account also impurities [52]. In agreement with HSQC-analyses, the first DTG peak during pyrolysis ranging from 200 to 300 °C can be referred to as the degradation of carbohydrates and contained extractives, i.e., fatty acids, pheromones, etc., as observed in the GC-MS headspace analysis, respectively. The second DTG peak located between 300 and 380 °C can be associated with the presence of mixed HGS structures as common in CL and SSL, followed by the third DTG peak that can be seen as reflecting G-only, or G-dominant lignin structures as typical in PL, SBAL and SBNL between 375 and 500 °C. The three DTG peaks indicate the development of three main components: A reactive carbon constituent, a carbon constituent with intermediate reactivity, and a less reactive carbon structure with reactivity that approaches that of commercial guaiacol [53]. The reactivities of lignin isolated from spruce bark with or without the use of the acid catalyst during the organosolv process were similar in terms of thermal characteristics, reflecting the basic structural similarity.

Data further underline the fact that both bark lignins have a similar composition. The first DTG peak was in absolute terms more visible for the bark lignins compared to other samples. This eventually reflects the greater amount of extractives remaining in the lignin after fractionation, as reported by the HSQC-analyses. SSL showed only a shoulder of the DTG peak with the maximum temperature shifted by 10 °C to higher temperatures compared to both bark lignins. Accumulative analysis of the curves indicate that lignins fractionated from herbaceous species are thermally less stable than those of softwood, which is in line with previous results [54,55]. Interestingly, the thermal data do not seem to reflect the ratio of aliphatic to aromatic OH-groups, or the impact of phenolic OH-group content. Both bark lignin, i.e., SBAL and SBNL, exhibiting significantly more phenolic than aliphatic OH-groups, show similar DTG curves compared to the other lignins. SBAL and SSL, representing the two lignins with the highest amount of phenolic OH, do not show drastic differences to the other samples.

4. Discussion

Thermogravimetric experiments demonstrated that the intrinsic reactivity of organosolv lignin under pyrolysis conditions or towards O_2 and CO_2 was similar due to the presence of DTG peaks in the same temperature range, with noteworthy exception of the PL sample. The PL sample was less reactive than other lignin samples. In principle, the reactivity of lignin can be affected by differences in organic composition and ash content [56,57]. However, the ash content of both spruce lignin SBAL and SBNL was below 0.6%, db, as shown in the proximate analysis (Table 2). Both bark lignin samples showed similar peak temperatures at 250 and 400 °C. Moreover, the first DTG peak of ash lean CL is shifted to the lower temperatures emphasizing that the degradation of organosolv lignin strongly depends, apart from fundamental structural differences, on the remaining carbohydrates and extractives after fractionation and less on the ash content. The results have shown that lignins richer in guaiacol unit are less reactive than herbaceous lignins containing additionally H- and S-type phenolics, as shown in Figure 2b,c. This could be explained by the formation of large condensed and polyaromatic structures from guaiacol during heating of softwood lignin, whereas the HGS and GS lignins contain a high number of methoxy groups leading to less thermal stability on the basis of an altered reactivity [58,59]. On the basis of the array of lignins studied in here it is not possible, however, to identify points in the curves that would clearly reflect differences in phenolic OH-group content or the ratio between aliphatic and aromatic OH-groups. Even when comparing G-lignin PL with structurally similar SBAL and SBNL, spectroscopically observed trends in terms of OH-group contents are not clearly delineable in thermogravimetric analysis.

The formation of levoglucosan found in case of SBNL indicates that an acid treatment is beneficial for facilitating removal of sugars in the generation of bark lignins. Simultaneously, the formation of phenols and lignin recondensation was enhanced during organosolv spruce bark fractionation using acid catalysts, as shown previously [60]. The presence of a catalyst promotes the cleavage of lignin aryl-ether bonds, releasing lignin fractions with lower molecular weight [61]. However, the reactivity of SBAL and SBNL samples in the thermal analysis was similar despite the differences in a polymer size and chemical composition in terms of impurities present. In addition, HSQC data suggest that the acid treatment in case of SBAL has no clear influence on the effectiveness of the hemicellulose removal. Chemical and HSQC-based analysis of SSL shows that sweet sorghum bagasse contains a higher concentration of polysaccharides impurities compared to the other lignin samples; however, headspace analysis did not reveal the formation of compounds that could be traced back to sugar residues. This finding is thus in line with the findings for carbohydrate-containing spruce bark lignins SBAL and SBNL. The literature reports a linkage between cellulose and lignin in spruce bark that is resistant to cleavage by sodium hydroxide [62]. Presence of sugar residues in the sample emphasizes, nevertheless, the more tedious process to extract the lignin from sweet sorghum bagasse. By means of purification techniques the residual polysaccharide content can be reduced by changing also the composition of the residual polysaccharides, e.g., glucose content decreases and the arabinose content increases due to the participation of this sugar in a lignin-polysaccharide linkage [63]. The compositional differences among lignins affected the distribution of weight losses during thermogravimetric analysis.

The results of this work demonstrate that the compositional differences among the lignin samples form the base for observed differences in the pyrolysis and O_2/CO_2 reactivity, as well as for the nature of released compounds during HSGC-MS analysis at different temperatures. The level of impurities in the organosolv lignin is generally low, and these low levels of remaining extractives and sugars in the lignins after organosolv fractionation do not affect the thermal stability of lignin; to a limited extend, however, they influence the formation of released products during HSGC-MS heating, leading to the formation of furfural-based products and the detection of fatty acids and derivatives.

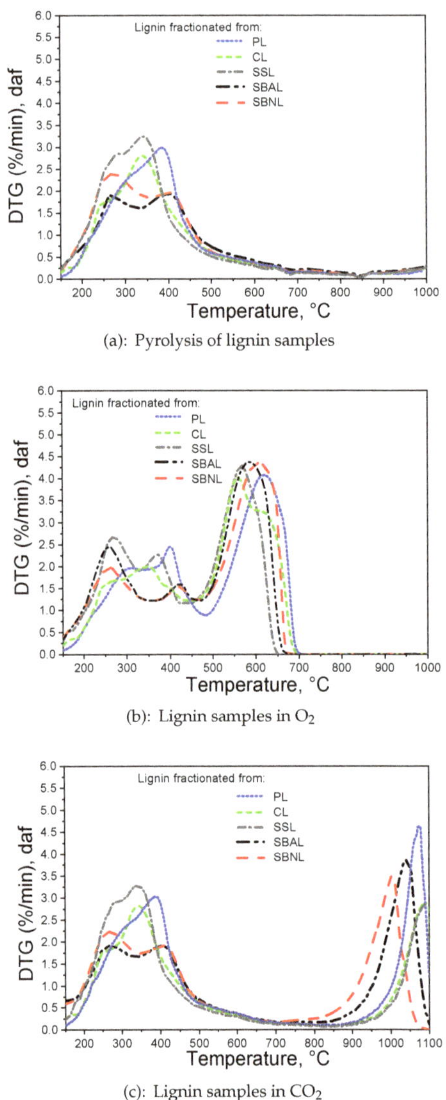

(a): Pyrolysis of lignin samples

(b): Lignin samples in O_2

(c): Lignin samples in CO_2

Figure 2. (**a–c**) DTG curves of lignin samples in 100 % N_2, O_2 (5 % volume fraction O_2 + 95 % volume fraction N_2) and CO_2 (20 % volume fraction CO_2 + 80 % volume fraction N_2). The abbreviations of lignin fractionated from pine sawdust (PL), cotton stalks (CL), sweet sorghum bagasse (SSL), spruce bark with (SBAL) and without acid addition (SBNL) were used.

5. Conclusions

Five organosolv lignins from two different classes of biomass were chemically and structurally analysed. Three softwood lignins and two herbaceous lignins were compared. Softwood lignin samples consisted of one main wood sample and two bark samples. These five lignins were analysed for impurities and structural features; the identified characteristics were correlated to data obtained in thermal treatment and gas phase analyses at various temperatures. The analysis of volatiles showed that the composition

Processes **2020**, *8*, 860

of the lignin backbones can be predicted in terms of a monomer type content. Presence of potential impurities, such as carbohydrates and fatty acids in original lignin can be identified using the head-space GC-MS analysis of volatiles. More interestingly, it was possible to correlate the general shape of the curves obtained in thermogravimetric analysis under various atmospheres to structural differences of lignins. These results suggest that G-type lignins can potentially be differentiated in HGS- and GS-lignins by comparison of mass loss curves, as long as they were isolated from the same biorefinery process, which yields lignins of comparable impurities. With respect to valorisation approaches, important differences in the relative amount of free phenolic OH-groups, and/or their ratio to aliphatic OH-groups in the side chains were not explicitly reflected, but contributed in a complex manner to the shape of mass loss curves.

Author Contributions: Conceptualization, A.T., L.M. and H.L.; methodology, H.L., C.C., A.T. and L.M.; software, B.B., A.B., A.T. and H.L.; validation, L.M. and A.B.; formal analysis, A.B., H.L. and A.T.; investigation, B.W., A.B., H.L., L.M. and A.T.; resources, U.R., P.C., C.C. and B.W.; data curation, A.B., B.W., H.L. and C.C.; writing–original draft preparation, A.B., A.T., H.H., J.J.L. and L.M.; writing–review and editing, U.R., P.C., C.C. and B.W.; visualization, A.T.; supervision, U.R. and P.C.; project administration, J.J.L. and L.M.; funding acquisition, B.W., A.T., U.R., P.C., C.C and J.J.L. All authors have read and agreed to the published version of the manuscript.

Funding: This research received no external funding.

Acknowledgments: Anna Trubetskaya, Bernd Wittgens and Anders Brunsvik gratefully acknowledge the financial support from the EU project Biomass Research Infrastructure for Sharing Knowledge (BRISK2). Heiko Lange acknowledges the MIUR Grant 'Dipartimento di Eccellenza 2018-2022' to the Department of Pharmacy of the University of Naples 'Federico II'. Ulrika Rova, Paul Christakopoulos and Leonidas Matsakas would like to thank Bio4Energy, a strategic research environment appointed by the Swedish government, for supporting this work.

Conflicts of Interest: The authors declare no conflict of interest. The funders had no role in the design of the study; in the collection, analyses, or interpretation of data; in the writing of the manuscript, or in the decision to publish the results.

Appendix A. FTIR Analysis

FTIR spectra of lignin samples are summarized in Table A1.

Table A1. Summary of FT-IR peak/band assignment for fractionated lignin samples. The abbreviations of lignin fractionated from pine sawdust (PL), cotton stalks (CL), sweet sorghum bagasse (SSL), spruce bark with (SBAL) and without acid addition (SBNL) were used.

| Band Position, cm^{-1} | | | | | |
SSL	CL	PL	SBNL	SBAL	Peak Assignment
		2960, 2933			C-H stretch [64]
		2953			C-H stretching in CH_2 and CH_3 [64]
1701			1694		C=O stretching, unconjugated [65]
		1597			aromatic skeletal vibrations (S > G) [38]
		1514			aromatic skeletal vibrations (G > S) [65]
1460			1450		C-H deformations in CH_2 and CH_3 [39]
1420					Aromatic skeletal vibrations [39]
		1372			C-H bending [66]
	1328				S ring breathing [39]
	1267		1272		G ring stretching [66]
1215			1202		C-C and C-O stretch [67]
1163			1155		C-O stretch in ester groups (HGS) [39]
		1115			C-O stretch, -OCH_3 (S) [68,69]
			1072		C-O stretching [70]
		1028			C-C, C-OH, C-H ring [39]
		990			HC=CH-out-of-plane deformation [45]
					CH-out-of-plane deformation in ethylenic double bonds [67]
828		820			CH-out-of-plane in positions 2, 5, and 6 of G units [45]

References

1. Umeki K.; Yamamoto K.; Namioka T.; Yoshikawa K. High temperature steam-only gasification of woody biomass. *Appl. Energy* **2010**, *87*, 791–798. [CrossRef]
2. Akin, D.E.; Benner R. Degradation of polysaccharides and lignin by ruminal bacteria and fungi. *Appl. Environ. Microbiol.* **1988**, *54*, 1117–1125. [CrossRef] [PubMed]
3. Dutta T.; Papa G.; Wang E.; Sun J.; Isern N.G.; Cort, J.R. Characterization of Lignin Streams during Bionic Liquid-Based Pretreatment from Grass, Hardwood, and Softwood. *ACS Sustain. Chem. Eng.* **2018**, *6*, 3079–3090. [CrossRef]
4. Doherty, W.O.S.; Mousavioun, P.; Fellows, C.M. Value-adding to cellulosic ethanol: Lignin polymers. *Ind. Crops. Prod.* **2011**, *33*, 259–276. [CrossRef]
5. Alvira P.; Tomas-Pejo E.; Ballesteros M.; Negro M.J. Pretreatment technologies for an efficient bioethanol production process based on enzymatic hydrolysis: A review. *Biores. Technol.* **2010**, *10*, 4851–4861. [CrossRef]
6. Trubetskaya, A.; Brown, A.; Tompsett, G.A.; Timko, M.T.; Kling, J.; Umeki, K. Characterization and reactivity of soot from fast pyrolysis of lignocellulosic compounds and monolignols. *Appl. Energy* **2018**, *212*, 1489–1500. [CrossRef]
7. Kalogiannis, K.G..; Matsakas, L.; Aspden, J.; Lappas, A.A.; Rova, U.; Christakopoulos, P. Acid Assisted Organosolv Delignification of Beechwood and Pulp Conversion towards High Concentrated Cellulosic Ethanol via High Gravity Enzymatic Hydrolysis and Fermentation. *Energies* **2018**, *23*, 1647. [CrossRef]
8. Zhao, X.; Cheng, K.; Liu, D. Organosolv pretreatment of lignocellulosic biomass for enzymatic hydrolysis. *Appl. Microbiol. Biotech.* **2009**, *82*, 815–827. [CrossRef] [PubMed]
9. Matsakas, L.; Nitsos, C.; Raghavendran, V.; Yakimenko, O.; Persson, G.; Christakopoulos, P. A novel hybrid organosolv: Steam explosion method for the efficient fractionation and pretreatment of birch biomass. *Biotech. Biofuels* **2018**, *11*, 1–14.
10. Zhang, J.; Tang, M.; Viikari, L. Xylans inhibit enzymatic hydrolysis of lignocellulosic materials by cellulases. *Biores. Technol.* **2012**, *121*, 8–12. [CrossRef]
11. Kristensen, J.B.; Thygesen, L.G.; Felby, C.; Jorgensen, H.; Elder, T. Cell wall structural changes in wheat straw pretreated for bioethanol production. *Biotech. Biofuels* **2008**, *1*, 1–9.
12. Sipponen, M.H.; Rahikainen, J.; Leskinen, T.; Pihlajaniemi, V.; Mattinen, M.L.; Lange, H. Structural changes of lignin in biorefinery pretreatments and consequences to enzyme-lignin interactions. *Nordic. Pulp. Paper Res. J.* **2018**, *32*, 550–571. [CrossRef]
13. Kalogiannis, K.G.; Matsakas, L.; Lappas, A.A.; Rova, U.; Christakopoulos, P. Aromatics from Beechwood Organosolv Lignin through Thermal and Catalytic Pyrolysis. *Energies* **2019**, *12*, 1606. [CrossRef]
14. Wyman, C.E.; Dale, B.E.; Elander, R.T.; Holtzapple, M.; Ladisch, M.R.; Lee, Y.Y. Coordinated development of leading biomass pretreatment technologies. *Biores. Technol.* **2005**, *96*, 1959–1966. [CrossRef] [PubMed]
15. Constant, S.; Wienk, H.L.J.; Frissen, A.E.; de Peinder, P.; Boelens, R.; van Es, D. New insights into the structure and composition of technical lignins: a comparative characterisation study. *Green Chem.* **2016**, *18*, 2651–2665. [CrossRef]
16. Nitsos, C.; Rova, U.; Christakopoulos, P. Organosolv Fractionation of Softwood Biomass for Biofuel and Biorefinery Applications. *Energies* **2018**, *11*, 1–23.
17. Huber GW.; Iborra S.; Corma A. Synthesis of Transportation Fuels from Biomass: Chemistry, Catalysts, and Engineering. *Energies* **2006**, *106*, 4044–4098.
18. Yang, H.; Yan, R.; Chen, H.; Lee, D.H.; Zheng, C. Characteristics of hemicellulose, cellulose and lignin pyrolysis. *Fuel* **2007**, *86*, 1781–1788. [CrossRef]
19. Stefanidis, S.D.; Kalogiannis, K.G.; Iliopoulou, E.F.; Michailof, C.M.; Pilavachi, P.A.; Lappas, A.A. A study of lignocellulosic biomass pyrolysis via the pyrolysis of cellulose, hemicellulose and lignin. *J. Anal. Appl. Pyrolysis.* **2014**, *105*, 143–150. [CrossRef]
20. Rinaldi, R.; Jastzebski, R.; Clough, M.T.; Ralph, J.; Kennema, M.; Bruijnincx, P.C.A. Paving the Way for Lignin Valorisation: Recent Advances in Bioengineering, Biorefining and Catalysis. *Angew. Chem.* **2016**, *55*, 8164–8215. [CrossRef]
21. Karnaouri, A.; Lange, H.; Crestini, C.; Rova, U.; Christakopoulos, P. Chemoenzymatic Fractionation and Characterization of Pretreated Birch Outer Bark. *ACS Sustain. Chem. Eng.* **2016**, *4*, 5289–5302. [CrossRef]

22. Lange, H.; Schiffels, P.; Sette, M.; Sevastyanova, O.; Crestini, C. Fractional Precipitation of Wheat Straw Organosolv Lignin: Macroscopic Properties and structural Insights. *ACS Sustain. Chem. Eng.* **2016**, *4*, 5136–5151. [CrossRef]

23. Nitsos, C.; Stoklosa, R.; Lange, H.; Hodge, D.; Rova, U.; Christakopoulos, P. Isolation and Characterization of Organosolv and Alkaline Lignins from Hardwood and Softwood Biomass. *ACS Sustain. Chem. Eng.* **2016**, *4*, 5171–5193. [CrossRef]

24. Matsakas, L.; Christakopoulos, P. Fermentation of liquefacted hydrothermally pretreated sweet sorghum bagasse to ethanol at high-solids content. *Biores. Technol.* **2013**, *127*, 202–208. [CrossRef]

25. Kemppainen, K.; Siika-aho, M.; Pattathil, S.; Giovando, S.; Kruus, K. Spruce bark as an industrial source of condensed tannins and non-cellulosic sugars. *Ind. Crops. Prod.* **2014**, *52*, 158–168. [CrossRef]

26. MS-SEARCH. NIST Mass Spectrometry Data Center: NIST/EPA/NIH Mass Spectral Database. Available online: http://chemdata.nist.gov (accessed on 1 January 2011).

27. Mu, L.; Matsakas, L.; Wu, J.; Chen, M.; Rova, U.; Christakopoulos, P.. Two important factors of selecting lignin as efficient lubricating additives in poly (ethylene glycol): Hydrogen bond and molecular weight. *Int. J. Biol. Macromol.* **2019**, *129*, 564–570. [CrossRef] [PubMed]

28. Asikkala, J.; Tamminen, T.; Argyropoulos, D.S. Accurate and Reproducible Determination of Lignin Molar Mass by Acetobromination. *J. Agric. Food Chem.* **2012**, *60*, 8968–8973. [CrossRef]

29. Meng, X.; Crestini, C.; Ben, H.; Hao, N.; Pu, Y.; Ragauskas, A.J.. Determination of hydroxyl groups in biorefinery resources via quantitative ^{31}P NMR spectroscopy. *Nat. Protoc.* **2019**, *14*, 2627–2647. [CrossRef] [PubMed]

30. Gordobil, O.; Moriana, R.; Zhang, L.; Labidi, J.; Sevastyanova, O. Assessment of technical lignins for uses in biofuels and biomaterials: Structure-related properties, proximate analysis and chemical modification. *Ind. Crops. Prod.* **2014**, *83*, 155–165. [CrossRef]

31. Guo, H.; Zhang, B.; Qi, Z.; Li, C.; Ji, J.; Dai, T.. Valorization of Lignin to Simple Phenolic Compounds over Tungsten Carbide: Impact of Lignin Structure. *ChemSusChem* **2016**, *10*, 523–532. [CrossRef]

32. Nadji, H.; Diouf, P.N.; Benaboura, A.; Bedard, Y.; Riedl, B.; Stevanovic, T. Comparative Study of Lignin Isolated from Alfa grass (Stipa Tenacissima L.). *Biores Tech* **2009**, *100*, 3585–3592. [CrossRef]

33. Sluiter, A.; Hames, B.; Ruiz, R.; Scarlata, C.; Sluiter, J.; Templeton, D. *Determination of Structural Carbohydrates and Lignin in Biomass. Golden (CO)*; Report No. NREL/TP-510-42618. Contract No.; DE-AC36-08-GO28308; National Renewable Energy LAboratory: Golden, CO, USA, 2011.

34. Thring, R.W.; Vanderlaan, M.N.; Griffin, S.L. Fractionation of ALCELL lignin by sequential solvent extraction. *J. Wood Chem. Technol.* **1996**, *16*, 139–154. [CrossRef]

35. Brodin, I.; Sjöholm, E.; Gellerstedt, G. The behavior of kraft lignin during thermal treatment. *J. Anal. Appl. Pyrolysis.* **2010**, *87*, 70–77. [CrossRef]

36. Cui, C.; Sun, R.; Argyropoulos, D.S. Fractional Precipitation of Softwood Kraft Lignin: Isolation of Narrow Fractions Common to a Variety of Lignins. *ACS Sustain. Chem. Eng.* **2014**, *2*, 959–968. [CrossRef]

37. Sun, S.L.; Wen, J.L.; Ma, M.G.; Li, M.F.; Sun, R.C. Revealing the Structural Inhomogeneity of Lignans from Sweet Sorghum Stem by Successive Alkali Extractions. *J. Agric. Food Chem.* **2013**, *61*, 4226–4235. [CrossRef]

38. El-Hendawy, A.N.A. Variation in the FTIR spectra of a biomass under impregnation, carbonization and oxidation conditions. *J. Anal. Appl. Pyrolysis.* **2004**, *75*, 159–166. [CrossRef]

39. Inkrod, C.; Raita, M.; Champreda, V.; Laosiripojana, N. Characteristics of Lignin Extracted from Different Lignocellulosic Materials via Organosolv Fractionation. *BioEnergy Res.* **2018**, *11*, 277–290. [CrossRef]

40. Abdelaziz, O.Y.; Brink, D.P.; Prothmann, J.; Ravi, K.; Sun, M.; Garcia-Didalgo, J. Biological valorization of low molecular weight lignin. *Biotechnol. Adv.* **2016**, *34*, 1318–1346. [CrossRef]

41. Pawliszyn, J. Theory of Solid-Phase Microextraction. *J. Chromatogr. Sci.* **2000**, *38*, 270–278. [CrossRef] [PubMed]

42. Rocha, S.M.; Gonçalves, V.; Evtuguin, D.; Delgadillo, I. Distinction and identification of lignins based on their volatile headspace composition. *Talanta* **2008**, *75*, 594–597. [CrossRef]

43. Tarasov, D.; Leitch, M.; Fatehi, P. Lignin-carbohydrate complexes: Properties, applications, analyses, and methods of extraction: A review. *Biotechnol. Biofuels* **2018**, *11*, 1–28.

44. Koller, J.; Baumer, U.; Kaup, Y.; Schmid, M.; Weser, U. Analysis of a pharaonic embalming tar. *Nature* **2003**, *425*, 784. [CrossRef] [PubMed]

45. Faix, O. Classification of Lignins from Different Botanical Origins by FT-IR Spectroscopy. *Holzforschung* **1991**, *45*, 21–28. [CrossRef]

46. Xue, L.; Zhao, Z.; Zhang, Y.; Chu, D.; Mu, J. Analysis of Gas Chromatography-Mass Spectrometry Coupled with Dynamic Headspace Sampling on Volatile Organic Compounds of Heat-Treated at High Temperatures. *Biores* **2016**, *11*, 3550–3560. [CrossRef]

47. Oliveira, D.; Finger-Teixeira, A.; Mota, T.; Salvador, V.H.; Moreira-Vilar, F.C.; Molinari, H.B.C. Ferilic Acid: A Key Component in Grass Lignocellulose Recalcitrance to Hydrolysis. *Plant Biotechnol. J.* **2015**, *13*, 1224–1232. [CrossRef]

48. Gutierrez, A.; Rodriguez, I.M.; Del Rio, J.C. Chemical Characterization of Lignin and Lipid Fractions in Industrial Hemp Bast Fibbers Used for Manufacturing High-Quality Paper Pulps. *J. Agric. Food Chem.* **2006**, *54*, 2138–2144. [CrossRef]

49. Marques, G.; Gutierrez, A.; del Rio, J.C. Chemical Characterization of Lignin and Lipid Fractions in Leaf Fibers of Curaua (*Ananas erectifolius*). *J. Agric. Food Chem.* **2007**, *55*, 1327–1336. [CrossRef]

50. Surup, G.R.; Foppe, M.; Schubert, D.; Deike, R.; Heidelmann, M.; Trubetskaya, A. The effect of feedstock origin and temperature on the structure and reactivity of char from pyrolysis at 1300–2800 °C. *Fuel* **2019**, *235*, 306–316. [CrossRef]

51. Surup, G.R.; Nielsen, H.K.; Heidelmann, M.; Trubetskaya, A. Characterization and reactivity of charcoal from high temperature pyrolysis (800–1600 °C). *Fuel* **2019**, *235*, 1544–1554. [CrossRef]

52. Du, S.; Valla, J.A.; Bollas, G.M. Characteristics and origin of char and coke from fast and slow, catalytic and thermal pyrolysis of biomass and relevant model compounds. *Green Chem.* **2013**, *15*, 3214–3229. [CrossRef]

53. Cen, K.; Cao, X.; Chen, D.; Zhou, J.; Chen, F.; Li, M. Leaching of alkali and alkaline earth metallic species (AAEMs) with phenolic substances in bio-oil and its effect on pyrolysis characteristics of moso bamboo. *Fuel Process. Technol.* **2020**, *200*, 106332. [CrossRef]

54. Yan, K.; Liu, F.; Chen, Q.; Ke, M.; Huang, X.; Hu, W. Pyrolysis characteristics and kientics of lignin derived from enzymatic hydrolysis residue of bamboo pretreated with white-rot fungus. *Biotechnol. Biofuels* **2016**, *9*, 1–11.

55. Cherif, M.F.; Trache, D.; Brosse, N.; Benaliouche, F.; Tarchoun, A.F. Comparison of the Physicochemical Properties and Thermal Stability of Organosolv and Kraft Lignins from Hardwood and Softwood Biomass for Their Potential Valorization. *Waste Biomass Valoriz* **2020**, *15*, 1–13.

56. Shimizu, S.; Yokoyama, T.; Akiyama, T.; Matsumoto, Y. Reactivity of Lignin with Different Composition of Aromatic Syringyl/Guaiacyl Structures and Erythro/Threo Side Chain Structures in β-O-4 Type during Alkaline Delignification: As a Basis for the Different Degradability of Hardwood and Softwood Lignin. *J. Agric. Food Chem.* **2012**, *60*, 6471–6476. [CrossRef] [PubMed]

57. Trubetskaya, A.; Jensen, P.A.; Jensen, A.D.; Umeki, K.; Gardini, D.; Kling, J.; Glarborg, P. Effects of several types of biomass fuels on the yield, nanostructure and reactivity of soot from fast pyrolysis at high temperatures. *Appl. Energy* **2016**, *171*, 468–482. [CrossRef]

58. Jiang, G.; Nowakowski, D.J.; Bridgwater, A.V. Effect of the temperature on the composition of lignin pyrolysis products. *Energy Fuel* **2010**, *24*, 4470–4475. [CrossRef]

59. Trubetskaya, A.; Larsen Andersen, M.; Talbro Barsberg, S. The Nature of Stable Char Radicals: An ESR and DFT Study of Structural and Hydrogen Bonding Requirements. *ChemPlusChem* **2018**, *83*, 780–786. [CrossRef] [PubMed]

60. Mood, S.H.; Golfeshan, A.H.; Tabatabaei, M.; Abbasalizadeh, S.; Ardjmand, M. Comparison of different ionic liquids pretretment for barley straw enzymatic saccharification. *Biotech* **2013**, *3*, 399–406.

61. Kumar, A.; Kumar, J.; Bhaskar, T. Utilization of lignin: A sustainable and eco-friendly approach. *J. Energy Inst.* **2020**, *93*, 235–271. [CrossRef]

62. Brauns, F.E.; Brauns, D.A. The Thermal Decomposition of Lignin. In *The Chemistry of Lignin*; Academic Press: Cambridge, MA, USA, 1960; p. 814.

63. Fengel, D.; Wegener, G. *Wood: Chemistry, Ultrastructure, Reactions*; Walter de Gruyter: Berlin, Germany, 1989; p. 612.

64. Arellano, O.; Flores, M.; Guerra, J.; Hidalgo, A.; Rojas, D.; Strubinger, A. Hydrothermal Carbonization of Corncob and Characterization of the Obtained Hydrochar. *Chem. Eng. Trans.* **2016**, *50*, 235–240.

65. Adapa, P.K.; Tabil, L.G.; Schoenau, G.J.; Canam, T.; Dumonceaux, T. Quantitative Analysis of Lignocellulosic Components of Non-Treated and Steam Exploded Barley, Canola, Oat and Wheat Straw Using Fourier Transform Infrared Spectroscopy. *J. Agric. Sci. Technol. B* **2011**, *1*, 177–188.
66. Fan, M.; Dai, D.; Huang, B. Chapter 3 Fourier Transform Infrared Spectroscopy for Natural Fibres. In Fourier Transform. *IntechOpen* **2012**, *3*, 45–68.
67. Lisperguer, J.; Perez, P.; Upizar, S. Structure and thermal properties of lignins: characterization by infrared spectroscopy and differential scanning calorimetry. *J. Chil. Chem. Soc.* **2009**, *54*, 460–463. [CrossRef]
68. Arafat, A.; Samad, S.A.; Masum, S.M.; Moniruzzaman, M. Preparation and Characterization of Chitosan from Shrimp shell waste. *IJSER* **2015**, *6*, 538–541.
69. Liu, Z.; Quek, A.; Hoekman, S.K.; Balasubramanian, R. Production of solid biochar fuel from waste biomass by hydrothermal carbonization. *Fuel* **2013**, *103*, 943–949. [CrossRef]
70. Lopes, J.O.; Garcia, R.A.; Souza, N.D. Infrared spectroscopy of the surface of thermally-modified teak juvenile wood. *Maderas Cienc. Technol.* **2018**, *20*, 737–746. [CrossRef]

Article

Lipid Production by Yeasts Growing on Commercial Xylose in Submerged Cultures with Process Water Being Partially Replaced by Olive Mill Wastewaters

Evangelos Xenopoulos [1], **Ioannis Giannikakis** [1], **Afroditi Chatzifragkou** [1,2], **Apostolis Koutinas** [1] **and Seraphim Papanikolaou** [2,*]

[1] Department of Food Science & Human Nutrition, Agricultural University of Athens, 75 Iera Odos, 11855 Athens, Greece; evansxeno@gmail.com (E.X.); jjb.giannis.007@hotmail.com (I.G.); a.chatzifragkou@reading.ac.uk (A.C.); akoutinas@aua.gr (A.K.)

[2] Department of Food and Nutritional Sciences, University of Reading, Whiteknights, P.O. Box 226, Reading RG6 6AD, UK

[*] Correspondence: spapanik@aua.gr; Tel.: +30-210-5294700

Received: 24 May 2020; Accepted: 8 July 2020; Published: 11 July 2020

Abstract: Six yeast strains belonging to *Rhodosporidium toruloides*, *Lipomyces starkeyi*, *Rhodotorula glutinis* and *Cryptococcus curvatus* were shake-flask cultured on xylose (initial sugar—S_0 = 70 ± 10 g/L) under nitrogen-limited conditions. *C. curvatus* ATCC 20509 and *L. starkeyi* DSM 70296 were further cultured in media where process waters were partially replaced by the phenol-containing olive mill wastewaters (OMWs). In flasks with $S_0 \approx 100$ g/L and OMWs added yielding to initial phenolic compounds concentration (PCC_0) between 0.0 g/L (blank experiment) and 2.0 g/L, *C. curvatus* presented maximum total dry cell weight—$TDCW_{max} \approx 27$ g/L, in all cases. The more the PCC_0 increased, the fewer lipids were produced. In OMW-enriched media with $PCC_0 \approx 1.2$ g/L, TDCW = 20.9 g/L containing ≈ 40% *w/w* of lipids was recorded. In *L. starkeyi* cultures, when $PCC_0 \approx 2.0$ g/L, TDCW ≈ 25 g/L was synthesized, whereas lipids in TDCW = 24–28% *w/w*, similar to the experiments without OMWs, were recorded. Non-negligible dephenolization and species-dependent decolorization of the wastewater occurred. A batch-bioreactor trial by *C. curvatus* only with xylose ($S_0 \approx 110$ g/L) was performed and TDCW = 35.1 g/L (lipids in TDCW = 44.3% *w/w*) was produced. Yeast total lipids were composed of oleic and palmitic and to lesser extent linoleic and stearic acids. *C. curvatus* lipids were mainly composed of nonpolar fractions (i.e., triacylglycerols).

Keywords: lignocellulosic sugars; microbial lipid; olive mill wastewater; *Cryptococcus curvatus*; *Lipomyces starkeyi*

---

1. Introduction

Lignocellulosic materials represent the largest and the most attractive biomass resource worldwide that can serve as cheap feedstock of monosaccharides in a remarkable plethora of microbial fermentations. These low-cost materials like woody biomass, grass, agricultural and forestry solid wastes, process waters rich (or potentially very rich) in lignocellulosic sugars (i.e., spent sulfite liquor, waste xylose mother liquid), municipal solid wastes, etc., are particularly abundant in nature and have a very important potential for various types of microbial bioconversions [1–8]. Commercial xylose, being one of the principal monomer sugars of these feedstocks, is produced after acid hydrolysis of various types of lignocellulosic biomass (i.e., corn cobs, sugarcane bagasse, etc.) [2,3,7], followed by subsequent condensation and crystallization [1,6]. The capability of microorganisms to grow on and produce biotechnological compounds during culture on C-5 sugars and specifically during fermentations on xylose, that, as stressed, is one of the most abundant sugars on the lignocellulosic biomass, presents continuously growing importance [3–5,7,8].

Olive mill wastewaters (OMWs) are the major effluent deriving from olive oil production process, specifically when traditional (viz. press extraction systems) or 3.0- and 2.5-phase centrifugation systems are employed. Despite the gradual utilization of 2.0-phase centrifugation systems (that generate lower OMW quantities than the 3.0- or the 2.5-phase systems) within the EU countries, OMWs are always considered as one of the most important agro-industrial wastes generated into the Mediterranean region; these residues are seasonally produced in very high quantities, whereas their strong odor and dark color as also their relatively high organic load have a direct negative impact on the environment if they are released without previous treatment. This important residue of the olive oil industry is one of the most difficult to treat wastes because of its high content in phenolic compounds [9,10]. The increased concentration of OMWs in organic matter and phenolic components results in reduction of the available concentration of oxygen when these residues are released without prior treatment and this upsets the balance of ecosystems and the soil porosity, resulting in contaminated aquifers and polluted environments [10–12]. In addition, the phenolic compounds found into the OMWs are, in general, quite instable and their polymerization during uncontrolled release often leads to the generation of high-molecular-weight, hardly degradable substances [10,13,14]. The annual production of OMWs is estimated to be > 15×10^6 m^3 [13] and this very high residue production together with the seasonal production of these wastewaters render their environmentally friendly disposal and management as very important priorities, specifically for the Mediterranean countries [9,13,14].

The last years a new trend has appeared in relation to the valorization various types of food-processing wastewaters or low-quality waters referring to their simultaneous utilization as substrate and as fermentation water implemented in various types of microbial-guided bioprocesses. Specifically, "concentrated" wastes and residues (like crude glycerol, molasses, etc.) and renewable low- or zero-cost hydrophilic (i.e., glucose syrups, low-purity sugars, white grape pomace, etc.) or hydrophobic (i.e., various low quality fatty compounds) carbon sources could be diluted; in this type of dilution, instead of the tap water, OMWs (or similar types of wastewaters like the table olive-processing wastewaters) can be used as process waters. Accordingly, added-value metabolites could be produced during these fermentation processes with simultaneous partial detoxification (i.e., decolorization and phenol removal) of the implicated wastewaters [11,12,15–21]. On the other hand—and in combination with the strategy previously mentioned—OMWs could initially be subjected to purification in order to recover useful materials that are found into the residue (i.e., antioxidant compounds) [14], and, thereafter, utilize the bulky remaining wastewater as process water and simultaneous substrate in fermentation processes.

Microbial lipids (single-cell oils—SCOs) are produced by the "oleaginous" microorganisms (these that can accumulate lipid to quantities ≥20% in DCW during growth on glucose in conditions favoring lipogenesis; [22,23]). These lipids possess similar fatty acid (FA) composition with various edible oils [24]. Therefore, these fatty materials can be implicated as perfect agents in the synthesis of the so-called "2nd generation" biodiesel (in fact, it is the biodiesel the production of which is not in competition with the arable land) [3,5,22,25]. SCOs can also be employed as starting materials for the synthesis of several added-value oleochemical compounds or can be implicated as high added-value fatty supplements in the food-processing industries (specifically the ones that contain rarely found poly-unsaturated FAs or these that have composition similarities with expensive exotic fats) [3–5,7,8,22,23,26,27].

In the current investigation, a number of yeasts belonging to the species *Rhodosporidium toruloides*, *Rhodotorula glutinis*, *Cryptococcus curvatus* and *Lipomyces starkeyi* (in total six strains) were tested as regards their potential to assimilate commercial, non-purified xylose, deriving from lignocellulosic biomass. Trials were carried out in shake-flask mode under nitrogen-limited conditions, favoring the production of SCO [22–24]. In the next step, the most promising among the previously screened yeasts, namely *Cryptococcus curvatus* ATCC 20509 and *Lipomyces starkeyi* DSM 70296, were cultivated in media presenting high initial xylose concentrations with initial nitrogen remaining constant, in order to further enhance the production of SCO. Moreover, according to a new trend recently appeared in the Industrial Biotechnology, OMWs were employed as liquid medium in order to partially replace tap water in the fermentation carried out. Finally, *C. curvatus* ATCC 20509 culture that presented very interesting total biomass and SCO production on high-xylose concentration media was scaled-up in bench top laboratory-scale bioreactor. Quantitative and kinetic considerations concerning the yeast physiological behavior were critically discussed.

2. Materials and Methods

2.1. Microorganism, Media and Cultures

The strains used in the present study belonged to the species *Rhodosporidium toruloides* (strains DSM 4444 and NRRL Y-27012), to the species *Rhodotorula glutinis* (strain NRRL YB-252), to the species *Cryptococcus curvatus* (strains ATCC 20509 and NRRL Y-1511) and to the species *Lipomyces starkeyi* (strain DSM 70296). The strain with the code nomination ATCC was given from the American Type Culture Collection, the strains with the code nomination DSM were provided by the German Collection of Microorganisms and Cell Cultures, while strains encoded NRRL Y-derive from the NRRL culture collection. With the exception of *R. toruloides* and *R. glutinis*, strains were maintained on YPDA slants (10-g/L glucose, 10-g/L yeast extract, 10-g/L peptone and 20-g/L agar) at 4 °C. *R. toruloides* and *R. glutinis* were maintained on YPDMA medium (10-g/L glucose, 10-g/L yeast extract, 10-g/L malt extract, 5-g/L peptone and 20-g/L agar) at 4 °C.

Most of the experiments were carried out in 250-mL Erlenmeyer flasks containing 50 ± 1 mL of growth medium, previously sterilized at T = 115 ± 1 °C for 20 min and inoculated with 1 mL of 24-h exponential preculture yeast incubated at 180 rpm at T = 28 ± 1 °C (use of an orbital shaker Zhicheng ZHWY 211C; PR of China). This type of shake-flask culture performed in the current investigation (utilization of 250-mL flasks filled with the ⅕ of their volume and agitated at 180 rpm) implicates full aerobic conditions (viz. dissolved oxygen tension—DOT ≥ 20% *v/v*) throughout the flask experiments performed [11,12,16]. The yeast preculture was carried out in YPD medium. Liquid cultures were performed in a medium in which the salt composition was as in Papanikolaou et al. [28]. Yeast extract at 3.0 g/L and peptone at 1.0 g/L were used as nitrogen sources. Yeast extract contained *c.* 7% *w/w* nitrogen, while peptone contained *c.* 14% *w/w* nitrogen, respectively. Commercial-type xylose (purity of *c.* 95%, *w/w*) implicated as starting material for the large-scale synthesis of xylitol, was used as the sole carbon source. In the first part of the study, a screening of all available microorganisms was carried out on media containing xylose at initial total sugar (S_0) concentration ≈70 g/L. In the nest step of the experimental procedure and in order to further enhance the production of cellular lipids, the most promising of the previously screened microorganisms (it was found that were mainly the strains *C. curvatus* ATCC 20509 and *L. starkeyi* DSM 70296) were cultured on media containing higher S_0 concentration (≈100 g/L) with initial nitrogen concentration into the medium remaining the same as in the previously mentioned screening experiment (yeast extract at 3.0 g/L and peptone at 1.0 g/L), therefore trials in media with higher initial C/N molar ratio were prepared.

In one case (trial using xylose at $S_0 \approx 100$ g/L, with initial nitrogen concentration as mentioned above), a batch bioreactor trial was performed by the strain *C. curvatus* ATCC 20509. The culture was carried out in a bench top 3-L bioreactor (New Brunswick, SC, USA), containing 1.9 L of fermentation medium. The reactor was aseptically inoculated with 100 mL of 24-h yeast preculture (see details in the previous section). The culture conditions in the bioreactor trials were as follows: incubation temperature 28 ± 1 °C; agitation rate 450 ± 5 rpm, aeration 0.1–1.5 vvm (the aeration was set on cascade mode, and the air-pump debit varied automatically within the above-mentioned vvm values in order to maintain a dissolved oxygen tension (DOT) value above 20%, *v/v*, in all culture phases); pH = 6.0 ± 0.1 regulated with automatic addition of KOH (2 M).

In the frame of sustainability and water-saving policies adopted, fermentation waters were partially replaced by olive mill wastewaters (OMWs) obtained from a three-phase decanter olive mill in the region Kalentzi (Corinthia, Peloponnisos, Greece). After their collection, OMWs were immediately frozen at–20 ± 1 °C and before use, these wastes were thawed and the solids were removed by centrifugation and subsequent filtration [17], to ensure the uniformity of the liquid material implicated in the fermentations. This procedure is in accordance with the current Greek legislation [29] regarding pretreatment of the waste before its disposal. Before fermentations, chemical analyses of the used OMWs that would replace process water were carried out; the initial pH of OMWs used was = 4.9 ± 0.1. Initial phenolic compounds concentration (PCC_0) in the residue was found to be ≈3.5 g/L (phenolic compounds (PC) were expressed as gallic acid equivalents). Moreover, OMWs contained a small concentration of total sugars (≈10.0 g/L). These sugars, after HPLC analysis conducted (see analysis details in the next paragraphs) were mainly composed of glucose (*c.* 80% of sugars) and fructose (*c.* 20% of sugars). Insignificant quantities of organic acids (citric acid at *c.* 2.0 g/L and acetic acid at *c.* 2.5 g/L—analysis equally performed through HPLC) were also detected. Protein quantity into the residue, as quantified through the Lowry method, was found to be ≈3.0 g/L. The quantity of ammonium ions, as determined with the aid of an ammonium selective electrode (SA 720, Orion, Boston, MA, USA) was found to be ≈10 mg/L. The electrical conductivity of the wastewater at 25 °C was 13.85 ms/cm (Hanna HI-8733 conductivity meter, Padova, Italy). Finally, triple extraction of OMWs by hexane revealed very small concentrations of fatty materials (0.4 ± 0.1 g/L of olive oil) in the volume of the residue. In the cases in which OMWs partially replaced the process fermentation waters in the cultures carried out, due to the indeed low concentrations of proteins, organic acids and oils, these compounds were not taken into consideration concerning their impact upon the quantitative physiological response of the microorganisms tested. In contrast, in the quantitative determination of total sugars in the growth medium, the contribution of sugars added from the OMWs, although almost negligible as compared with the initial concentration of xylose added, was taken into consideration.

The initial pH in the culture media after the addition of the salts was found to be = 6.0 ± 0.1 while pH value during all trials was always maintained between the ranges of 5.2–5.8, due to the buffer capacity of the salts of the employed medium. Since microorganisms did not produce appreciable quantities of organic acids, no external addition of base was requested in order to maintain the pH value within this range.

2.2. Determinations and Analyses

For the flask cultures, flasks were periodically removed from the incubator while for the bioreactor experiments *c.* 10 mL of culture was aseptically collected. In the flask trials, pH was measured in a SevenCompact™ pH S210 pH meter (Mettler-Toledo, Columbus, OH, USA). The whole content of the 250-mL flasks or the sample collected from the bioreactor experiment were subjected to centrifugation at 9000× *g*/10 min at 4 °C (Universal 320R-Hettich centrifuge, Tuttlingen, Germany). Wet biomass collected after the first centrifugation, was then extensively washed with distilled water and centrifugation was applied once more at the same conditions. Wet yeast biomass was dried at T = 90 ± 5 °C for 24 h to obtain the total dry cell weight (TDCW—X, in g/L) [28]. Biomass yield ($Y_{X/S}$, in g/g), was expressed as the grams of cell dry weight (X) produced, per grams of total sugar (mostly or completely xylose; S) consumed (g TDCW/g total sugars consumed).

Total cellular lipid (L) extracted from the TDCW with a chloroform-methanol mixture (2/1, *v/v*) according to Gardeli et al. [30], was determined gravimetrically and was expressed as g/L. Lipid in TDCW (%, *w/w*; $Y_{L/X}$), was calculated based on percentage of the accumulated lipid (L, g/L) per produced TDCW (X, g/L). Extracted intracellular lipids were converted to their fatty acid methyl esters (FAMEs) and analyzed in a gas chromatography (GC) according to Fakas et al. [31]. Specifically, a quantity of total lipids (<100 mg) was converted to the respective methyl esters in a two-step reaction using $CH_3O^-Na^+$ and CH_3OH/HCl as methylation agents [31]. The identification of the analyzed methyl esters was based on the comparison of retention times with known FAMEs standards. In the representations done, the relative area of the FAMEs is presented. Moreover, in some instances, crude lipid of *C. curvatus* ATCC 20509 was washed with 0.88%-*w/v* KCl solution and was subsequently fractionated to its principal lipid fractions (viz. neutral lipids (NLs), glycolipids plus sphingolipids

(G + S) and phospholipids (PLs), respectively) according to Fakas et al. [31] and Gardeli et al. [30]. Lipid fractions were trans-methylated according to the previously mentioned protocol. Thin layer chromatography (TLC) analysis in this crude lipid ("Folch" extract) was performed on glass pre-coated silica gel G plates (Merck, Darmstadt, Germany) (20 cm × 20 cm; thickness = 0.25 mm) as previously described [30]. Plates were pre-washed and activated according to Papanikolaou et al. [32] and separation of crude total lipid was carried out with petroleum ether/diethyl ether/glacial acetic acid (80:20:1, *v/v/v*). Plate application of the sample and the lipid standards and visualization after the development were conducted according to Papanikolaou et al. [32]. Glyceryl trioleate (TAG), cholesterol (CL), cholesteryl linoleate (CE), oleic acid (FFA), and monononadecanoin (MAG) [33] were employed as lipid standards for the identification of the main bands on TLC plates.

Determination of total intracellular polysaccharides (IPS) was carried out using a modified protocol described by Diamantopoulou et al. [34]. In brief, a precisely weighted quantity of TDCW (≤100 mg) was subjected to boiling (100 °C) with 20 mL of 2.5-N HCl for 1 h, and afterwards the whole was neutralized to pH = 7.0 with 2.5-N NaOH. Then, IPS were quantified as glucose equivalents with the DNS method [35] and were expressed in both absolute (g/L) and relative ($Y_{IPS/X}$, % of total polysaccharides in TDCW) values.

Total reducing sugars in the fermentation medium (viz. xylose for the trials with no OMWs, mostly xylose and to lesser extent glucose and fructose for the trials in which OMWs were added) were quantified according to the DNS method [35]. Pure xylose (purity = 99% *w/w*) was employed in order to trace the calibration curve. In some instances, and in order to perform cross-validation of the results, the concentration of the remaining xylose was also determined during the fermentation using HPLC. Xylitol, produced in small concentrations and organic acids and sugars found in the OMWs, were also determined through HPLC analysis, performed according to Papanikolaou et al. [16]. The area of each compound was determined according its retention time and the concentration of each compound was determined using reference curves and expressed as g/L. Free amino nitrogen (FAN) concentration into the liquid samples was determined at 570 nm using the ninhydrin colorimetric method with glycine employed in order to trace the calibration curve, according to Kachrimanidou et al. [36].

Determination of total phenol compounds concentration (PCC) into the medium was carried out according to the method described by the Folin–Ciocâlteu protocol modified according to Aggelis et al. [37] in the sample supernatant. Absorbencies were measured at 750 nm in a Hitachi *U*-2000 Spectrophotometer (Hitachi High Technologies Corp., Fukuoka, Japan). The concentration of phenolic compounds was expressed in equivalence of gallic acid according to a reference curve. Moreover, in order to determine the decolorization efficiency of the fermentations, 0.5-mL samples were mixed well with 14.5 mL distilled water and the absorbance was measured at 395 nm (same spectrophotometer as previously) according to Sayadi and Ellouz [38]. The decolorization percentage was calculated using the equation: % A = $[(A_0 - A_1)/A_0] \times 100$, where, A_0 is the absorbance at time 0 and A_1 is the absorbance at each experimental point during the fermentation.

2.3. Data Analysis

Each experimental point of all the kinetics presented in tables and figures is the mean value of two independent determinations; in fact, for each experiment presented, two lots of independent cultures using different inocula were conducted; the standard error (SE) was < 15%. Data were plotted using Kaleidagraph 4.0 Version 2005 showing the mean values with the standard error mean. Throughout the text, indices 0 and max represent the initial and the maximum quantity of the elements in each kinetics presented.

3. Results and Discussion

3.1. Initial Screening on Commercial-Type Xylose

In the first part of this investigation, all available strains were cultivated on commercial-type xylose at initial total sugar (xylose) (S_0) concentration $\approx$70 g/L under nitrogen-limited conditions. The achieved kinetic results are illustrated in Table 1. All strains presented appreciable TDCW production (X_{max} ranging between 15.6 and 19.0 g/L, respective total biomass yield on sugar consumed $Y_{X/S} \approx$ 0.22–0.28 g/g). $Y_{X/S}$ values seemed to be slightly higher in the middle of the cultures and before X_{max} values were recorded. Moreover, variable quantities of SCO were produced by the microorganisms tested; the lower lipid in TDCW quantities ($Y_{L/X}$, % *w/w*) were recorded for the strains *Rhodosporidium toruloides* NRRL Y-27012 ($Y_{L/X} <$ 20% *w/w*) and *Rhodotorula glutinis* NRRL YB-252 ($Y_{L/X} \approx$ 10% *w/w*) despite the fact that growth was conducted in media favoring the production of SCO. This result is in disagreement with recent investigations demonstrating the strain *R. toruloides* NRRL Y-27012 produced huge quantities of SCO ($Y_{L/X} >$ 40% *w/w*, in some trials this value was = 54.3% *w/w*, with an L_{max} value $\approx$ 12 g/L) on media composed of crude glycerol, the principal waste stream deriving from biodiesel production process [39]. Equally, significant SCO quantities were produced by the strain *R. glutinis* NRRL YB-252 ($Y_{L/X}$ = 38.2% *w/w*, corresponding to $L_{max} \approx$ 7.2 g/L) in similar types of media (crude glycerol) [40], demonstrating that in the above-mentioned yeast, as well as in a plethora of other yeast species and genera, glycerol is a very competitive substrate related with the production of TDCW and added-value extracellular and intracellular metabolites [33,40–43]. It is not clearly understandable why such important discrepancies exist between the lipid production process for *R. toruloides* NRRL Y-27012 and *R. glutinis* NRRL YB-252 growing on xylose and glycerol, but in any case—and in accordance with the literature [4,5,22,23,44,45]—it appears that the carbon source seems to play a very crucial role in the de novo lipid production process, even if implicated substrates present important similarities among them in molecular and metabolic level. In this section, though, it must be indicated that while the intracellular metabolism of glycerol and glucose (or other hexoses) theoretically presents important similarities (these compounds are mostly metabolized through the EMP pathway) [5,22,23], the intracellular metabolism of xylose could present some differences compared with the above-mentioned compounds. These differences could, finally, reflect in the quantity of cellular lipids produced by the microorganisms growing on these substrates; for instance, xylose metabolism implicates in many instances the pentose–phosphate pathway [5,22,23], that is somehow less efficient as regards biomass production compared with the typical EMP glycolysis utilized for the catabolism of hexoses or glycerol [5,22]. On the other hand, xylose metabolism implicates in its first step, the reduction of xylose into xylitol, that in many cases in which oleaginous microorganisms are involved [30,33] is secreted into the medium and it is not re-utilized in order to be converted into SCO. Therefore, there would be a carbon "loss" related with the production of microbial lipids when xylose is used as microbial substrate amenable to be converted into cellular lipid compared with utilization of hexoses or glycerol [22,30,33].

The remaining "red" yeast strain implicated in the current investigation, namely *R. toruloides* DSM 4444, presented efficient cell growth and non-negligible lipid accumulation ($Y_{L/X} \approx$ 30% *w/w*) on media composed of commercial-type xylose (Table 1). The above-mentioned strain has been revealed capable to produce interesting SCO quantities during growth of glycerol and, to lesser extent glycerol/xylose blends [33], whereas it accumulated indeed impressive quantities of storage lipid during growth on flour-rich waste stream hydrolysates, mostly composed of glucose, during growth in shake-flask or bioreactor experiments [46].

C. curvatus NRRL Y-1511 presented the highest TDCW production among all tested strains cultivated on commercial-type xylose (X_{max} = 19 g/L) even though this strain did not produce significant SCO production under the given culture conditions ($Y_{L/Xmax} <$ 25% *w/w* for all growth steps) (Table 1). The very same has been reported to produce a SCO quantity = 4.3 g/L (respective $Y_{L/Xmax}$ value $\approx$ 30% *w/w*) during growth on lactose, in shake-flask fermentations [47]. Harde et al. [48] reported a SCO produced quantity = 5.1 g/L (respective $Y_{L/X} \approx 38\%$ *w/w*) during growth of the above-mentioned yeast strain on xylose-based shake-flask fermentations. On the other hand, the same strain was not capable to produce significant SCO quantities ($Y_{L/X}$ values always < 20% *w/w*) during growth on media composed of saccharose (i.e., molasses, crude sucrose, orange peel waste extracts) [47,49]. Moreover, in the current submission, *L. starkeyi* DSM 70296 was proved to be a robust TDCW-producing strain during growth on commercial-type xylose, in shake-flask experiments under nitrogen-limited conditions, although total lipid quantities produced ($Y_{L/Xmax} \approx 34\%$ *w/w*) were somewhat or remarkably lower compared with the ones achieved on glycerol [39] or (mostly glucose-based) flour-rich waste stream hydrolysates [50] in shake-flask and bioreactor experiments. All these results and their comparisons with the literature indicate, once more, that important differentiations in the lipid production bioprocesses may exist even when carbon sources presenting biochemical similarities are used as substrates for the given oleaginous microorganisms employed.

Table 1. Quantitative data of yeast strains deriving from experiments performed on commercial-type xylose, in nitrogen-limited shake-flask cultures, in which the initial sugar concentration was adjusted to *c.* 70 g/L. Four different points in the fermentations are represented: (a) when the maximum quantity of total dry cell weight (TDCW—*X*, g/L) was observed; (b) when the maximum quantity of total cellular lipids (*L*, g/L) was observed; (c) when the maximum quantity of xylitol (Xyl, g/L) was observed; (d) when the maximum quantity of total intracellular polysaccharides (IPS, g/L) was observed. Fermentation time (h), quantities of TDCW (*X*, g/L), total lipid (*L*, g/L), intracellular polysaccharides (IPS, g/L), total sugar (xylose) consumed (S_{cons}, g/L), lipid in TDCW ($Y_{L/X}$,% *w/w*) and polysaccharides in DCW ($Y_{IPS/X}$,% *w/w*) are depicted for all the above-mentioned fermentation points. Culture conditions: growth on 250-mL conical flasks filled with the ⅕ of their volume at 180 ± 5 rpm, initial pH = 6.0 ± 0.1, pH ranging between 5.2 and 5.8, incubation temperature 28 °C. Each experimental point is the mean value of two independent measurements (SE < 15%).

Yeasts		Time (h)	S_{cons} (g/L)	*X* (g/L)	*L* (g/L)	Xyl (g/L)	$Y_{L/X}$ (%, *w/w*)	IPS (g/L)	$Y_{IPS/X}$ (%, *w/w*)
Rhodosporidium	c	72	26.8	5.6	0.6	4.3	10.7	1.3	23.2
toruloides	d	120	37.2	10.7	2.8	4.0	26.2	2.3	21.4
DSM 4444	a, b	264	57.8	16.5	4.8	3.1	29.1	1.7	10.3
Rhodosporidium	d	120	40.1	11.9	1.3	Tr.*	10.9	3.2	26.9
Toruloides NRRL Y-27012	a, b, c	264	59.0	15.6	3.0	Tr.*	19.5	2.7	17.3
Rhodotorula glutinis	b	168	41.7	10.1	1.1	0.2	9.9	1.9	12.6
NRRL YB-252	a, c, d	264	59.0	15.1	1.0	2.0	6.6	5.0	33.1
Cryptococcus curvatus	c, d	192	50.9	16.6	2.9	4.5	17.5	5.7	34.3
NRRL Y-1511	a, b	220	63.0	19.0	4.4	2.6	23.2	5.4	28.4
Lipomyces starkeyi	c	144	46.4	15.6	4.0	1.0	25.6	5.8	37.1
DSM 70296	a, b, d	170	56.1	17.5	5.9	Tr. *	33.7	6.3	36.0
Cryptococcus curvatus	c	144	40.4	13.5	5.6	4.5	41.5	2.8	20.7
ATCC 20509	a, b, d	216	64.0	17.4	8.1	3.2	46.6	3.2	18.4

* Tr. < 0.1 g/L.

With the exception of the strain *R. glutinis* NRRL YB-252 that gradually and throughout the culture seemed to accumulate total intracellular polysaccharides (IPS) (in accordance with the results achieved during growth on glycerol under nitrogen-limited conditions; see Filippousi et al. [40]), all other yeast strains used seemed to present higher IPS in TDCW values ($Y_{IPS/X}$, % *w/w*) at the relative earlier growth steps, and these values seemed to decrease as growth proceeded. Moreover, in several of the tested microorganisms (i.e., strains *C. curvatus* NRRL Y-1511, *R. toruloides* NRRL Y-27012, *R. toruloides* DSM 4444, *L. starkeyi* DSM 70296), appreciable $Y_{IPS/X}$ quantities (i.e., ≥ 35% *w/w*) have been reported even at the very early growth steps (i.e., in fermentation time, $t \leq 60$ h) where assimilable nitrogen was found into the medium—or it had barely been exhausted, in accordance with results in which other low-molecular weight hydrophilic carbon sources had been used as substrates under nitrogen-limited conditions (i.e., crude sucrose, lactose, biodiesel-derived glycerol, etc.; see: [33,40,47]). In contrast, in other yeast strains reported in the literature (yeast species/genera like *Apiotrichum curvatum*, *Yarrowia lipolytica*, other *R. toruloides* strains, *Metschnikowia* sp.) that had been cultivated on hydrophilic carbon sources under nitrogen-limited conditions, it has been seen that accumulation of intracellular polysaccharides inside the cells occurred continuously and without interruption after nitrogen deprivation from the medium, that in some cases occurred simultaneously with lipid accumulation [33,40,51,52].

Most of the yeast strains cultivated on xylose produced small quantities of xylitol (Xyl), which in some of the performed trials was reconsumed in order for TDCW to be synthesized. The concentration of the secreted xylitol was not significant (Xyl $\leq$ 4.5 g/L), and in any case it was drastically lower than the one reported for other eukaryotic microorganisms cultivated on media based on xylose like *Thamnidium elegans* [53], *Ashbya gossypii* [54] and *Mortierella isabellina* [30]. Given that Xyl concentration was not high under the present culture conditions, the biosynthesis of this metabolite was not further studied in the current investigation.

The most promising among the microorganisms tested in the "screening" part of the work was the yeast strain *C. curvatus* ATCC 20509, and the kinetic profile of growth of this strain on commercial-type xylose in shake-flask cultures is depicted in Figure 1a–c. The microorganism presented efficient microbial growth with almost not any lag phase at all occurring during growth. The μ_{max} value, calculated by the formula $\mu_{max} = \frac{\ln(X_1/X_0)}{t_1-t_0}$ (where X_1 was the first experimental point after inoculation and t_1 was the respective time; X_0 was the initial TDCW quantity and $t_0 = 0$ h) was found to be $= 0.19$ h^{-1}. Moreover, cellular lipids in appreciable quantities were accumulated only after virtual depletion of nitrogen from the growth medium, which occurred at $t \approx 60$ h after inoculation (see and compare evolutions of FAN in Figure 1c and $Y_{L/X}$ in Figure 1b). Interestingly, xylose consumption rate seemed to be uninterrupted despite nitrogen limitation that rapidly occurred in the medium (Figure 1c), in disagreement with results reported for several types of oleaginous microorganisms cultivated on sugars or similarly metabolized compounds under nitrogen-limited conditions, where sugar uptake rate was (much) higher in the balanced growth phase in which nitrogen in significant quantities was found into the medium, compared with that recorded in the lipid-accumulating phase in which nitrogen had been exhausted [30,55–59]. On the other hand, IPS in non-negligible concentrations (i.e., $Y_{IPS/X}$ > 35% *w*/*w*) were recorded at the early growth steps during the presence of assimilable nitrogen into the medium. $Y_{IPS/X}$ values significantly decreased as the fermentation proceeded (Figure 1b,c). This sequential biosynthesis of intracellular polysaccharides and storage lipids that finally becomes an "interplay" between their productions has been originally observed in another *C. curvatus* strain (namely NRRL Y-1511) during its flask cultures on commercial-type lactose. This physiological feature seems to be quite characteristic for the species *C. curvatus*, and it has also been observed, nevertheless less clearly than in the case of *C. curvatus*, for other oleaginous species like *M. isabellina* growing on sugars [30] or *R. toruloides* growing on glycerol [33].

Finally, at the late growth steps and when the concentration of xylose seemed not capable to saturate the microbial metabolic activities, cellular lipids were mobilized in favor of energy creation, in accordance with several reports indicated in the literature for *Mucor circinelloides* [60], *Y. lipolytica* [28,32,41], *M. isabellina* [30,32,44], *Cunninghamella echinulata* [57], *Trichoderma viride* [61], *Candida curvata* [62,63] and other oleaginous strains, suggesting that lipid mobilization (turnover) in the oleaginous microorganisms is a quite common feature observed regardless of the carbon source that was used in order for cellular lipids to be created [22].

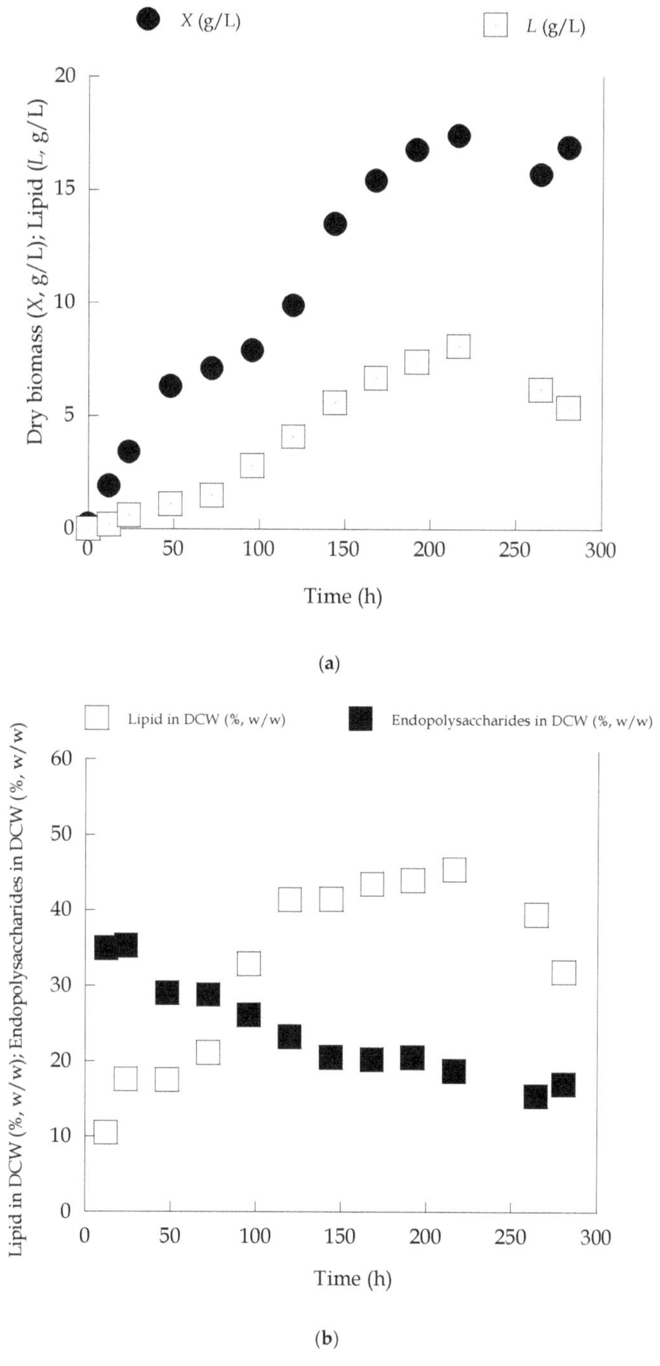

(**a**)

(**b**)

Figure 1. *Cont.*

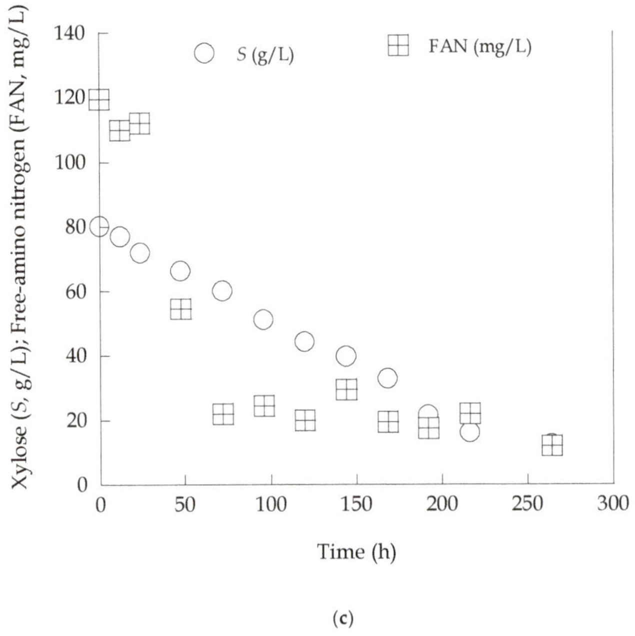

(c)

Figure 1. (**a**) Kinetics of total dry cell weight (TDCW; X, g/L) and lipid (L, g/L) production, (**b**) lipid in TDCW (%, w/w) and endopolysaccharides in TDCW (%, w/w) evolution and (**c**) xylose (S, g/L) and free-amino nitrogen (FAN, mg/L) assimilation by *Cryptococcus curvatus* ATCC 20509, during growth on commercial-type xylose in shake-flask experiments under nitrogen-limited conditions. Culture conditions: growth on 250-mL conical flasks filled with the ⅕ of their volume at 180 ± 5 rpm, initial pH = 6.0 ± 0.1, pH ranging between 5.2 and 5.8, incubation temperature 28 °C, initial sugar concentration ≈70 ± 10 g/L. Each experimental point is the mean value of two independent measurements (SE < 15%).

3.2. Lipid Production in High-Xylose Concentration Media Partially Diluted with Olive Mill Wastewaters

In the second part of the current investigation, the most promising of the previously screened strains as regards their lipid-producing capabilities (namely *C. curvatus* ATCC 20509 and *L. starkeyi* DSM 70296), were cultured on media composed of higher S_0 concentrations (adjusted to ≈ 100 g/L), in which the concentration of extracellular nitrogen remained constant (as in the "screening" experiment). The increase of the S_0 concentration in media in which nitrogen availability remained stable, would be attainable to further "boost" the production of SCOs, since higher initial C/N molar ratio media would have been prepared and typically and according to the theory lipid accumulation would have further been stimulated [5,7,8,23]. In these media, OMWs were added in various quantities, in order to partially replace tap water in the fermentations carried out. As previously indicated, this concept was realized within the frame of zero-waste release and water-saving policies that are currently adopted, in a dual scope: To partially replace water by this highly polluted and toxic wastewater, and to potentially perform partial detoxification (i.e., removal of phenolic compounds and color) during lipid production bioprocesses performed. Moreover, it must be stressed that the phenolic compounds that are presented into the OMWs, present significant chemical similarities with recalcitrant phenol-type compounds that are found in various abundant xylose-rich lignocellulose-type wastewaters like spent-sulfite liquor [64–66] and waste xylose mother liquid [1,6]. The composition itself of the media that were currently prepared in order to carry out the lipid production bioprocesses (viz. media containing initial xylose at c. 100 g/L, insignificant glucose and fructose quantities (<5 g/L) and variable PC_0 concentrations ranging between 0.0 g/L (no OMWs added) and c. 2.0 g/L (60% OMWs and 40% water)) present noticeable similarities with the above-mentioned liquid waste-streams. Therefore, in this part of the work it can be considered among other issues, that lipid production was studied in media mimicking the composition of various types of abundant xylose-containing lignocellulosic-type wastewaters.

The obtained results for *C. curvatus* ATCC 20509 and *L. starkeyi* DSM 70296 cultures performed on xylose-based media blended with OMWs are depicted in Table 2. Concerning the cases of *C. curvatus*, the obtained results indicated that within the range of PCC_0 added into the medium (up to *c.* 2.0 g/L) significant TDCW occurred regardless of the addition of OMWs (and, hence, phenolic compounds) into these media. Moreover, interestingly, it appears that the more the OMWs were added into the medium, the more the $TDCW_{max}$ concentrations increased. In a similar trend, as far as the case of *L. starkeyi* is concerned, addition of OMWs in significant concentrations (viz. PCC_0 up to *c.* 2.0 g/L) did not seem to have serious impact upon the TDCW synthesis by this microorganism, that always remained in high levels ($TDCW_{max} \approx 25$–26 g/L irrespective of the addition of OMWs), demonstrating its suitability related with biomass production on these types of media, in accordance with results reported for other strains of this species [67,68].

Table 2. Quantitative data of *Cryptococcus curvatus* ATCC 20509 and *Lipomyces starkeyi* DSM 702096 originated from kinetics performed on commercial-type xylose, in nitrogen-limited shake-flask cultures, in which OMWs were added in various concentrations into the medium and the initial sugar (mostly xylose) concentration was adjusted to *c.* 100 g/L. The initial phenolic compounds concentration (PCC_0) is illustrated for all runs carried out. Two different points in the fermentations are represented: (a) when the maximum quantity of total dry cell weight (TDCW—X, g/L) was observed; (b) when the maximum quantity of total cellular lipids (L, g/L) was observed. Fermentation time (h), quantities of TDCW (X, g/L), total lipid (L, g/L), total sugar (mostly xylose) consumed (S_{cons}, g/L) and lipid in TDCW ($Y_{L/X}$,% w/w) are depicted for all the above-mentioned fermentation points. Culture conditions: growth on 250-mL conical flasks filled with the ⅕ of their volume at 180 ± 5 rpm, initial pH = 6.0 ± 0.1, pH ranging between 5.2 and 5.8, incubation temperature 28 °C. Each experimental point is the mean value of two independent measurements (SE < 15%).

PCC_0 (g/L)		Time (h)	S_{cons} (g/L)	X (g/L)	L (g/L)	$Y_{L/X}$ (%, w/w)
		Cryptococcus curvatus ATCC 20509				
0.00 *	b	168	87.3	21.0	10.0	47.6
	a	215	89.9	23.0	6.0	26.1
1.16	b	144	77.6	20.9	8.4	40.2
	a	215	89.7	24.4	3.6	14.8
1.65	b	240	94.8	24.7	5.0	20.2
	a	336	97.0	25.7	1.8	7.0
1.91	b	192	92.1	23.8	2.5	10.5
	a	216	94.0	27.0	1.6	5.9
		Lipomyces starkeyi DSM 702096				
0.00 *	b	216	84.1	25.3	6.0	23.7
	a	239	92.3	26.2	4.4	16.7
1.96	b	239	92.8	21.1	5.9	27.9
	a	255	96.0	25.0	4.0	16.0

*: No olive mill wastewaters (OMWs) added.

It is not a simple task to explain this type of feature of the microbial "resistance"—or, even more, the "boost" of TDCW production in the presence of OMWs in the fermentations carried out. It is widely known that these wastewaters contain phenolic compounds that are recalcitrant and highly toxic substances, the presence of which constitutes the main problem of the safe remediation and disposal of OMWs and similar types of wastewaters [9,20,37,38,69]. On the other hand, it must be pointed out that within the range of PCC_0 found in the current investigation (viz. 0.0 up to 2.0 or even 3.0 g/L), similar results concerning the "resistance" of TDCW production—or even the "boost" of growth in the presence of OMWs for other yeast strains—has been reported [11,12,16–18,70]. Specifically, the increment of TDCW production by the strain *Saccharomyces cerevisiae* MAK-1 in which OMWs were added in various initial concentrations (additions up to $PCC_0 \approx 3.0$ g/L) was impressive as compared with the blank experiment ($PCC_0 = 0.0$ g/L—no OMWs added) [18]. Likewise, various strains of the nonconventional yeast *Y. lipolytica* noticeably resisted despite the presence of phenolic compounds in significant concentrations (i.e., PCC_0 up to 5.5 g/L) in the media [11,12,17,70,71]. In accordance with the achieved in the present study results, other yeasts belonging to the species *Candida cylindracea*, *C. tropicalis*, *C. albidus* and *Trichosporon cutaneum* seemed to demonstrate remarkable resistance upon the phenolic compounds of OMWs [15,68,72–74]. Likewise, some addition of OMWs into glucose-based media (i.e., OMWs added to 20% or 30% *v/v*) seemed to enhance TDCW production not only in yeast cultures but also in other cultures of microbial species like the edible fungi *Lentinula edodes* and *Pleurotus ostreatus* growing on OMW-based media [69,75].

The addition of phenolic compounds into the medium significantly altered the cellular metabolism of *C. curvatus* shifting the intracellular carbon flow towards the synthesis of lipid-free biomass, and significantly decreasing the accumulation of lipid inside the yeast cells (Table 2). When PCC_0 was = 0.0 g/L (no OMWs added), *C. curvatus* produced noticeable SCO quantities ($L_{max} = 10.0$ g/L corresponding to $Y_{L/X} \approx 48\%$ *w/w*) that were reduced when increasing PCC_0 and thus, increasing OMWs amounts were added into the medium. When PCC_0 was adjusted to ≈ 1.2 g/L (viz. trials in which *c.* 100 g/L of xylose were diluted to *c.* 35% *v/v* of OMWs and *c.* 65% *v/v* of tap water), again the production of SCO was quite satisfactory ($L_{max} = 8.4$ g/L corresponding to $Y_{L/X} = 40.2\%$ *w/w*). Further increase of OMWs addition into the medium (i.e., PCC_0 up to ≈ 2.0 g/L meaning that xylose was diluted to *c.* 60% *v/v* of OMW and *c.* 40% of tap water) noticeably decreased both the L_{max} and $Y_{L/X}$ values for *C. curvatus*. On the other hand, SCO production bioprocess for *L. starkeyi* growing on phenol-containing wastewaters enriched with commercial-type xylose seemed to be more "robust" compared with the one of *C. curvatus*; as it was demonstrated, at least for the range of PCC_0 tested in the current submission (PCC_0 up to *c.* 2.0 g/L), lipid production seemed unaffected by the addition of toxic phenolic compounds into the medium (see Table 2). The above-mentioned physiological feature of *L. starkeyi* (meaning, in fact, that lipid production seemed unaffected by the addition of OMWs into the medium, at least until $PCC_0 \approx 2.0$ g/L) together with the fact that $TDCW_{max}$ production was equally not negatively influenced by the addition of the same range of OMWs into the medium, were the main reasons for which the growth of *L. starkeyi* was not tested in intermediate PCC_0 values (i.e., 1.2 or 1.7 g/L), as it happened with *C. curvatus*. Moreover, another interesting result achieved in this second set of experiments, was related with the fact that the highest concentrations of TDCW and total lipids did not occur at the same fermentation time (see Table 2); apparently, the carbon flow that occurred during the fermentation steps after nitrogen limitation led to significant lipid accumulation (that was higher compared with the results reported in Table 1, as it has been anticipated), but finally, cellular lipids were subjected to degradation despite the fact that xylose remained in some non-negligible concentrations into the medium. At the latter growth steps, the low xylose concentration seemed incapable to saturate the metabolic requirements of *C. curvatus* and *L. starkeyi*, leading to consumption of the cellular lipids in favor of lipid-free material, and for this reason TDCW concentration peaked at the end of the cultures (see Table 2).

The addition of OMWs into the medium seemed to have contradictory results in relation to several types of oleaginous microorganisms; for instance, addition of PC increased, in some cases noticeably, the quantity of lipids in TDCW for some strains of the yeast *Y. lipolytica* [11,17]. Equally, addition of OMWs in glucose-based cultures of oleaginous Zygomycetes, seemed to enhance the $Y_{L/Xmax}$ values of several oleaginous Zygomycetes, and due to this feature, OMWs had been characterized as a "lipogenic" medium [26]. In the above-mentioned studies, as in the current investigation, OMWs deriving from 3.0-phase extraction systems and containing comparable (and, in general, relatively low) concentrations of sugars were employed as microbial substrates and fermentation waters. Moreover, these OMWs contained indeed negligible concentrations of residual olive oil, as is the case of the current investigation. This fact excludes any ex novo lipid accumulation process that could be carried out from olive oil found into the growth medium; in the above-mentioned studies as in the current investigation, physiological differences concerning the microbial behavior seemed to be attributed mostly to the PC presence into the medium. Contradictory to these results, the addition of OMWs negatively affected the production of SCO in the strain *L. starkeyi* NRRL Y-11557, than the control experiment (trial on glucose in which no OMWs were added) [68]. Generally, the addition of OMWs can significantly change the spectrum of the final products synthesized by various types of yeasts, compared with the control trials (no OMWs added) [11,12,15,17,70,73]. The kinetics of TDCW and lipid production and total extracellular sugars (mostly xylose) and FAN evolution during growth of *C. curvatus* in media in which PCC_0 was adjusted to ≈ 1.2 g/L is depicted in Figure 2a,b.

The μ_{max} value, calculated as previously was found to be slightly lower than in the previous trial (=0.13 h^{-1}) and this was due to the potential inhibiting effect of the phenol-type wastewater found into the medium. As in the previous set of experiments (see Figure 1a,b) cellular lipids were subjected to biodegradation when somehow low quantities of xylose, incapable to saturate the microbial metabolic requirements, were found into the medium.

Given that *C. curvatus* presented an efficient growth and a quite satisfactory lipid-accumulation in shake-flask cultures in which high initial xylose concentrations were added, whereas the addition of OMWs into the medium negatively affected the process of lipid accumulation, in the next part of the present investigation it was decided to scale-up the culture in a bench top laboratory-scale bioreactor (active volume = 2 L). Commercial xylose ($S_0 \approx 100$ g/L) and nitrogen compounds (yeast extract at 3.0 g/L and peptone at 1.0 g/L) were added as in the previous part of the investigation, whereas no OMWs were added into the medium, in order to enhance the maximum the quantity of cellular lipids produced by *C. curvatus*. The obtained results are shown in Figure 3a,b. Compared with the equivalent trial performed in shake-flask mode (see Table 2, entry in which $PCC_0 = 0.0$ g/L) it may be assumed that xylose uptake rate seemed to be slightly lower in the bioreactor experiment than in the shake-flask trial. This finding is in agreement with results in which filamentous fungi (like *T. elegans* and *M. isabellina*) have been reported to consume more rapidly sugars (including xylose) in shake-flask experiments than in batch bioreactor experiments [44,53,59]. On the other hand—and in accordance with the literature [5,22,23]—the batch bioreactor experiment of *C. curvatus* cultivated on commercial-type xylose revealed $TDCW_{max}$ and L_{max} values that were significantly higher than in the respective shake-flask trial (after 340 h of incubation X_{max} and SCO_{max} quantities achieved were 37.0 and 16.4 g/L with respective $Y_{L/X}$ value = 44.3% *w/w*). Moreover, the global yield of SCO produced per quantity of xylose consumed ($Y_{L/S}$) was ≈ 0.17 g/g (Figure 3b), that is higher than the respective shake-flask experiment (≈ 0.11 g/g– see results in Table 2, entry in which $PCC_0 = 0.0$ g/L).

In any case, the production of TDCW and lipids for *C. curvatus* and *L. starkeyi* cultivated on commercial-type xylose and blends of commercial-type xylose and OMWs seemed promising. Various bioreactor experiment studies using batch, fed-batch and continuous strategies have appeared in a number of times in the literature [50,76–81], including in various cases the strains *C. curvatus* ATCC 20509 and *L. starkeyi* DSM 70,296. Comparisons of TDCW and lipid production with data collected from the relevant literature concerning *C. curvatus* and *L. starkeyi* strains are depicted in Table 3. From this table it can be deduced that strains of the species *L. starkeyi* (and also the currently used DSM 70296 strain) can present excellent SCO production in bioreactor experiments [50,76], therefore potentially high lipid production can be achieved by this microorganism cultivated on xylose-based media in bioreactor trials. On the other hand, the reported in the current investigation results by *C. curvatus* ATCC 20509, especially the ones achieved in batch bioreactor trials during growth on commercial-type xylose, were revealed as quite competitive and promising, demonstrating the potential of the microorganism in this type of bioprocesses.

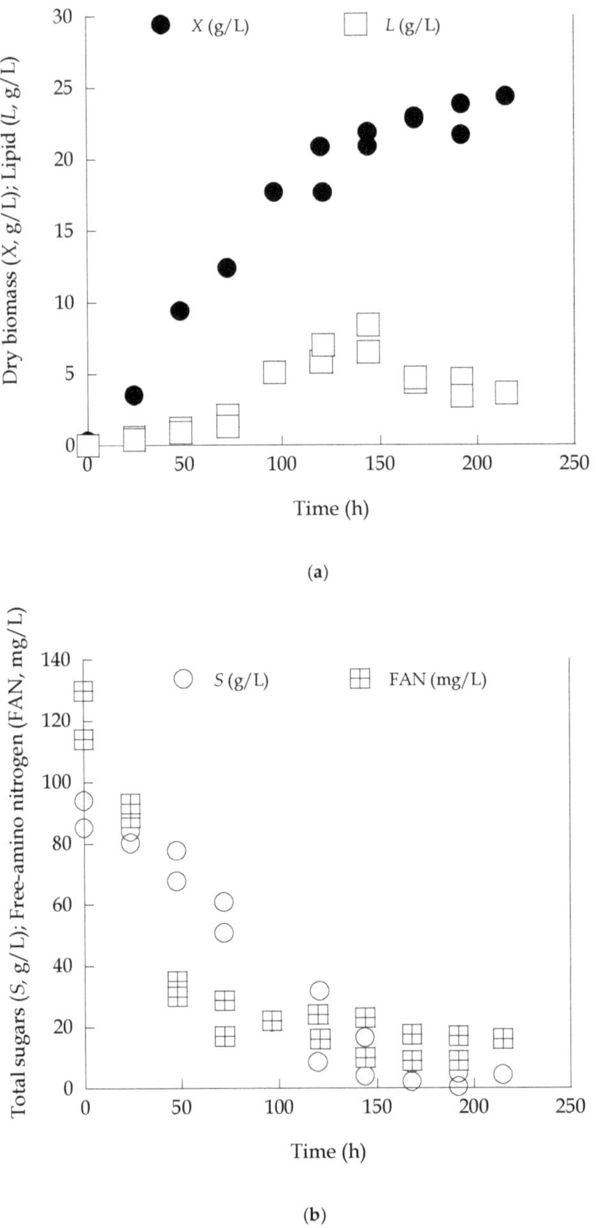

Figure 2. (a) Kinetics of total dry cell weight (TDCW; *X*, g/L) and lipid (*L*, g/L) production and (b) xylose (*S*, g/L) and free-amino nitrogen (FAN, mg/L) assimilation by *Cryptococcus curvatus* ATCC 20509, during growth on commercial-type xylose in media diluted with olive mill wastewaters, at initial phenol-content concentration ≈1.2 g/L in shake-flask experiments under nitrogen-limited conditions. Culture conditions: growth on 250-mL conical flasks filled with the ⅕ of their volume at 180 ± 5 rpm, initial pH = 6.0 ± 0.1, pH ranging between 5.2 and 5.8, incubation temperature 28 °C, initial sugar concentration ≈100 ± 10 g/L. Each experimental point is the mean value of two independent measurements (SE < 15%).

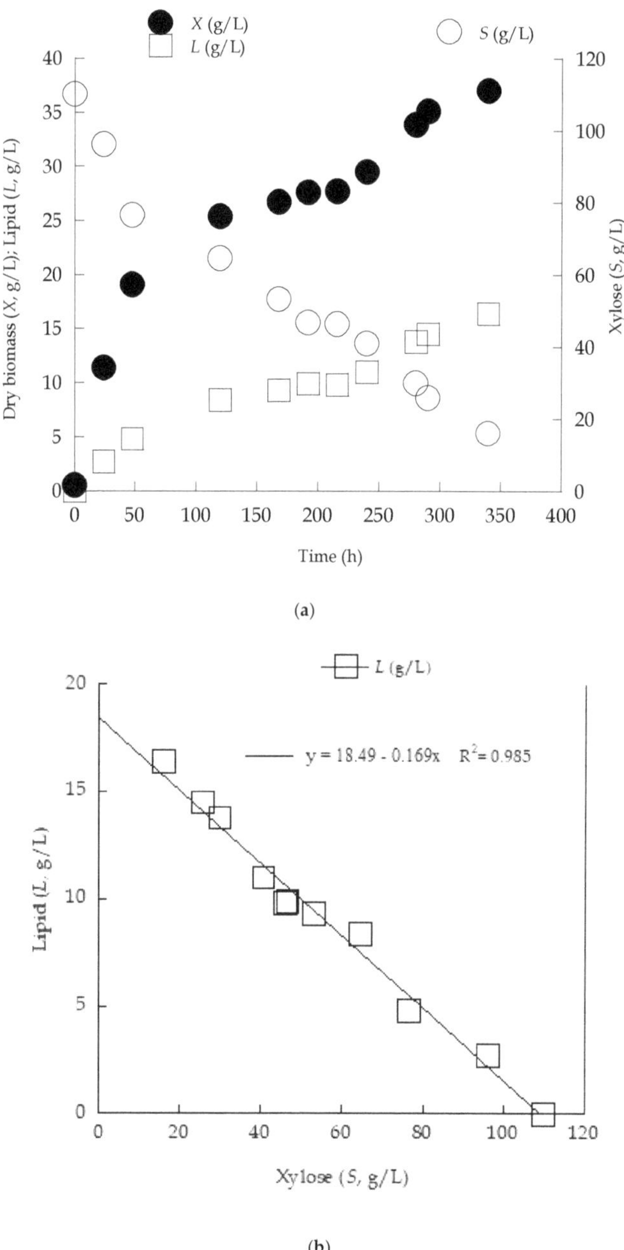

(**a**)

(**b**)

Figure 3. (**a**) Kinetics of evolution of total dry cell weight (TDCW; *X*, g/L), lipid (*L*, g/L) and xylose (*S*, g/L) and (**b**) lipid produced (*L*, g/L) vs. remaining xylose (*S*, g/L) by *Cryptococcus curvatus* ATCC 20509, during growth on commercial-type xylose in batch bioreactor experiment under nitrogen-limited conditions. Culture conditions: growth on laboratory-scale bioreactor (active volume 2.0 L), agitation rate 450 ± 5 rpm, incubation temperature 28 °C, aeration rate up to 1.5 vvm, initial xylose concentration (S_0) ≈100 g/L, pH ranging between 5.9 and 6.1, oxygen saturation ≥ 20% (*v/v*) for all growth phases. Each experimental point is the mean value of two independent measurements (SE < 15%).

Table 3. Total dry cell weight (TDCW, g/L) and lipid in TDCW ($Y_{L/X}$,% *w/w*) values obtained by (a) *Cryptococcus curvatus* and (b) *Lipomyces starkeyi* strains growing on several carbon sources and fermentation configurations.

Strain	Culture Mode	Carbon Source	TDCW (g/L)	$Y_{L/X}$ (%, *w/w*)	Reference
(a)					
Cryptococcus curvatus					
D	Continuous bioreactor	Glucose	13.5	29.0	Evans and Ratledge [77]
		Xylose	15.0	37.0	
ATCC 20509	Batch bioreactor / Continuous recycling	Whey permeate	21.6 / 85.0	36.0 / 35.0	Ykema et al. [78]
	Batch bioreactor	Prickly pear juice	10.9	45.8	Hassan et al. [79]
	Fed-batch bioreactor	Glycerol (pure)	118.0	25.0	Meesters et al. [80]
	Fed-batch bioreactor	Glycerol (crude)	32.9	52.9	Liang et al. [81]
	Fed-batch bioreactor	Glycerol (crude)	44.9	49.0	Cui et al. [82]
	Single-stage continuous	Acetic acid	5.1	66.4	Gong et al. [83]
NRRL Y-1511	Shake flasks	Lactose (commercial)	14.5	29.7	Tchakouteu et al. [47]
	Shake flasks / Shake flasks	Glucose / Xylose (pure)	13.2 / 13.3	19.7 / 38.3	Harde et al. [48]
ATCC 20509	Continuous bioreactor	Acetic acid	26.7	48–53	Béligon et al. [84]
	Shake flasks	WPHL	17.3	52.5	Zhou et al. [85]
	Fed-batch bioreactor	Cheese whey	66.8	49.6	Kopsahelis et al. [86]
ATCC 20509	Shake flasks	Xylose (commercial) / Xylose/OMWs	21.0 / 20.9	47.6 / 40.2	Present study
	Batch bioreactor	Xylose (commercial)	37.0	44.3	
(b)					
Lipomyces starkeyi					
DSM 70295	Shake flasks	Glucose / Glucose/sewage sludge	9.4 / 9.3	68.0 / 72.3	Angerbauer et al. [87]
AS 2.1560	Batch bioreactor	Glucose	30.0	46.0	Liu et al. [88]
CBS 187	Shake flask	Xylose (pure)	12.5	80.0	Oguri et al. [89]
DSM 70296	Fed-batch bioreactor / Batch bioreactor	Glucose/xylose / SCBHL	82.4 / 13.9	46.9 / 26.7	Anschau et al. [76]
AS 2.1560	Fed-batch bioreactor	Xylose (pure)	94.7	65.5	Lin et al. [90]
CBS 1807	Shake flasks	Glucose/fructose / SSJc	12.3 / 21.7	47.3 / 29.5	Matsakas et al. [91]
DSM 70296	Fed-batch bioreactor / Shake flasks	FRWHL/glucose / FRWHL	109.8 / 30.5	57.8 / 40.4	Tsakona et al. [50]
	Shake flasks	Glycerol (crude)	34.4	35.9	Tchakouteu et al. [39]
NRRL Y-11557	Shake flasks	Glucose/OMWs	9.5	24.5	Dourou et al. [68]
DSM 70296	Shake flasks	Xylose (commercial) / Xylose/OMWs	25.3 / 21.1	23.7 / 27.9	Present study

WPHL—waste paper hydrolysate; SCBH—sugarcane bagasse hydrolysate; SSJc—sweet sorghum juice; FRWH—flour-rich waste hydrolysate; OMWs—olive mill wastewaters.

Concerning the trials in OMW-based media, irrespective of the PCC_0 imposed into the medium, *C. curvatus* proceeded to a gradual dephenolization of the medium, reaching to a total phenolic content removal of ≈25–28% *w/w* (Figure 4a). On the other hand, decolorization process was not "parallel" with that of dephenolization (Figure 4b), reaching much more rapidly to a color removal "plateau" (*c.* 25% of decolorization), whereas, interestingly, at the late fermentation steps (i.e., t > 200 h), although dephenolization constantly occurred, the color into the medium seemed to be more "dense" than in the earlier growth steps. As far as the detoxification process led by *L. starkeyi* was concerned, the kinetic profile of phenol removal was almost identical with the ones recorded for *C. curvatus* (see Figure 3a). In contrast—and despite the non-negligible dephenolization that occurred—*L. starkeyi* did not remove at all color from the wastewater. It may be assumed, therefore, that color and phenol removal from phenol-containing wastewaters, seem processes that are not obligatorily implicated to each other, in agreement with results recorded for the detoxification of wastewaters performed by edible and medicinal mushrooms [37,69], several types of yeasts like strains of *Y. lipolytica* and *S. cerevisiae* [11,16,18,70] or even crude enzymes deriving from mushroom cultivations [92].

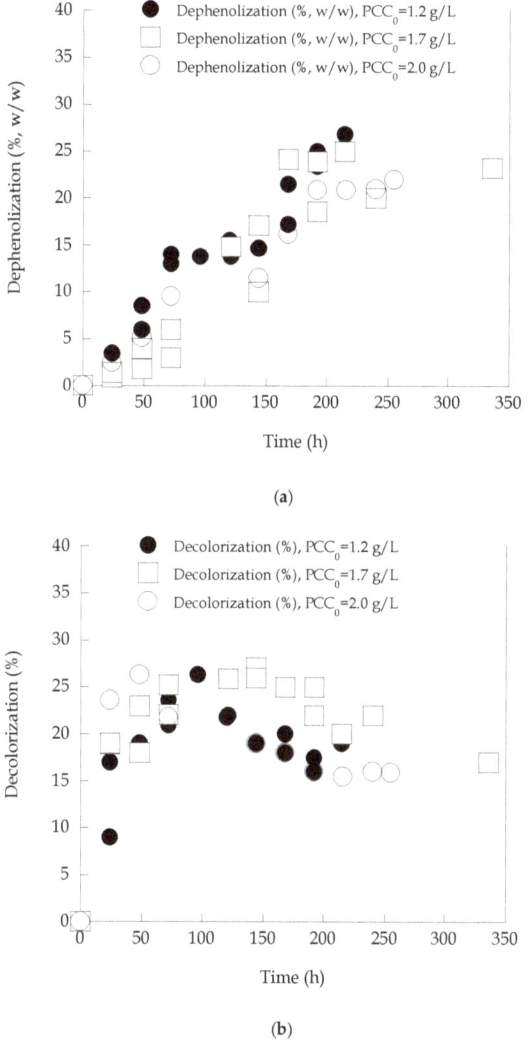

(**a**)

(**b**)

Figure 4. Kinetics of (**a**) dephenolization (%, *w/w*) and (**b**) decolorization (%) performed by *Cryptococcus curvatus* ATCC 20509, during growth on commercial-type xylose in media diluted with olive mill wastewaters, at initial phenol-content concentrations ≈1.2 g/L, ≈ 1.7 g/L and ≈ 2.0 g/L, in shake-flask experiments under nitrogen-limited conditions. Culture conditions: growth on 250-mL conical flasks at 180 ± 5 rpm, initial pH = 6.0 ± 0.1, pH ranging between 5.2 and 5.8, incubation temperature T = 28 °C, initial sugar concentration ≈100 ± 10 g/L. Each experimental point is the mean value of two measurements (SE < 15%).

Yeast strains compared to higher fungi, do not possess the mechanisms of producing the appropriate extracellular oxidases [10,13,92] to break down phenolic compounds that are found in several phenol-containing wastewaters [37,38,69,92]. Oxidative enzymes produced by higher fungi in these processes include but are not limited to lignin peroxidase (*LiP*, E.C. 1.11.1.14), manganese-dependent peroxidase (*MnP*, E.C. 1.11.1.13), phenol oxidase (laccase) (*Lac*, E.C. 1.10.3.2) [37,38,69,92]. On the other hand, besides OMWs, phenol-containing wastes are nowadays produced in significant quantities from several industrial processes (e.g., coal conversion, petroleum refining, sugar refining, paper production processes, etc.) [7,10,37], therefore these "detoxification" and "remediation" bioprocesses present a constantly increasing scientific and industrial importance. The last years, in an increasing number of reports, a non-negligible reduction of phenolic-type compounds has been observed in cultures performed by yeast species like *C. cylindracea*, *C. tropicalis*, *L. starkeyi* and *T. cutaneum* [15,67,68,73,74,93]. Since, as previously indicated, yeasts lack the existence of phenol-oxidizing enzymes [10,13], the OMW decolorization and removal of phenol compounds by yeasts should not be achieved by such mechanisms. Potentially, adsorption of phenolic compounds in the yeast cells may exist [11]. It could be also supposed that partial utilization of phenolic compounds as carbon source by the microorganism may occur [67,73,74].

3.3. Cellular Lipid Analysis

All screened strains were analyzed concerning the fatty acid (FA) composition of their total lipids at the stationary growth phase (FA composition analysis performed at t = 120–150 h after inoculation) (Table 4). In agreement with the literature (for critical reviews see: [3,5,22,23,25]) the principal FAs found in variable quantities were mainly oleic acid (Δ9C18:1) and palmitic acid (C16:0). Other FAs like stearic (C18:0) palmitoleic (Δ9C16:1) and linoleic (Δ9,12C18:2) were found in lower concentrations, whereas poly-unsaturated FAs containing >2 double bonds (i.e., α- and γ-linolenic acid, arachidonic acid, etc.) were not detected, since these compounds are the principal storage lipophilic compounds in oleaginous fungi and algae [25–27]. The FA compositions of lipids produced by other *C. curvatus*—or generally, other nonconventional yeast species/genera employed as cell factories amenable to produce SCOs (like *Y. lipolytica*, *Rhodosporidium* sp., *Lipomyces* sp., etc.) during growth on sugars or other hydrophilic substrates (like glycerol)—can present differences that reflect on the microbial growth stage, the incubation temperature, the initial molar ratio C/N and the initial concentration of substrate used without any systematic common effect on the modification of cellular FAs by the above-mentioned parameters [33,39,40,47,48,58,67,68,81–87]. On the other hand, SCOs containing increased concentrations of the FA Δ9C18:1 and non-negligible ones of the FAs C16:0 and Δ9,12C18:2 (as is the case of the yeast lipids produced in the current investigation; see Table 4) constitute perfect fatty materials amenable to produce high-quality "2nd generation" biodiesel [3–5,7,8,22,25].

In the present investigation, the addition of OMWs into the culture medium, did not seem to bring significant modifications in the FA composition of the total cellular lipids produced by *C. curvatus* (Table 5a) and *L. starkeyi* (Table 5b). As previously, the main cellular FAs were the Δ9C18:1 and C16:0, therefore, SCOs produced by *C. curvatus* and *L. starkeyi* cultivated on commercial xylose/OMW blends were suitable materials to be converted into nonconventional biodiesel. The FA C16:0 seemed to be presented in slightly higher concentrations in the cellular lipids of *L. starkeyi* compared with these of *C. curvatus*. In disagreement with the results indicated in the current investigation, addition of OMWs, as indicated by the PCC_0 that was comparable with the current investigation, seemed to significantly increase the quantity of the cellular FA Δ9C18:1 decreasing that of C18:0 and Δ9C16:1 for various strains of the yeasts *S. cerevisiae* and *Y. lipolytica* [11,12,16–18]. It had been postulated that the presence of microbial inhibitors like gallic acid, caffeic acid, polyphenols, etc., could perform potential activation of the cellular Δ9 desaturase (catalyzing the reaction of conversion of the FA C18:0 to the FA Δ9C18:1) [11,18]. However, as shown in the current investigation (Table 5a,b), the mentioned feature of the yeasts *S. cerevisiae* and *Y. lipolytica* in not observed for *L. starkeyi* and *C. curvatus*.

Table 4. Fatty-acid composition of the cellular lipids produced by yeast strains cultivated on commercial-type xylose employed as sole carbon source in shake-flask experiments (S_0 concentration $\approx$ 70 g/L). Time of fermentation for the determination of the fatty-acid composition was at the stationary growth phase (between 120 and 150 h after inoculation). Culture conditions as in Table 1.

Yeast Strain	Fatty-Acid Composition of Yeast Lipids (%, *w/w*)					
	C16:0	Δ9C16:1	C18:0	Δ9C18:1	Δ9,12C18:2	Others
Rhodosporidium toruloides DSM 4444	29.5	0.6	7.0	51.6	7.7	3.6
Rhodosporidium toruloides NRRL Y-27012	28.4	0.5	6.9	53.7	7.5	3.0
Rhodotorula glutinis NRRL YB-252	20.5	1.5	2.0	55.5	9.0	11.5
Cryptococcus curvatus NRRL Y-1511	22.5	0.5	14.0	49.9	13.0	0.1
Lipomyces starkeyi DSM 702096	28.9	5.0	5.4	50.1	5.5	5.1
Cryptococcus curvatus ATCC 20509	22.5	0.9	9.0	51.1	10.8	5.7

Table 5. Fatty-acid composition of cellular lipids of *Cryptococcus curvatus* ATCC 20509 (a) and *Lipomyces starkeyi* DSM 702096 (b) during growth on commercial-type xylose, in nitrogen-limited shake-flask cultures, in which OMWs were added in various concentrations into the medium and the initial sugar (mostly xylose) concentration was adjusted to *c.* 100 g/L. The initial phenolic compounds concentration (PCC_0) is also illustrated for all runs carried out. LE is the late exponential phase ($t \approx 48$ h) and S is the stationary phase ($t \approx 150$–180 h for *C. curvatus* and t = 215 h for *L. starkeyi*). Culture conditions as in Table 2.

(a) *Cryptococcus curvatus*		Fatty-Acid Composition of Cellular Lipids (%, *w/w*)				
PCC_0 (g/L)	Fermentation Period	C16:0	C18:0	Δ9C18:1	Δ9,12C18:2	Others
0.0	LE	28.0	8.6	50.0	10.5	2.9
	S	24.0	11.0	54.3	7.0	3.7
1.16	LE	25.3	11.4	46.3	5.4	11.6
	S	21.5	13.0	54.0	7.5	4.0
1.65	LE	25.0	12.0	49.0	6.1	7.9
	S	19.7	13.7	54.4	8.2	4.0
1.91	LE	24.8	12.2	51.2	6.8	5.0
	S	20.1	13.9	54.0	8.9	3.1
(b) *Lipomyces starkeyi*		Fatty-Acid Composition of Cellular Lipids (%, *w/w*)				
PCC_0 (g/L)	Fermentation Period	C16:0	C18:0	Δ9C18:1	Δ9,12C18:2	Others
0.0	LE	31.5	3.9	49.0	5.8	9.8
	S	32.0	5.0	53.0	2.7	7.3
1.96	LE	32.3	4.7	49.7	6.6	6.7
	S	33.4	6.3	52.5	2.9	4.9

C. curvatus lipid was mainly composed of neutral lipids (NLs) (85–88% *w/w* of cellular lipids) whereas smaller quantities of glycolipids plus sphingolipids (G + S) (10–13% *w/w* of cellular lipids) principally phospholipids (PLs) (*c.* 2.0% *w/w* of cellular lipids) were quantified, in an analysis performed at the stationary (and, therefore, "oleaginous") phase (t $\approx$ 170–190 h after inoculation) (Table 6a,b). In most instances, in the case of oleaginous microorganisms, NLs are mainly composed of triacylglycerols (TAGs) and to lesser extent steryl esters, while low concentrations of mono- and diacylglycerols (MAGs and DAGs) are identified [7,8,23]. It is also noted that NLs have been revealed to be the major component of the cellular lipids irrespective of the fact of significant or insignificant accumulation of lipids in many yeast species like *R. toruloides*, *Y. lipolytica*, *Pichia membranifaciens*, etc., when growth is performed on sugars or glycerol under nitrogen-limited conditions [32,41,58,94].

(Finally, *C. curvatus* ATCC 20509 stored total lipids (extracted with the aid of chloroform/methanol 2/1 *v/v* mixture) that were mainly composed of triacylglycerols (TAGs) (Figure 5), in accordance with many results reported for various oleaginous yeasts [4,33,94] and fungi [32,59].

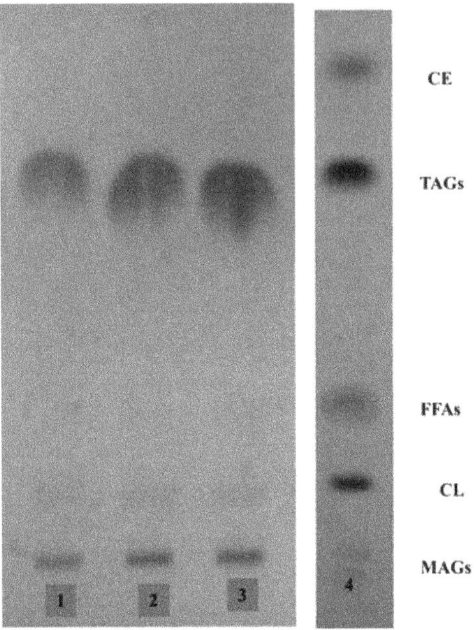

Figure 5. TLC analysis of the "crude" "Folch" extract (lanes 1, 2 and 3) of *Cryptococcus curvatus* ATCC 20509 lipid produced during growth on xylose in shake-flask experiments (S_0 concentration $\approx$100 g/L), *c.* 170 h after inoculation. Lane 1 corresponds to deposited sample of 4 µL, lane 2 of 6 µL and lane 3 of 8 µL. Total lipids were diluted to chloroform/methanol 2/1 *v/v* solution, at concentration $\approx$10 mg/mL. Lane 4 corresponds to a mixture of neutral lipid standards (mononoadecanoin—MAG, cholesterol—CL, oleic acid—FFA, glyceryl trioleate—TAG and cholesteryl linoleate—CE). Solvents: petroleum ether/diethyl ether/glacial acetic acid (70:30:1, *v/v/v*); time of run 45 min (16 cm above the origin); intermediate drying at room temperature for 15–30 min. Detection system: H_2SO_4:H_2O 1:1 (*v/v*) at 110–120 °C for 30 min. Culture conditions as in Table 2.

Table 6. Quantities (in %, *w/w*) of neutral lipid (NL), sphingolipid and glycolipid (G + S) and phospholipid (PL) fractions during lipid accumulation phase of *Cryptococcus curvatus* ATCC 20509 cultivated on commercial xylose at $S_0 \approx$ 100 g/L without OMWs added (PCC$_0$ = 0.0 g/L) in shake-flask experiment, *c.* 170 h after inoculation (a) and batch bioreactor experiment, *c.* 190 h after inoculation (b) and fatty-acid composition (in%, *w/w*) of the mentioned lipid fractions as compared with the fatty-acid composition of total lipid for the relevant fermentation point. Culture conditions for the shake-flask trials as in Table 2, for the batch bioreactor trial as in Figure 3.

(a) Flask Experiment					
	% *w/w*	C16:0	C18:0	Δ9C18:1	Δ9,12C18:2
Total lipid		26.0	9.7	53.2	7.2
NLs	85.2	29.0	6.5	50.8	10.8
G+S	12.6	26.0	7.2	53.4	6.1
PLs	2.2	22.1	8.0	59.0	7.0
(b) Bioreactor Experiment					
	% *w/w*	C16:0	C18:0	Δ9C18:1	Δ9,12C18:2
Total lipid		32.1	6.0	54.4	2.9
NLs	88.1	40.2	4.8	49.4	3.0
G+S	10.0	37.3	8.8	49.3	2.1
PLs	1.9	26.4	5.1	61.4	4.0

4. Conclusions

Wild-type yeast strains were screened towards their ability to convert commercial xylose, a low-cost carbon source deriving from lignocellulosic biomass, into SCO. Trials were carried out with initial xylose at *c.* 70 g/L using six yeast strains of the species *Rhodosporidium toruloides*, *Lipomyces starkeyi*, *Rhodotorula glutinis* and *Cryptococcus curvatus* and the most promising results as regards lipid production were recorded for the strains *C. curvatus* ATCC 20509 and *L. starkeyi* DSM 70296. In the next step, media presenting high xylose quantities ($\approx$100 g/L) in which OMWs were added in various concentrations, were employed as fermentation media. These media mimic various abundant xylose-rich lignocellulose-type wastewaters like spent-sulfite liquor and waste xylose mother liquid. Both *C. curvatus* and *L. starkeyi* presented interesting TDCW production. In the media without OMWs added, *C. curvatus* produced SCO = 10 g/L (lipid in TDCW $\approx$ 48% *w/w*. Scale-up in laboratory-scale bioreactor clearly increased the production of TDCW and SCO synthesized by *C. curvatus*; the values for both biomass and lipids produced in bioreactor experiments (37.0 g/L and 16.4 g/L respectively) were among the very high ones achieved in the international literature for this type of conversion (commercial xylose towards SCO) demonstrating the high potential of this strain on the mentioned bioprocess.

In media presenting increasing phenolic compounds, lipid in TDCW values decreased whereas shift towards the creation of lipid-free material was observed. In contrast, lipid production bioprocess was significantly unaltered by the presence of phenolic compounds into the medium for *L. starkeyi* (in all cases lipids in TDCW quantities 24–28% *w/w*, similar to the blank experiment were recorded). Interesting dephenolization occurred for both yeast strains, irrespective of the initial concentration of the phenolic compounds found into the medium. On the contrary, decolorization of the wastewater occurred only by *C. curvatus*.

Total lipids of all yeasts tested were mainly composed of the FAs oleic and palmitic and to lesser extent linoleic and stearic, found in variable concentrations and constituting perfect fatty materials amenable for the synthesis of high-quality biodiesel. In *C. curvatus*, lipids were mainly composed of nonpolar fractions (i.e., triacylglycerols) whereas polar lipids (i.e., phospholipids, glyco + sphingolipids) were found at drastically lower concentrations.

Concluding, both *C. curvatus* and *L. starkeyi* are robust TDCW- and SCO-producing microorganisms during growth on xylose and xylose/OMWs blends. The current submission has demonstrated the potential of the bioconversion of these compounds or also residues mimicking these substrates (i.e., abundant xylose-containing wastewaters) into yeast biomass and lipid with satisfactory productions and conversion yields.

Author Contributions: Conceptualization, S.P. and A.K.; methodology, S.P.; software, S.P.; validation, S.P., A.K. and E.X.; formal analysis, E.X.; investigation, E.X and I.G.; resources, S.P.; data curation, S.P.; writing—original draft preparation, E.X.; writing—review and editing, S.P and A.C.; visualization, E.X.; supervision, S.P.; project administration, S.P.; funding acquisition, S.P. and A.K. All authors have read and agreed to the published version of the manuscript.

Funding: This research was financially supported by the project entitled "Research Infrastructure for Waste Valorization and Sustainable Management of Resources—INVALOR" (MIS 5002495) which is implemented under the Action "Reinforcement of the Research and Innovation Infrastructure", funded by the Operational Program "Competitiveness, Entrepreneurship and Innovation" (NSRF 2014–2020) and co-financed by Greece and the European Union (European Regional Development Fund).

Conflicts of Interest: The authors declare no conflicts of interest.

References

1. Cheng, H.; Wang, B.; Lv, J.; Jiang, M.; Lin, S.; Deng, Z. Xylitol production from xylose mother liquor: A novel strategy that combines the use of recombinant *Bacillus subtilis* and *Candida maltose*. *Microb. Cell Fact.* **2011**, *10*, 5. [CrossRef] [PubMed]

2. Sarris, D.; Papanikolaou, S. Biotechnological production of ethanol: Biochemistry, processes and technologies. *Eng. Life Sci.* **2016**, *16*, 307–329. [CrossRef]

3. Patel, A.; Arora, N.; Sartaj, K.; Pruthi, V.; Pruthi, P.A. Sustainable biodiesel production from oleaginous yeasts utilizing hydrolysates of various non-edible lignocellulosic biomasses. *Renew. Sustain. Energy Rev.* **2016**, *62*, 836–855. [CrossRef]

4. Patel, A.; Mikes, F.; Bühler, S.; Matsakas, L. Valorization of brewers' spent grain for the production of lipids by oleaginous yeast. *Molecules* **2018**, *23*, 3052. [CrossRef] [PubMed]

5. Patel, A.; Karageorgou, D.; Rova, E.; Katapodis, P.; Rova, U.; Christakopoulos, P.; Matsakas, L. An Overview of potential oleaginous microorganisms and their role in biodiesel and omega-3 fatty acid-based industries. *Microorganisms* **2020**, *8*, 434. [CrossRef] [PubMed]

6. Wang, H.; Li, L.; Zhang, L.; An, J.; Chang, H.; Deng, Z. Xylitol production from waste xylose mother liquor containing miscellaneous sugars and inhibitors: One-pot biotransformation by *Candida tropicalis* and recombinant *Bacillus subtilis*. *Microb. Cell Fact.* **2016**, *15*, 82. [CrossRef] [PubMed]

7. Qin, L.; Liu, L.; Zeng, A.-P.; Wei, D. From low-cost substrates to single cell oils synthesized by oleaginous yeasts. *Bioresour. Technol.* **2017**, *245*, 1507–1519. [CrossRef]

8. Diwan, B.; Parkhey, P.; Gupta, P. From agro-industrial wastes to single cell oils: A step towards prospective biorefinery. *Folia Microbiol.* **2018**, *63*, 547–568. [CrossRef]

9. Mantzavinos, D.; Kalogerakis, N. Treatment of olive mill effluents. *Environ. Int.* **2005**, *31*, 289–295. [CrossRef]

10. Crognale, S.; D'Annibale, A.; Federici, F.; Fenice, M.; Quaratino, D.; Petruccioli, M. Olive oil mill wastewater valorisation by fungi. *J. Chem. Technol. Biotechnol.* **2006**, *81*, 1547–1555. [CrossRef]

11. Sarris, D.; Stoforos, N.G.; Mallouchos, A.; Kookos, I.K.; Koutinas, A.A.; Aggelis, G.; Papanikolaou, S. Production of added-value metabolites by *Yarrowia lipolytica* growing in olive mill wastewater-based media under aseptic and non-aseptic conditions. *Eng. Life Sci.* **2017**, *17*, 695–709. [CrossRef] [PubMed]

12. Sarris, D.; Rapti, A.; Papafotis, N.; Koutinas, A.A.; Papanikolaou, S. Production of added-value chemical compounds through bioconversions of olive-mill wastewaters blended with crude glycerol by a *Yarrowia lipolytica* strain. *Molecules* **2019**, *24*, 222. [CrossRef] [PubMed]

13. Ahmed, P.A.; Fernández, P.M.; de Figueroa, L.I.C.; Pajot, H.F. Exploitation alternatives of olive mill wastewater: Production of value-added compounds useful for industry and agriculture. *Biofuel Res. J.* **2019**, *6*, 980–994. [CrossRef]

14. Alfano, A.; Corsuto, L.; Finamore, R.; Savarese, M.; Ferrara, F.; Falco, S.; Santabarbara, G.; De Rosa, M.; Schiraldi, C. Valorization of olive mill wastewater by membrane processes to recover natural antioxidant compounds for cosmeceutical and nutraceutical applications or functional foods. *Antioxidants* **2018**, *7*, 72. [CrossRef]

15. D'Annibale, A.; Sermanni, G.G.; Federici, F.; Petruccioli, M. Olive-mill wastewaters: A promising substrate for microbial lipase production. *Bioresour. Technol.* **2006**, *97*, 1828–1833. [CrossRef]

16. Papanikolaou, S.; Galiotou-Panayotou, M.; Fakas, S.; Komaitis, M.; Aggelis, G. Citric acid production by *Yarrowia lipolytica* cultivated on olive-mill wastewater-based media. *Bioresour. Technol.* **2008**, *99*, 2419–2428. [CrossRef]

17. Sarris, D.; Galiotou-Panayotou, M.; Koutinas, A.A.; Komaitis, M.; Papanikolaou, S. Citric acid, biomass and cellular lipid production by *Yarrowia lipolytica* strains cultivated on olive mill wastewater-based media. *J. Chem. Technol. Biotechnol.* **2011**, *86*, 1439–1448. [CrossRef]

18. Sarris, D.; Giannakis, M.; Philippoussis, A.; Komaitis, M.; Koutinas, A.A.; Papanikolaou, S. Conversions of olive mill wastewater-based media by *Saccharomyces cerevisiae* through sterile and non-sterile bioprocesses: Bioconversions of olive-mill wastewaters by *Saccharomyces cerevisiae*. *J. Chem. Technol. Biotechnol.* **2013**, *88*, 958–969. [CrossRef]

19. Papadaki, E.; Tsimidou, M.Z.; Mantzouridou, F.T. Changes in phenolic compounds and phytotoxicity of the Spanish-style green olive processing wastewaters by *Aspergillus niger* B60. *J. Agric. Food Chem.* **2018**, *66*, 4891–4901. [CrossRef]

20. Papadaki, E.; Mantzouridou, F.T. Current status and future challenges of table olive processing wastewater valorization. *Biochem. Eng. J.* **2016**, *112*, 103–113. [CrossRef]

21. Papadaki, E.; Mantzouridou, F.T. Citric acid production from the integration of spanish-style green olive processing wastewaters with white grape pomace by *Aspergillus niger*. *Bioresour. Technol.* **2019**, *280*, 59–69. [CrossRef] [PubMed]

22. Papanikolaou, S.; Aggelis, G. Lipids of oleaginous yeasts. Part II: Technology and potential applications. *Eur. J. Lipid Sci. Technol.* **2011**, *113*, 1052–1073. [CrossRef]

23. Papanikolaou, S.; Aggelis, G. Sources of microbial oils with emphasis to *Mortierella (Umbelopsis) isabellina* fungus. *World J. Microbiol. Biotechnol.* **2019**, *35*, 63. [CrossRef] [PubMed]

24. Sutanto, S.; Zullaikah, S.; Tran-Nguyen, P.L.; Ismadji, S.; Ju, Y.-H. *Lipomyces starkeyi*: Its current status as a potential oil producer. *Fuel Proc. Technol.* **2018**, *177*, 39–55. [CrossRef]

25. Athenaki, M.; Gardeli, C.; Diamantopoulou, P.; Tchakouteu, S.S.; Sarris, D.; Philippoussis, A.; Papanikolaou, S. Lipids from yeasts and fungi: Physiology, production and analytical considerations. *J Appl. Microbiol.* **2018**, *124*, 336–367. [CrossRef] [PubMed]

26. Bellou, S.; Makri, A.; Sarris, D.; Michos, K.; Rentoumi, P.; Celik, A.; Papanikolaou, S.; Aggelis, G. The olive mill wastewater as substrate for single cell oil production by Zygomycetes. *J. Biotechnol.* **2014**, *170*, 50–59. [CrossRef]

27. Bellou, S.; Triantaphyllidou, I.-E.; Aggeli, D.; Elazzazy, A.M.; Baeshen, M.N.; Aggelis, G. Microbial oils as food additives: Recent approaches for improving microbial oil production and its polyunsaturated fatty acid content. *Curr. Opin. Biotechnol.* **2016**, *37*, 24–35. [CrossRef]

28. Papanikolaou, S.; Chevalot, I.; Komaitis, M.; Aggelis, G.; Marc, I. Kinetic profile of the cellular lipid composition in an oleaginous *Yarrowia lipolytica* capable of producing a cocoa-butter substitute from industrial fats. *Antonie Van Leeuwenhoek* **2001**, *80*, 215–224. [CrossRef]

29. Valta, K.; Kosanovic, T.; Malamis, D.; Moustakas, K.; Loizidou, M. Overview of water usage and wastewater management in the food and beverage industry. *Desalin. Water Treat.* **2015**, *53*, 3335–3347. [CrossRef]

30. Gardeli, C.; Athenaki, M.; Xenopoulos, E.; Mallouchos, A.; Koutinas, A.A.; Aggelis, G.; Papanikolaou, S. Lipid production and characterization by Mortierella (Umbelopsis) isabellina cultivated on lignocellulosic sugars. *J. Appl. Microbiol.* **2017**, *123*, 1461–1477. [CrossRef]

31. Fakas, S.; Papanikolaou, S.; Galiotou-Panayotou, M.; Komaitis, M.; Aggelis, G. Lipids of *Cunninghamella echinulata* with emphasis to γ-linolenic acid distribution among lipid classes. *Appl. Microbiol. Biotechnol.* **2006**, *73*, 676–683. [CrossRef] [PubMed]

32. Papanikolaou, S.; Rontou, M.; Belka, A.; Athenaki, M.; Gardeli, C.; Mallouchos, A.; Kalantzi, O.; Koutinas, A.A.; Kookos, I.K.; Zeng, A.-P.; et al. Conversion of biodiesel-derived glycerol into biotechnological products of industrial significance by yeast and fungal strains. *Eng. Life Sci.* **2017**, *17*, 262–281. [CrossRef] [PubMed]

33. Diamantopoulou, P.; Filippousi, R.; Antoniou, D.; Varfi, E.; Xenopoulos, E.; Sarris, D.; Papanikolaou, S. Production of added-value microbial metabolites during growth of yeast strains on media composed of biodiesel-derived crude glycerol and glycerol/xylose blends. *FEMS Microbiol. Lett.* **2020**, *367*, fnaa063. [CrossRef] [PubMed]

34. Diamantopoulou, P.; Papanikolaou, S.; Katsarou, E.; Komaitis, M.; Aggelis, G.; Philippoussis, A. Mushroom polysaccharides and lipids synthesized in liquid agitated and static cultures. Part II: Study of *Volvariella volvacea*. *Appl. Biochem. Biotechnol.* **2012**, *167*, 1890–1906. [CrossRef] [PubMed]

35. Miller, G.L. Use of Dinitrosalicylic acid reagent for determination of reducing sugar. *Anal. Chem.* **1959**, *31*, 426–428. [CrossRef]

36. Kachrimanidou, V.; Kopsahelis, N.; Chatzifragkou, A.; Papanikolaou, S.; Yanniotis, S.; Kookos, I.; Koutinas, A.A. Utilisation of by-products from sunflower-based biodiesel production processes for the production of fermentation feedstock. *Waste Biomass Valor.* **2013**, *4*, 529–537. [CrossRef]

37. Aggelis, G.; Iconomou, D.; Christou, M.; Bokas, D.; Kotzailias, S.; Christou, G.; Tsagou, V.; Papanikolaou, S. Phenolic Removal in a model olive oil mill wastewater using *Pleurotus ostreatus* in bioreactor cultures and biological evaluation of the process. *Water Res.* **2003**, *37*, 3897–3904. [CrossRef]

38. Sayadi, S.; Ellouz, R. Decolourization of olive mill waste-waters by the white-rot fungus *Phanerochaete chrysosporium*: Involvement of the lignin-degrading system. *Appl. Microbiol. Biotechnol.* **1992**, *37*, 813–817. [CrossRef]

39. Tchakouteu, S.S.; Kalantzi, O.; Gardeli, C.; Koutinas, A.A.; Aggelis, G.; Papanikolaou, S. Lipid production by yeasts growing on biodiesel-derived crude glycerol: Strain selection and impact of substrate concentration on the fermentation efficiency. *J. Appl. Microbiol.* **2015**, *118*, 911–927. [CrossRef]

40. Filippousi, R.; Antoniou, D.; Tryfinopoulou, P.; Nisiotou, A.A.; Nychas, G.-J.; Koutinas, A.A.; Papanikolaou, S. Isolation, identification and screening of yeasts towards their ability to assimilate biodiesel-derived crude glycerol: Microbial production of polyols, endopolysaccharides and lipid. *J. Appl. Microbiol.* **2019**, *127*, 1080–1100. [CrossRef]

41. Makri, A.; Fakas, S.; Aggelis, G. Metabolic activities of biotechnological interest in *Yarrowia lipolytica* grown on glycerol in repeated batch cultures. *Bioresour. Technol.* **2010**, *101*, 2351–2358. [CrossRef] [PubMed]

42. Morgunov, I.G.; Kamzolova, S.V.; Lunina, J.N. The citric acid production from raw glycerol by *Yarrowia lipolytica* yeast and its regulation. *Appl. Microbiol. Biotechnol.* **2013**, *97*, 7387–7397. [CrossRef] [PubMed]

43. Canonico, L.; Ashoor, S.; Taccari, M.; Comitini, F.; Antonucci, M.; Truzzi, C.; Scarponi, G.; Ciani, M. Conversion of raw glycerol to microbial lipids by new *Metschnikowia* and *Yarrowia lipolytica* strains. *Ann. Microbiol.* **2016**, *66*, 1409–1418. [CrossRef]

44. Chatzifragkou, A.; Fakas, S.; Galiotou-Panayotou, M.; Komaitis, M.; Aggelis, G.; Papanikolaou, S. Commercial sugars as substrates for lipid accumulation in *Cunninghamella echinulata* and *Mortierella isabellina* fungi. *Eur. J. Lipid Sci. Technol.* **2010**, *112*, 1048–1057. [CrossRef]

45. Kothri, M.; Mavrommati, M.; Elazzazy, A.M.; Baeshen, M.N.; Moussa, T.A.A.; Aggelis, G. Microbial sources of polyunsaturated fatty acids (PUFAs) and the prospect of organic residues and wastes as growth media for PUFA-producing microorganisms. *FEMS Microbiol. Lett.* **2020**, *367*, fnaa028. [CrossRef]

46. Tsakona, S.; Skiadaresis, A.G.; Kopsahelis, N.; Chatzifragkou, A.; Papanikolaou, S.; Kookos, I.K.; Koutinas, A.A. Valorisation of side streams from wheat milling and confectionery industries for consolidated production and extraction of microbial lipids. *Food Chem.* **2016**, *198*, 85–92. [CrossRef]

47. Tchakouteu, S.S.; Chatzifragkou, A.; Kalantzi, O.; Koutinas, A.A.; Aggelis, G.; Papanikolaou, S. Oleaginous yeast *Cryptococcus curvatus* exhibits interplay between biosynthesis of intracellular sugars and lipids: Biochemical features of *Cryptococcus curvatus*. *Eur. J. Lipid Sci. Technol.* **2015**, *117*, 657–672. [CrossRef]

48. Harde, S.M.; Wang, Z.; Horne, M.; Zhu, J.Y.; Pan, X. Microbial lipid production from SPORL-pretreated douglas fir by *Mortierella isabellina*. *Fuel* **2016**, *175*, 64–74. [CrossRef]

49. Carota, E.; Petruccioli, M.; D'Annibale, A.; Gallo, A.M.; Crognale, S. Orange peel waste-based liquid medium for biodiesel production by oleaginous yeasts. *Appl. Microbiol. Biotechnol.* **2020**, *104*, 4617–4628. [CrossRef]

50. Tsakona, S.; Kopsahelis, N.; Chatzifragkou, A.; Papanikolaou, S.; Kookos, I.K.; Koutinas, A.A. Formulation of fermentation media from flour-rich waste streams for microbial lipid production by *Lipomyces starkeyi*. *J. Biotechnol.* **2014**, *189*, 36–45. [CrossRef]

51. Evans, C.T.; Ratledge, C. Phosphofructokinase and the regulation of the flux of carbon from glucose to lipid in the oleaginous yeast Rhodosporidium toruloides. *J. Gen. Microbiol.* **1984**, *130*, 3251–3264. [CrossRef]

52. Park, W.-S.; Murphy, P.A.; Glatz, B.A. Lipid metabolism and cell composition of the oleaginous yeast *Apiotrichum curvatum* grown at different carbon to nitrogen ratios. *Can. J. Microbiol.* **1990**, *36*, 318–326. [CrossRef] [PubMed]

53. Zikou, E.; Chatzifragkou, A.; Koutinas, A.A.; Papanikolaou, S. Evaluating glucose and xylose as cosubstrates for lipid accumulation and γ-linolenic acid biosynthesis of *Thamnidium elegans*. *J. Appl. Microbiol.* **2013**, *114*, 1020–1032. [CrossRef] [PubMed]

54. Díaz-Fernández, D.; Aguiar, T.Q.; Martín, V.I.; Romaní, A.; Silva, R.; Domingues, L.; Revuelta, J.L.; Jiménez, A. Microbial lipids from industrial wastes using xylose-utilizing *Ashbya gossypii* strains. *Bioresour. Technol.* **2019**, *293*, 122054. [CrossRef] [PubMed]

55. Papanikolaou, S.; Sarantou, S.; Komaitis, M.; Aggelis, G. Repression of reserve lipid turnover in *Cunninghamella echinulata* and *Mortierella isabellina* cultivated in multiple-limited media. *J. Appl. Microbiol.* **2004**, *97*, 867–875. [CrossRef] [PubMed]

56. Papanikolaou, S.; Diamantopoulou, P.; Chatzifragkou, A.; Philippoussis, A.; Aggelis, G. Suitability of low-cost sugars as substrates for lipid production by the fungus *Thamnidium elegans*. *Energy Fuel.* **2010**, *24*, 4078–4086. [CrossRef]

57. Fakas, S.; Galiotou-Panayotou, M.; Papanikolaou, S.; Komaitis, M.; Aggelis, G. Compositional shifts in lipid fractions during lipid turnover in *Cunninghamella echinulata*. *Enzyme Microb. Technol.* **2007**, *40*, 1321–1327. [CrossRef]

58. Chatzifragkou, A.; Makri, A.; Belka, A.; Bellou, S.; Mavrou, M.; Mastoridou, M.; Mystrioti, P.; Onjaro, G.; Aggelis, G.; Papanikolaou, S. Biotechnological conversions of biodiesel derived waste glycerol by yeast and fungal species. *Energy* **2011**, *36*, 1097–1108. [CrossRef]

59. Bellou, S.; Moustogianni, A.; Makri, A.; Aggelis, G. Lipids containing polyunsaturated fatty acids synthesized by Zygomycetes grown on glycerol. *Appl. Biochem. Biotechnol.* **2012**, *166*, 146–158. [CrossRef]

60. Aggelis, G.; Sourdis, J. Prediction of lipid accumulation-degradation in oleaginous micro-organisms growing on vegetable oils. *Antonie Van Leeuwenhoek* **1997**, *72*, 159–165. [CrossRef]

61. Serrano-Carreon, L.; Hathout, Y.; Bensoussan, M.; Belin, J.-M. Lipid accumulation in *Trichoderma* species. *FEMS Microbiol. Lett.* **1992**, *93*, 181–187. [CrossRef]

62. Holdsworth, J.E.; Veenhuis, M.; Ratledge, C. Enzyme activities in oleaginous yeasts accumulating and utilizing exogenous or endogenous lipids. *J. Gen. Microbiol.* **1988**, *134*, 2907–2915. [CrossRef] [PubMed]

63. Holdsworth, J.E.; Ratledge, C. Lipid turnover in oleaginous yeasts. *J. Gen. Microbiol.* **1988**, *134*, 339–346. [CrossRef]

64. Koutinas, A.A.; Vlysidis, A.; Pleissner, D.; Kopsahelis, N.; Lopez Garcia, I.; Kookos, I.K.; Papanikolaou, S.; Kwan, T.H.; Lin, C.S.K. Valorization of industrial waste and by-product streams via fermentation for the production of chemicals and biopolymers. *Chem. Soc. Rev.* **2014**, *43*, 2587–2627. [CrossRef] [PubMed]

65. Alexandri, M.; Papapostolou, H.; Vlysidis, A.; Gardeli, C.; Komaitis, M.; Papanikolaou, S.; Koutinas, A.A. Extraction of phenolic compounds and succinic acid production from spent sulphite liquor. *J. Chem. Technol. Biotechnol.* **2016**, *91*, 2751–2760. [CrossRef]

66. Alexandri, M.; Papapostolou, H.; Komaitis, M.; Stragier, L.; Verstraete, W.; Danezis, G.P.; Georgiou, C.A.; Papanikolaou, S.; Koutinas, A.A. Evaluation of an integrated biorefinery based on fractionation of spent sulphite liquor for the production of an antioxidant-rich extract, lignosulphonates and succinic acid. *Bioresour. Technol.* **2016**, *214*, 504–513. [CrossRef]

67. Yousuf, A.; Sannino, F.; Addorisio, V.; Pirozzi, D. Microbial conversion of olive oil mill wastewaters into lipids suitable for biodiesel production. *J. Agric. Food Chem.* **2010**, *58*, 8630–8635. [CrossRef]

68. Dourou, M.; Kancelista, A.; Juszczyk, P.; Sarris, D.; Bellou, S.; Triantaphyllidou, I.-E.; Rywinska, A.; Papanikolaou, S.; Aggelis, G. Bioconversion of olive mill wastewater into high-added value products. *J. Clean. Prod.* **2016**, *139*, 957–969. [CrossRef]

69. Tsioulpas, A.; Dimou, D.; Iconomou, D.; Aggelis, G. Phenolic removal in olive oil mill wastewater by strains of *Pleurotus* spp. in respect to their phenol oxidase (laccase) activity. *Bioresour. Technol.* **2002**, *84*, 251–257. [CrossRef]

70. Tzirita, M.; Kremmyda, M.; Sarris, D.; Koutinas, A.A.; Papanikolaou, S. Effect of salt addition upon the production of metabolic compounds by *Yarrowia lipolytica* cultivated on biodiesel-derived glycerol diluted with olive-mill wastewaters. *Energies* **2019**, *12*, 3649. [CrossRef]

71. Lanciotti, R.; Gianotti, A.; Baldi, D.; Angrisani, R.; Suzzi, G.; Mastrocola, D.; Guerzoni, M.E. Use of *Yarrowia lipolytica* strains for the treatment of olive mill wastewater. *Bioresour. Technol.* **2005**, *96*, 317–322. [CrossRef] [PubMed]

72. Federici, F.; Montedoro, G.; Servili, M.; Petruccioli, M. Pectic enzyme production by *Cryptococcus albidus* var. *albidus* on olive vegetation waters enriched with sunflower calathide meal. *Biol. Wastes* **1988**, *25*, 291–301. [CrossRef]

73. Ettayebi, K.; Errachidi, F.; Jamai, L.; Tahri-Jouti, M.A.; Sendide, K.; Ettayebi, M. Biodegradation of polyphenols with immobilized *Candida tropicalis* under metabolic induction. *FEMS Microbiol. Lett.* **2003**, *223*, 215–219. [CrossRef]

74. Chtourou, M.; Ammar, E.; Nasri, M.; Medhioub, K. Isolation of a yeast, *Trichosporon cutaneum*, able to use low molecular weight phenolic compounds: Application to olive mill waste water treatment. *J. Chem. Technol. Biotechnol.* **2004**, *79*, 869–878. [CrossRef]

75. Lakhtar, H.; Ismaili-Alaoui, M.; Philippoussis, A.; Perraud-Gaime, I.; Roussos, S. Screening of strains of *Lentinula edodes* grown on model olive mill wastewater in solid and liquid state culture for polyphenol biodegradation. *Int. Biodeterior. Biodegrad.* **2010**, *64*, 167–172. [CrossRef]

76. Anschau, A.; Xavier, M.C.A.; Hernalsteens, S.; Franco, T.T. Effect of feeding strategies on lipid production by *Lipomyces starkeyi*. *Bioresour. Technol.* **2014**, *157*, 214–222. [CrossRef]

77. Evans, C.T.; Ratledge, C. A comparison of the oleaginous yeast, Candida curvata, grown on different carbon sources in continuous and batch culture. *Lipids* **1983**, *18*, 623–629. [CrossRef]

78. Ykema, A.; Verbree, E.C.; Kater, M.M.; Smit, H. Optimization of lipid production in the oleaginous yeast *Apiotrichum curvatum* in whey permeate. *Appl. Microbiol. Biotechnol.* **1988**, *29*, 211–218. [CrossRef]

79. Hassan, M.; Blanc, P.J.; Pareilleux, A.; Goma, G. Production of single-cell oil from prickly-pear juice fermentation by *Cryptococcus curvatus* grown in batch culture. *World J. Microbiol. Biotechnol.* **1994**, *10*, 534–537. [CrossRef]

80. Meesters, P.A.E.P.; Huijberts, G.N.M.; Eggink, G. High-cell-density cultivation of the lipid accumulating yeast *Cryptococcus curvatus* using glycerol as a carbon source. *Appl. Microbiol. Biotechnol.* **1996**, *45*, 575–579. [CrossRef]

81. Liang, Y.; Cui, Y.; Trushenski, J.; Blackburn, J.W. Converting crude glycerol derived from yellow grease to lipids through yeast fermentation. *Bioresour. Technol.* **2010**, *101*, 7581–7586. [CrossRef] [PubMed]

82. Cui, Y.; Blackburn, J.W.; Liang, Y. Fermentation optimization for the production of lipid by *Cryptococcus curvatus*: Use of response surface methodology. *Biomass Bioenergy* **2012**, *47*, 410–417. [CrossRef]

83. Gong, Z.; Shen, H.; Zhou, W.; Wang, Y.; Yang, X.; Zhao, Z.K. Efficient conversion of acetate into lipids by the oleaginous yeast *Cryptococcus curvatus*. *Biotechnol. Biofuels* **2015**, *8*, 189. [CrossRef] [PubMed]

84. Béligon, V.; Poughon, L.; Christophe, G.; Lebert, A.; Larroche, C.; Fontanille, P. Validation of a predictive model for fed-batch and continuous lipids production processes from acetic acid using the oleaginous yeast *Cryptococcus curvatus*. *Biochem. Eng. J.* **2016**, *111*, 117–128. [CrossRef]

85. Zhou, W.; Gong, Z.; Zhang, L.; Liu, Y.; Yan, J.; Zhao, M. Feasibility of lipid production from waste paper by the oleaginous yeast *Cryptococcus curvatus*. *BioResources* **2017**, *12*, 5249–5263. [CrossRef]

86. Kopsahelis, N.; Dimou, C.; Papadaki, A.; Xenopoulos, E.; Kyraleou, M.; Kallithraka, S.; Kotseridis, Y.; Papanikolaou, S.; Koutinas, A.A. Refining of wine lees and cheese whey for the production of microbial oil, polyphenol-rich extracts and value-added co-products. *J. Chem. Technol. Biotechnol.* **2018**, *93*, 257–268. [CrossRef]

87. Angerbauer, C.; Siebenhofer, M.; Mittelbach, M.; Guebitz, G.M. Conversion of sewage sludge into lipids by Lipomyces starkeyi for biodiesel production. *Bioresour. Technol.* **2008**, *99*, 3051–3056. [CrossRef]

88. Liu, H.; Zhao, X.; Wang, F.; Jiang, X.; Zhang, S.; Ye, M.; Zhao, Z.K.; Zou, H. The proteome analysis of oleaginous yeast *Lipomyces starkeyi*: Comparative proteomic analysis of *Lipomyces starkeyi*. *FEMS Yeast Res.* **2011**, *11*, 42–51. [CrossRef]

89. Oguri, E.; Masaki, K.; Naganuma, T.; Iefuji, H. Phylogenetic and biochemical characterization of the oil-producing yeast *Lipomyces starkeyi*. *Antonie Van Leeuwenhoek* **2012**, *101*, 359–368. [CrossRef]

90. Lin, J.; Li, S.; Sun, M.; Zhang, C.; Yang, W.; Zhang, Z.; Li, X.; Li, S. Microbial lipid production by oleaginous yeast in D-xylose solution using a two-stage culture mode. *RSC Adv.* **2014**, *4*, 34944–34949. [CrossRef]

91. Matsakas, L.; Sterioti, A.-A.; Rova, U.; Christakopoulos, P. Use of dried sweet sorghum for the efficient production of lipids by the yeast *Lipomyces starkeyi* CBS 1807. *Ind. Crop. Prod.* **2014**, *62*, 367–372. [CrossRef]

92. Economou, C.N.; Philippoussis, A.N.; Diamantopoulou, P.A. Spent mushroom substrate for a second cultivation cycle of Pleurotus mushrooms and dephenolization of agro-industrial wastewaters. *FEMS Microbiol. Lett.* **2020**, *367*, fnaa060. [CrossRef] [PubMed]

93. Lopes, M.; Araújo, C.; Aguedo, M.; Gomes, N.; Gonçalves, C.; Teixeira, J.A.; Belo, I. The use of olive mill wastewater by wild type *Yarrowia lipolytica* strains: Medium supplementation and surfactant presence effect. *J. Chem. Technol. Biotechnol.* **2009**, *84*, 533–537. [CrossRef]

94. Tiukova, I.A.; Brandenburg, J.; Blomqvist, J.; Sampels, S.; Mikkelsen, N.; Skaugen, M.; Arntzen, M.Ø.; Nielsen, J.; Sandgren, M.; Kerkhoven, E.J. Proteome analysis of xylose metabolism in *Rhodotorula toruloides* during lipid production. *Biotechnol. Biofuels* **2019**, *12*, 137. [CrossRef] [PubMed]

Article

Evaluation of the Interactive Effect Pretreatment Parameters via Three Types of Microwave-Assisted Pretreatment and Enzymatic Hydrolysis on Sugar Yield

Saleem Ethaib [1,2], Rozita Omar [1,*], Mustapa Kamal Siti Mazlina [3] and Awang Biak Dayang Radiah [2]

[1]   Department of Chemical and Environmental Engineering, Faculty of Engineering, University Putra Malaysia, Serdang 43400, Selangor, Malaysia; dr.saleem@utq.edu.iq

[2]   College of Engineering, University of Thi-Qar, Al Nassirya 64001, Iraq; dradiah@upm.edu.my

[3]   Department of Process and Food Engineering, Faculty of Engineering, University Putra Malaysia, Serdang 43400, Selangor, Malaysia; smazlina@upm.edu.my

*   Correspondence: rozitaom@upm.edu.my

Received: 12 May 2020; Accepted: 10 June 2020; Published: 6 July 2020

Abstract: This study aims to evaluate the sugar yield from enzymatic hydrolysis and the interactive effect pretreatment parameters of microwave-assisted pretreatment on glucose and xylose. Three types of microwave-assisted pretreatments of sago palm bark (SPB) were conducted for enzymatic hydrolysis, namely: microwave-sulphuric acid pretreatment (MSA), microwave-sodium hydroxide pretreatment (MSH), and microwave-sodium bicarbonate (MSB). The experimental design was done using a response surface methodology (RSM) and Box–Behenken Design (BBD). The pretreatment parameters ranged from 5–15% solid loading (SL), 5–15 min of exposure time (ET), and 80–800 W of microwave power (MP). The results indicated that the maximum total reducing sugar was 386 mg/g, obtained by MSA pretreatment. The results also illustrated that the higher glucose yield, 44.3 mg/g, was found using MSH pretreatment, while the higher xylose yield, 43.1 mg/g, resulted from MSA pretreatment. The pretreatment parameters MP, ET, and SL showed different patterns of influence on glucose and xylose yield via enzymatic hydrolysis for MSA, MSH, and MSB pretreatments. The analyses of the interactive effect of the pretreatment parameters MP, ET, and SL on the glucose yield from SPB showed that it increased with the high MP and longer ET, but this was limited by low SL values. However, the analysis of the interactive effect of the pretreatment parameters on xylose yields revealed that MP had the most influence on the xylose yield for MSA, MSH, and MSB pretreatments.

Keywords: microwave-assisted pretreatment; pretreatment parameters; enzymatic hydrolysis; glucose; xylose

---

1. Introduction

The utilization of solid waste, such as lignocellulosic biomass, as raw materials for the fuel, food, and pharmaceutical component industries is a global concern [1]. Forestry, agricultural, and agro-industrial residues are the main sources of these useful materials [2,3], such as sago palm bark (SPB) which is a by-product generated by the sago starch industry. Due to the presence of cellulose and hemicellulose contents in sago palm bark (> 63%) [4], SPB can provide a sustainable resource for sugar platform-based chemicals and organic fuels because of its availability in enormous quantities at low cost. The conversion of the lignocellulosic materials into different valuable products faces the complex structure of lignocellulosic materials, which make these materials resistant to some conversion possessing stages, such as enzymatic hydrolysis. Hence, the challenge in the hydrolysis stage is to achieve a high sugar yield from lignocellulosic biomass using limited amounts of energy and chemicals during pretreatment to reduce the investment cost. A pretreatment stage is a key to the use of lignocellulose materials in bio-alcohol compound production [5]. Various techniques have been developed for the pretreatment of lignocellulosic compounds, including physical and chemical pretreatment methods, such as steam explosion [6], diluted acid [7], alkali [8], and hydrothermal pretreatments [9]. Most of these methods of pretreatment involve high processing costs, due to harsh operating conditions, such as high pressure and/or temperature. Furthermore, highly concentrated chemicals, such as acids, are toxic to the enzymes or fermentative microorganisms and, therefore, require an additional processing step [10].

In recent years, microwave heating has received more recognition. The key benefit of microwave heating is the short amount of time it takes, relative to traditional heating; minutes versus hours [11]. This is because of the fundamental difference in the heat transfer mechanism between microwave and conventional heating [12]. Conventional heating requires surface heating until conduction, convection, or radiation transfers the heat inwards. However, during microwave heating, the microwave energy not only interacts with the surface material but, at the same time, penetrates the surface that comes into contact with the material's core [13]. Because of its high heating rate and easy operation, microwave heating is, therefore, a viable alternative to conventional heating methods, which have been widely used in many fields. The microwave-assisted pretreatment of various lignocellulosic biomass substrates was used in many studies [14–16]. These studies reported that microwave heating has a positive effect on cellulosic material digestion for downstream processes. Despite the large number of microwave-assisted application studies, microwave technology has not completely replaced conventional heating in industry. The problems associated with the processing of waste materials with microwaves include inherent difficulties with microwaves themselves and those inherent with processed materials. Microwave radiation can be applied via an applicator; therefore, it can be placed remotely and heating can be done in a clean environment. However, not all materials (e.g. transparent materials) are easily heated via microwave heating. Another characteristic of microwave processing is the differential coupling of materials, which enables selective heating. Adding absorbers to transparent material could help to increase the reaction temperature. On the other hand, using additives (absorbers) may result in unwanted impurities. At a temperature below the critical temperature of a material, microwave processing is self-limiting and, therefore, heating can cease after the process or phase is completed. For this reason, the efficacy of microwave-assisted pretreatment relies ultimately on the the pretreatment parameters. The factors that have most influence on sugar recovery in microwave pretreatments are solvent type, solid loading (solvent to feed ratio), exposure time, and microwave power [16–20].

Microwave heating-based processes may achieve a green and low-energy pretreatment cycle. The minimal use of energy requirements (short heating time), chemical auxiliaries (use of extreme diluted solvents), and the recycling of biomass wastes meet with the principles of green-extraction and can provide more sustainable and feasible routes for commodity production [21,22]. Therefore, this study investigates the use of low concentration solvents in microwave-assisted pretreatment and evaluates their effect on fermentable sugar yield from sago palm bark wastes through enzymatic hydrolysis. Extremely diluted solvents, such as sulphuric acid, sodium hydroxide, and sodium bicarbonate were applied in the microwave-assisted pretreatment prior to the enzymatic hydrolysis process. The microwave-assisted pretreatment methods were, namely, microwave-sulphuric acid pretreatment (MSA), microwave-sodium hydroxide pretreatment (MSH), and microwave-sodium bicarbonate pretreatment (MSB). Experimentally, the response surface methodology (RSM) was used for the design of experiment (DOE), to construct an empirical model based on the collected experimental data, and to highlights the interactions among the microwave-assisted pretreatment parameters and their effect on the enzymatic hydrolysis process. Therefore, this study aims to evaluate the sugar yield and the interactive effect of the key operating parameters of acidic and alkali microwave-assisted pretreatments. An emphasis is placed on how MSA, MSH, and MSB pretreatment processes effect glucose and xylose yield from the enzymatic hydrolysis of SPB.

2. Material and Methods

2.1. Substrate

The purchased sago palm trunks from a plantation in Melaka, Malaysia were used to prepare the experiment feedstock. These trunks were debarked to obtain the bark fraction (the outer layer). The collected bark was dried in an oven until reached a constant weight (temperature of 105 °C for ~24 h). Then, a woodchipper (Woodchipper from Pallmann Maschinenfabrik Gmb & Co KG in Zweibrücken, Germany type PZ 8) was used to chop the dried bark. The chopped matter was sieved by a chip classifier to get particle size (2–3 cm). Finally, these materials were stored in plastic bags at −20 °C until the experiments were carried out.

2.2. Enzymes and Chemicals

The cellulase enzymes from *Trichoderma reesei* (E.C. 3.2.1.4), xylanase from *Trichoderma viride* (E.C. 3.2.1.8) and β-glucosidase from Almond (E.C. 3.2.1.21) were purchased from Sigma-Aldrich (St. Louis, MO, USA) and used in enzymatic hydrolysis. In addition, the monosaccharides glucose, xylose analytical standards, and 3,5 dinitrosalicylic acid, as well as sodium azide were and used in the qualitative and quantitative analysis for sugar. Concentrated sulphuric acid, sodium hydroxide pellets, and sodium bicarbonate powder were purchased from R&M (Selangor, Malaysia) and were utilized in the pretreatment process. Citric acid monohydrate and sodium citrate were obtained from R&M (Selangor, Malaysia) to prepare the sodium citrate buffer.

2.3. Microwave-Assisted Pretreatment

The microwave-assisted pretreatment was performed in a single-phase stainless steel domestic microwave oven with a 206 mm (H) × 315 mm (W) × 353 mm (D) microwave oven cavity from Panasonic (NN-ST340M, Panasonic, Kadoma, Osaka Prefecture, Japan). This microwave has a 2.45 GHz magnetron. This magnetron was mounted at the side of the casing and with a maximal operation power of 800 W and five discrete settings. The microwave oven was modified by making a 25 mm round hole at the center of upper side to facilitate the connection between the reaction flask and the reflux condenser and the upper end of the reflex condenser was sealed by aluminum paper to prevent evaporation. The reflux unit is often included to control pressure by condensing the vaporized sample mixture, and this system operates at atmospheric conditions. The dried samples were transferred to a 1000 mL round-bottom flask that contained 100 mL of the pretreatment solvent. Then, the flask was placed inside the microwave cavity. After the pretreatment, the slurry was filtered through filter paper (0.45 μm) (Double Ring filter paper 102, China) to separate the solid residue and liquid faction (liquor). The filtered solid fraction was washed with distilled water to remove the pretreatment solvent and was dried at 60 °C for 48–72 h to get a constant weight. Then, the dried materials were stored at −20 °C in plastic bags for the subsequent enzymatic hydrolysis process.

Three solvents—0.05 M H_2SO_4, 0.1 M NaOH, and 0.01 M $NaHCO_3$—were used for the microwave-sulphuric acid pretreatment (MSA), microwave-sodium hydroxide pretreatment (MSH), and microwave-sodium bicarbonate pretreatment (MSB) of SPB, respectively. The pretreatments were performed at various solid loading, exposure time, and microwave power which resulted from the experimental design (DOE). The design of experiment (DOE) was done using Design Expert software (Version 7.1, Stat-EaseInc., Silicon Valley, CA, USA) a RSM approach. The Box–Behnken factorial design (BBD) with three independent variables and three levels was employed to plan experiments with consideration for the interactive effects among the variables during the pretreatments and their responses. Since there is no accurate procedure to directly measure the exact temperature and pressure of pretreatment in a domestic microwave oven, pretreatment was expressed in terms of the microwave power output that can be set on the instrument. The case study in this design involves the interaction effects of the pretreatment variables, microwave power (MP) (X_1), exposure time (ET) (X_2), and solid loading (SL) (X_3) on enzymatic hydrolysis from biomass. Table 1 shows the 17 experimental runs which were generated in terms of coded and actual variables.

Table 1. Number and Conditions of Experiments According RSM and BBD.

RUN	X_1	X_2	X_3	SL (%)	ET (min)	MP (W)
1	0	−1	1	10	5	800
2	0	1	1	10	15	800
3	0	1	−1	10	15	80
4	−1	0	1	5	10	800
5	−1	1	0	5	15	440
6	−1	−1	0	5	5	440
7	1	0	1	15	10	800
8	0	0	0	10	10	440
9	0	0	0	10	10	440
10	1	0	−1	15	10	80
11	0	0	0	10	10	440
12	1	−1	0	15	5	440
13	−1	0	−1	5	10	80
14	0	0	0	10	10	440
15	0	0	0	10	10	440
16	1	1	0	15	15	440
17	0	−1	−1	10	5	80

2.4. Enzymatic Hydrolysis

Enzymatic hydrolysis was performed in an incubator shaker at 55 °C and 150 rpm for 72 h. A total of 1.0 g of pretreated biomass, on a dry matter basis, was immersed in 30 mL of 50 mM sodium citrate buffer (pH 4.8) in a 250 mL Erlenmeyer flask. Cellulase was supplemented by 24 FPU/g, 2 UN/g of xylanase and β-glucosidase at an enzyme loading of 50 U/g. U and UN refer to the activity of β-glucosidase and xylanase, respectively, as reported by the manufacturer: "One U of β-glucosidase corresponds to the amount of enzyme which liberates 1 μmol of glucose per minute at pH 5.0 and 37 °C (salicin as substrate)" and "one UN will liberate 1 μmole of reducing sugar measured as xylose equivalents from xylan per minute at pH 4.5 at 30 °C. Additional β-glucosidase was essential to alleviate the cellobiose inhibition of cellulase. Prior to use, the cellulase activity assay for determining filter paper cellulase units (FPU) was performed as outlined by the NREL LAP-006 procedure [23] which was found to be 67 FPU/mL. A dose of 0.3% (*w/v*) sodium azide was added to avoid microbial contamination. Following hydrolysis, the samples were immediately transferred to a boiling water bath for 10 min to avoid further reaction and the denaturing of the enzymes. They were then cooled in an ice bath. Samples of the slurries were collected and filtered through a 0.22 μm nylon membrane, neutralized and kept at −30 °C for further sugar analysis. The sugar analysis was performed to identify the total reducing sugar using the dinitrosalicylic acid method (DNS) and to estimate monomeric sugar content via HPLC analysis.

2.5. Sugar Analysis

The total reducing sugar analysis was performed according to the DNS method of Miller (1959) [24]. The colored samples were then measured with a UV–VIS spectrophotometer (UV-2700, Shimadzu, Japan) at 540 nm using a standard curve of glucose. The monomeric sugars of glucose and xylose were analyzed by a high-performance liquid chromatography (HPLC) system (Alltech 2000, East Lyme, CT, USA) equipped with a RI detector and a Rezex RPM-Monosaccharide Pb^{+2} column, 300 mm in length and 7.5 mm in internal diameter, particle size: 8 (μm), Max back pressure: 1000 (PSI) (Phenomenex Inc., Torrance, CA, USA). The column was conditioned for 30 to 60 mins to reach a steady state using a mobile phase deionized water HPLC grade, which was already sonicated to deaerate the system at a flow rate of 0.6 mL/min. The column and detector temperatures were set at 85 and 40 °C, respectively and the injection volume of sample was 20 μL. The liquid samples were filtered using a 0.22 μm disposable nylon membrane syringe filter (Phenex Inc., England, UK) prior to HPLC analysis.

3. Results and Discussion

3.1. Effect of Microwave-Assisted Pretreatment Type on Sugar Yield

The effects of MSA, MSH, and MSB pretreatment on the physical and chemical characteristics of SPB were reported and discussed in a previous study by Ethaib et al. [4], and the composition of the pretreated solids is shown in Table 2.

Table 2. The chemical composition of solids before and after pretreatment [4].

Component % w/w	SPB			
	Untreated	MSA	MSH	MSB
Cellulose	40.79	47.23	47.1	44.92
Hemicellulose	22.32	19.55	24.21	27.18
Lignin (Removal)	25.85	17.68 (31.6%)	20.21 (21.8%)	18.83 (27.1%)
others	11.04	15.47	8.48	9.07

In the current study, the enzymatic hydrolysis of pretreated SPB using MSA, MSH, and MSB was carried out to evaluate the sugar yield and the interactive effect of microwave-assisted pretreatment parameters on sugar yield. In the present section, the sugar yield (mg/g of pretreated solids) and enzymatic hydrolysis efficiency of pretreated SPB were evaluated based on the total reducing sugar content using the DNS method on both glucose and xylose yields. Figure 1 depicts the levels of the total reducing sugar content that was obtained via MSA, MSH, MSB, and enzymatic hydrolysis for the seventeen experiment runs of the design of experiment. The examination of glucose and xylose yields using HPLC analysis are tabulated in Table 3 for MSA, MSH, and MSB based on BBD experiments. The results show that the sugar yield from enzymatic hydrolysis, which was calculated using the DNS method, was higher than the HPLC results. From a chemical standpoint, the DNS reagent reacts with all of the reducing sugars that contain aldehyde groups, which include all monosaccharaides, along with some disaccharides, such as cellobiose, oligosaccharides, and some polysaccharides [25]. In general, the HPLC can be used to detect the individual components of monosaccharaides, in this study, it was limited for glucose and xylose detection.

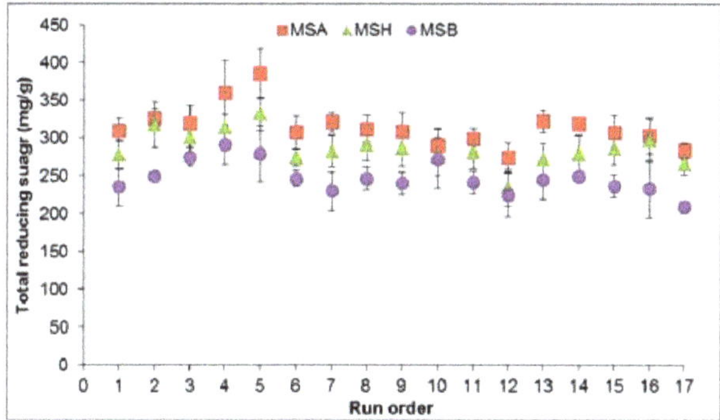

Figure 1. Total reducing sugar measured after enzymatic hydrolysis of sago palm bark for MSA, MSH, and MSB pretreatments.

Table 3. Glucose and xylose yield after enzymatic hydrolysis of sago palm bark for MSA, MSH, and MSB pretreatments.

Run	Pretreatment Conditions			MSA Pretreatment		MSH Pretreatment		MSB Pretreatment	
	SL(%)	ET (min)	MP (W)	Glucose (mg/g)	Xylose (mg/g)	Glucose (mg/g)	Xylose (mg/g)	Glucose (mg/g)	Xylose (mg/g)
1	10	5	800	25.5 ± 5.8	32.7 ± 2.8	32.9 ± 1.5	24.5 ± 0.0	16.0 ± 2.0	17.9 ± 1.5
2	10	15	800	20.8 ± 1.4	33.6 ± 5.8	35.8 ± 7.9	25.5 ± 0.2	19.6 ± 1.8	18.9 ± 0.3
3	10	15	80	27.1 ± 0.7	32.1 ± 7.6	32.5 ± 4.7	22.4 ± 1.4	18.8 ± 2.0	14.6 ± 2.3
4	5	10	800	24.4 ± 3.3	37.5 ± 5.7	44.3 ± 4.8	24.6 ± 0.7	20.1 ± 1.5	17.2 ± 2.4
5	5	15	440	37.5 ± 5.7	30.6 ± 3.2	40.9 ±3.4	24.7 ± 0.0	20.6 ± 0.3	14.7 ± 0.1
6	5	5	440	23.6 ± 0.8	31.4 ± 3.3	37.0 ± 1.0	15.8 ± 3.8	13.6 ± 1.9	16.8 ± 2.4
7	15	10	800	20.4 ± 2.1	43.1 ± 4.2	39.3 ± 1.0	22.9 ± 1.4	16.6 ± 1.0	17.2 ± 1.3
8	10	10	440	28.3 ± 2.4	33.8 ± 0.4	32.2 ± 1.4	19.6 ± 1.0	17.1 ± 2.0	21.2 ± 3.4
9	10	10	440	27.6 ± 2.2	37.1 ± 5.5	33.2 ± 1.7	18.3 ± 1.7	16.2 ± 1.3	19.9 ± 0.3
10	15	10	80	22.4 ± 2.3	34.5 ± 4.9	30.9 ± 5.3	17.7 ± 0.3	13.0 ± 2.6	16.3 ± 0.8
11	10	10	440	27.6 ± 0.8	31.1 ± 0.6	30.2 ± 1.4	18.2 ± 1.4	16.2 ± 1.3	19.0 ± 0.2
12	15	5	440	29.3 ± 3.9	33.0 ± 1.3	28.0 ± 1.4	14.7 ± 1.5	14.4 ± 0.7	16.5 ± 0.8
13	5	10	80	29.3 ± 1.5	38.9 ± 7.5	35.9 ± 3.8	19.7 ± 1.6	16.5 ± 0.3	19.8 ± 0.2
14	10	10	440	27.7 ± 3.7	35.5 ± 2.0	31.2 ± 2.6	19.6 ± 1.0	16.5 ± 0.3	19.8 ± 0.6
15	10	10	440	28.3 ± 3.0	32.0 ± 2.9	32.2 ± 1.3	18.8 ± 1.4	16.6 ± 1.8	18.9 ± 0.6
16	15	15	440	27.7 ± 3.9	35.5 ± 2.0	28.9 ± 0.9	17.3 ± 0.9	17.1 ± 1.0	13.9 ± 0.3
17	10	5	80	23.1 ± 1.2	30.1 ± 0.6	29.7 ± 1.0	14.9 ± 1.0	11.5 ± 2.4	17.4 ± 1.9

The results show that, overall, microwave-acid pretreatment (MSA), followed by the enzymatic hydrolysis of SPB gave a higher sugar yield in comparison to both types of alkaline pretreatments (MSH and MSB), as shown in Figure 1. The maximum total reducing sugar was found to be 386 mg/g when SPB was loaded at 5% and soaked in 0.05 M H_2SO_4 under 440 W for 15 mins. With 0.1 M NaOH and 0.01 M $NaCHO_3$ treatments under the same conditions, the yields were 332 mg/g and 279 mg/g, respectively. The results in Table 3 indicate that the highest glucose yield was found through MSH pretreatment, with an average glucose yield of 1.27—times the pretreatment yields of MSA and MSB, respectively. The highest glucose yield was found at 44.3 ± 4.8 mg/g in run 4 when the SPB underwent MSH pretreatment, loaded at 5% solid and subjected to 800 W for 5 mins. SPB, pretreated under the same conditions with sulphuric acid and bicarbonate sodium, only yielded 24.4 ± 3.3 mg/g and 20.1 ± 1.5 mg/g after the enzyme hydrolysis, respectively. Zhang and his co-workers [26] reported that NaOH solutions can cause the swelling and dissolution of cellulose. The swelling of cellulose causes significant changes in physical properties and an increase in cellulose volume. Moreover, the dissolution of cellulose can destroy the supramolecular structure of cellulose. Therefore, cellulose swelling and the increase in the cellulose content and its volume and the disruption of its structure, caused by the NaOH solution, may lead to an increase in the surface area available for the cellulase enzyme to react, resulting in increased glucose release.

Xylose yield in MSH pretreatment was lower than that of MSA pretreatment; however, a higher xylose yield was found at 43.1 ± 4.2 mg/g in run 7 using MSA pretreatment. The xylose yields for the MSH and MSB pretreatments under the same pretreatment conditions were 22.9 ± 1.4 mg/g and 17.2 ± 1.3 mg/g, respectively. This can be attributed to the ability of sulphuric acid to provide hydrogen ions to breakdown long hemicellulose chains and form shorter chain oligomers, facilitating the liberation of monomeric sugars C5 and C6 [24]. The average xylose yield in SPB hydrolysate via MSA was 1.72, which was 1.95 times the xylose yields of MSH and MSB, respectively.

Table 3 details the glucose and xylose yields for MSA, MSH, and MSB pretreatments, based on the variables SL, ET, and MP. These variables have a significant effect on glucose and xylose yields. For instance, in run 4, the glucose yield was 44.3 ± 2.8 mg/g when SPB underwent MSH pretreatment and was loaded at 5% (SL) and subjected to 800 W (MP) for 5 min. In run 2, increasing the solid loading to 10% at the same microwave power led to a decrease in glucose yield to 35.8 ± 3.9 mg/g, even though the time was increased to 10 min. This means that the SL has a significant effect on sugar yield compared to ET. A similar trend was found with MSA. For more details on the combined effects of these variables on glucose and xylose yield, statistical analysis was performed and a 3D contour plot was created for discussion in the following sections.

3.2. Analysis Effect of Microwave-Assisted Pretreatment Parameters Using ANOVA

In order to understand the effect of the parameters in microwave-assisted pretreatment on glucose and xylose yields, analyses of variance (ANOVA) for the BBD, as shown in Table 3, were performed using Design Expert software. The BBD was used to evaluate the effect of the three microwave-assisted pretreatment independent variables, SL, ET, and MP, on the response variables, glucose and xylose yields, via enzymatic hydrolysis to create models between these variables. Second-order multi regression models were constructed as a function of the three microwave-assisted pretreatment variables, SL (X_1), ET (X_2), and MP (X_3), on the predicted response of glucose yield (Y_1) and xylose yield (Y_2). The quality of fit of the polynomial model equations and their parameters were evaluated by determining the R^2 coefficient. Statistical and regression coefficient significances were checked against the probability (*p*-value). *p*-values of less than 0.05 were applied to validate the significant of the models and each of the variables, which, in turn, is necessary to understand the pattern of the mutual interactions between the test variables.

The glucose yield (Y_1) and xylose yield (Y_2) models for MSA pretreatments and the enzymatic hydrolysis of SPB are illustrated using second-order polynomial equations as follows:

$$Y_1 = 11.01211 + 0.35256X_1 + 1.71473X_2 + 0.035514X_3 - 0.15532X_1X_2 + 0.000394581X_1X_3 - 0.00120793X_2X_3 + 0.032591X_1{}^2 + 0.032933X_2{}^2 - 0.0000354291X_3{}^2 \tag{1}$$

$$Y_2 = 35.76288 - 2.78708X_1 + 2.91092X_2 - 0.022262X_3 + 0.033276X_1X_2 + 0.00138219X_1X_3 - 0.000152006X_2X_3 + 0.10210X_1{}^2 - 0.15315X_2{}^2 + 0.0000157648X_3{}^2 \tag{2}$$

The glucose yield (Y_1) and xylose yield (Y_2) models for MSH pretreatment and enzymatic hydrolysis are illustrated in Equations (3) and (4), respectively:

$$Y_1 = 41.5865 - 3.18614X_1 + 1.75493X_2 - 0.00832057X_3 - 0.028987X_1X_2 - 0.0000126471X_1X_3 + 0.0000142204X_2X_3 + 0.13534X_1{}^2 - 0.060397X_2{}^2 + 0.0000186437X_3{}^2 \tag{3}$$

$$Y_2 = 6.19687 + 0.39759X_1 + 1.65842X_2 - 0.00238738X_3 - 0.073023X_1X_2 - 0.000906216X_1X_3 + 0.0000219542X_3{}^2 \tag{4}$$

The regression model for the glucose yield (Y_1) and xylose yield (Y_2) of MSB pretreatment and the enzymatic hydrolysis of SPB are presented in Equations (5) and (6), respectively:

$$Y_1 = 9.14903 - 0.15517X_1 + 0.91215X_2 + 0.00852174X_3 - 0.022048X_1X_2 + 0.000128086X_1X_3 - 0.000514792X_2X_3 \tag{5}$$

$$Y_2 = 8.48181 - 1.27393X_1 + 1.55549X_2 - 0.00771416X_3 - 0.00612675X_1X_2 + 0.000498938X_1X_3 + 0.000525542X_2X_3 - 0.077275X_1{}^2 - 0.094553X_2{}^2 + 0.00000165916X_3{}^2 \tag{6}$$

Tables 3 and 4 display the ANOVA for glucose yields (Y_1) and xylose yields (Y_2) for MSA, MSH, and MSB pretreatments.

Table 4. ANOVA for glucose of MSA, MSH, and MSB pretreatments.

Pretreatment.	Source of Variations	Sum of Squares	Degree of Freedom	Mean Square	F-Value	*p*-Value	R^2
MSA	Model	232.76	9	25.86	6.30	0.0120	0.8901
	X_1 (SL)	28.16	1	28.16	6.86	0.0345	
	X_2 (ET)	16.67	1	16.67	4.06	0.0837	
	X_3 (MP)	14.95	1	14.95	3.64	0.0980	
	X_1X_2	60.31	1	60.31	14.69	0.0064	
	X_1X_3	2.02	1	2.02	0.49	0.5058	
	X_2X_3	18.91	1	18.91	4.61	0.0690	
	$X_1{}^2$	2.80	1	2.80	0.68	0.4365	
	$X_2{}^2$	2.85	1	2.85	0.70	0.4319	
	$X_3{}^2$	88.77	1	88.77	21.63	0.0023	
MSH	Model	286.65	9	31.85	6.59	0.0106	0.8944
	X_1 (SL)	120.08	1	120.08	24.85	0.0016	
	X_2 (ET)	13.82	1	13.82	2.86	0.1347	
	X_3 (MP)	68.05	1	68.05	14.08	0.0071	
	X_1X_2	2.10	1	2.10	0.43	0.5308	
	X_1X_3	0.02073	1	0.02073	0.004289	0.9841	
	X_2X_3	0.02621	1	0.02621	0.05422	0.9821	
	$X_1{}^2$	48.20	1	48.20	9.97	0.0160	
	$X_2{}^2$	9.60	1	9.60	1.99	0.2016	
	$X_3{}^2$	24.58	1	24.58	5.09	0.0587	
MSB	Model	90.99	6	15.17	30.11	<0.0001	0.9476
	X_1 (SL)	20.39	1	20.39	40.49	<0.0001	
	X_2 (ET)	43.28	1	43.28	85.93	<0.0001	
	X_3 (MP)	22.46	1	22.46	44.61	<0.0001	
	X_1X_2	1.22	1	1.22	2.41	0.1514	
	X_1X_3	0.21	1	0.21	0.42	0.5305	
	X_2X_3	3.43	1	3.43	6.82	0.0260	

The ANOVA for the glucose models (Y_1), pertaining to MSA, MSH, and MSB pretreatments, shows that the *p*-values are significant, as shown in Table 3. The results show that the *p*-value for the Y_1 of the MSA pretreatment model was 0.012, which means model is significant at a linear regression of R^2 0.8901 and a F-value of 6.3. The *p*-values also revealed that the independent variables, X_1 and X_2, and the quadratic term X_3 had significant effects on glucose yields during MSA pretreatment, as shown in Table 4.

The R^2 and associated *p*-values for the MSH pretreatment of Y_1 were 0.8944 and 0.0106, respectively at F-value 6.59, showing that this model is significant. The variables X_1 and X_2 and the quadratic term of X_1 illustrate their significant effect on glucose yields during MSH pretreatment, as shown in Table 4.

For MSB pretreatment, the *p*-value of the Y_1 model is very low (less than 0.0001) reflecting a high R^2 value (0.9476) and F-value 30.11. Based on the *p*-values of the model variables, the variables X_1, X_2, and X_3, and the interaction between X_2 and X_3, had significant effects on glucose yields during MSB pretreatment, as shown in Table 3.

An ANOVA was also executed for the xylose yield (Y_2) models of the three types of pretreatments. The *p*-values of the generated models are significant, as shown in Table 5. The *p*-value of 0.0363 for the Y_2 model for MSA pretreatment is significant where R^2 = 0.8431. This suggests that the interaction between the independent variables X_1 and X_3, and quadratic terms X_1 and X_2, had a significant effect on xylose yields during MSA pretreatment.

Table 5. ANOVA for xylose models of MSA, MSH, and MSB pretreatments.

Pretreatment	Source	Sum of Squares	Degree of Freedom	Mean Square	F-Value	*p*-Value	R^2
MSA Pretreatment	Model	155.51	9	17.28	4.18	0.0363	0.8431
	X_1 (SL)	7.68	1	7.68	1.86	0.2152	
	X_2 (ET)	2.59	1	2.59	0.63	0.4545	
	X_3 (MP)	15.88	1	15.88	3.84	0.0909	
	X_1X_2	2.77	1	2.77	0.67	0.4402	
	X_1X_3	24.76	1	24.76	5.99	0.0443	
	X_2X_3	0.30	1	0.30	0.072	0.7956	
	$X_1^{\,2}$	27.43	1	27.43	6.63	0.0367	
	$X_2^{\,2}$	61.72	1	61.72	14.93	0.0062	
	$X_3^{\,2}$	17.58	1	17.58	4.25	0.0782	
MSH Pretreatment	Model	201.90	9	22.43	17.89	0.0005	0.9583
	X_1 (SL)	22.13	1	22.13	17.65	0.0040	
	X_2 (ET)	56.07	1	56.07	44.72	0.0003	
	X_3 (MP)	64.22	1	64.22	51.22	0.0002	
	X_1X_2	13.33	1	13.33	10.63	0.0138	
	X_1X_3	0.024	1	0.024	0.019	0.8931	
	X_2X_3	10.64	1	10.64	8.49	0.0225	
	$X_1^{\,2}$	1.20	1	1.20	0.96	0.3605	
	$X_2^{\,2}$	0.011	1	0.011	0.08855	0.9277	
	$X_3^{\,2}$	34.73	1	34.73	27.70	0.0012	
MSB Pretreatment	Model	58.39	9	6.49	4.31	0.0335	0.8472
	X_1 (SL)	2.57	1	2.57	1.71	0.2328	
	X_2 (ET)	5.49	1	5.49	3.64	0.0979	
	X_3 (MP)	1.19	1	1.19	0.79	0.4037	
	X_1X_2	0.094	1	0.094	0.062	0.8100	
	X_1X_3	3.23	1	3.23	2.14	0.1866	
	X_2X_3	3.58	1	3.58	2.38	0.1669	
	$X_1^{\,2}$	15.71	1	15.71	10.44	0.0144	
	$X_2^{\,2}$	23.53	1	23.53	15.63	0.0055	
	$X_3^{\,2}$	0.19	1	0.19	0.13	0.7297	

For MSH pretreatment, the *p*-value of the Y_2 model was 0.0005, which means that this model is highly significant where $R^2 = 0.9583$ and the F-value = 17.89. The variables X_1, X_2, and X_3 are significant, as are the interactions between X_1 and X_2, and between X_2 and X_3. The quadratic term X_3 has a significant effect on xylose yields during MSH pretreatment, as shown Table 4.

Finally, Table 4 also shows the significance of the Y_2 model for MSB pretreatment, where: $R^2 = 0.8472$, *p*-value = 0.0335. In this table, only the quadratic terms of X_1 and X_2 are observed as having a significant effect on xylose yield under MSB pretreatment with *p*-values of 0.01440 and 0.0055, respectively.

3.3. Analysis Effect of Microwave-Assisted Pretreatment Parameters Using RSM Plots

In order to understand the main and interactive effects between the pretreatment parameters on glucose and xylose yields, 3D response surface plots and contour lines were generated for glucose and xylose yields, according to the pretreatment condition (SL, ET and MP). This allows for the examination of the interactive effect of the operating parameters. The shape and color of the surface plot and the contour lines provide information about the relationship between the pretreatment variables and related response. Contour plots are the projection of the response surface on a two-dimensional plane, while 3D surface plots are the projection of the response surface on a three-dimensional plane [27].

3.3.1. Pretreatment Parameters on Glucose Yield

Figure 2 displays the effect of the pretreatment parameters—solid loading (SL), exposure time (ET), and microwave power (MP)—on glucose yields during microwave-sulphuric acid (MSA) pretreatment. Figure 2a illustrates the response surfaces of the combined effects of MP and ET with a constant SL. The glucose yields gradually increased with an increase in MP and ET. However, after a certain point, further increases in MP caused a decrease in glucose yield. This might be related to the interactive effect of the third parameter (SL) whose effect on MP and ET can be clearly seen in the previous subsection in Table 3 regarding run 2 and run 7. It was reported that the increase in SL level could cause a decrease in the saccharification process. This might be attributed to the different "energy effect" with different SL. In other words, the samples with a high SL (and thus relatively low pretreatment solvent loading) receive less energy absorbed by the pretreatment solvent, resulting in a decrease in the internal heating and the oscillation of the pretreatment solvent molecules [17].

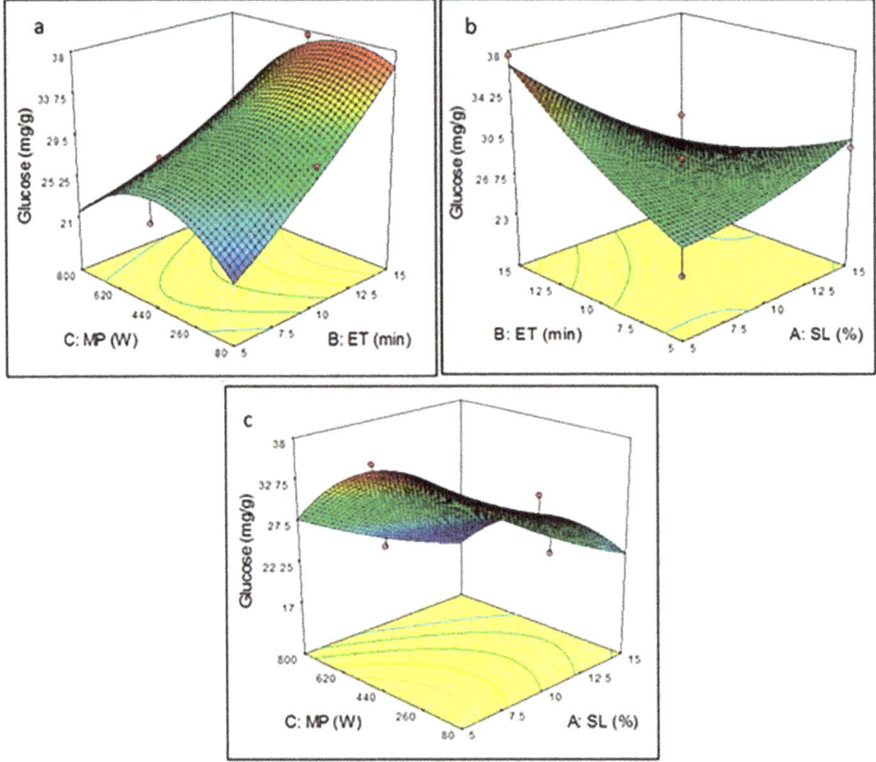

Figure 2. The effect of pretreatment parameters in MSA pretreatment on the glucose yield (**a**) with MP and ET at a constant SL, (**b**) with ET and SL at a constant MP, and (**c**) with MP and SL at a constant ET.

Figure 2b shows the relationship between ET and SL at a fixed MP. In this figure, the surface plot shows an increase in the glucose yield with an increase in ET, reaching a peak at the lowest SL (5%). An increase in ET allowed the sample (solvent + biomass) to absorb more microwave energy. This, in turn, generated more heat in the sample-breaking cellulose chains, which led to more enzyme accessibility during enzymatic hydrolysis, resulting in increased glucose yields [28].

Figure 2c shows a similar trend between MP and SL, where an increase in solid loading levels led to a gradual decrease in glucose yield. The higher glucose yields were found at the surface plot regions with low and medium SL. This confirms that high solid loadings have a negative effect on glucose yield because more energy is needed to produce a higher yield. A low solid loading allows the sample particles to receive more microwave energy per gram of the solid substrate. Moisture is heated inside the sample particles, evaporates, and generates tremendous pressure on the plant cell walls because of plant cell swelling. The pressure pushes the cell wall from inside, stretching and ultimately rupturing it, which facilitates a leaching out of the active constituents from the ruptured cell to the surrounding solvent, thus improving the yield of sugar [29].

Regarding MSH pretreatment, Figure 3 shows the combined effect of the pretreatment parameters on glucose yield. Figure 3a illustrates that a continuous increase in ET with a lower level of MP does not affect glucose yield, while increasing ET in combination with a high level of MP causes a steady increase in glucose yield due to the increase in heat generation within the sample. This may lead to a high delignification process because of disruptions in the lignin structure during the microwave-alkaline pretreatment [14]. Removing the lignin reduces the mechanical strength of the plant cell which, in turn, helps enzymes to access the cellulosic compounds inside the cell. Figure 3b shows that only lower levels of SL (less than 10%) result in a higher glucose yield at various ET levels. Figure 3c demonstrates that a higher level of microwave power leads to a higher glucose yield at lower levels of SL (5%), while an increase in SL to 15% causes a decrease in glucose yield mg/g (at a fixed ET). This is because samples which have low solid loading absorb more microwave energy, which helps to generate more heat within the sample. This reflects positively on enzymatic hydrolysis, as mentioned in an earlier study by Manaso et al. [28].

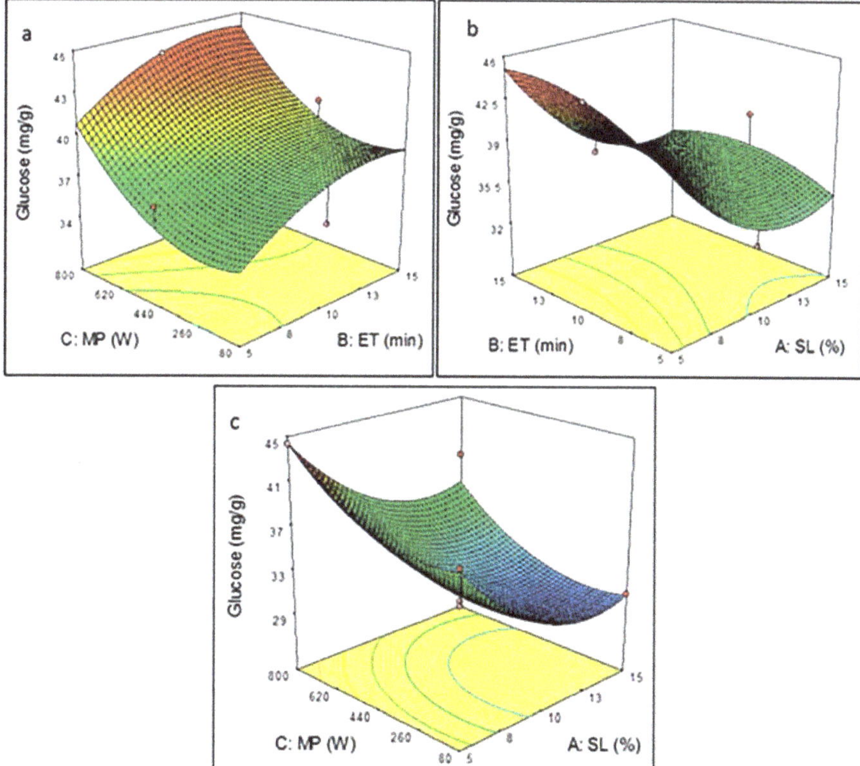

Figure 3. The effect of pretreatment parameters of MSH pretreatment on the glucose yield (**a**) with MP and ET at a constant SL, (**b**) with ET and SL at a constant MP, and (**c**) with MP and SL at a constant ET.

Regarding MSB pretreatment, Figure 4 depicts the effect of the operating parameters of this pretreatment on glucose yield. The response surface plot shape and the contour lines in Figure 4a indicate that there is a linear relationship among MP, ET, and glucose yield. An increase in both microwave power and exposure time intensified the generation of heat, enhancing the delignification process, making the lignocellulose more accessible for enzyme action. Figure 4b shows the interaction between ET and SL, where it can be seen that there is a contrasting relationship between the latter parameters and response. Similarly, Figure 4c shows the same relationship among MP, SL, and glucose yield as a result of the "energy effect", where low solid loaded samples received more microwave energy [30].

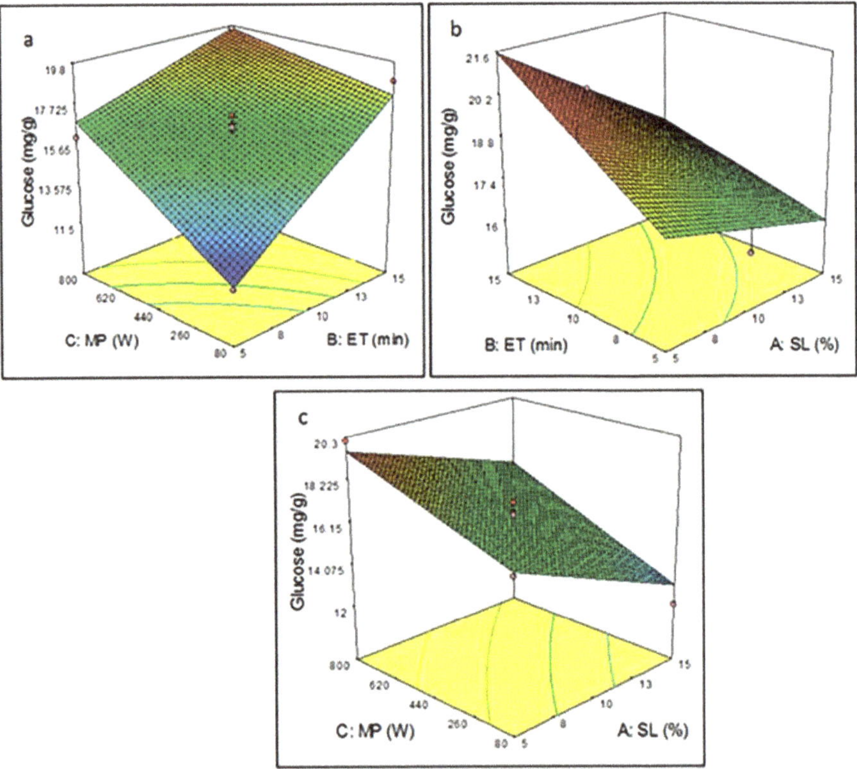

Figure 4. The effect of pretreatment parameters of MSB pretreatment on glucose yield (**a**) with MP and ET at a constant SL, (**b**) with ET and SL at a constant MP, and (**c**) with MP and SL at a constant ET.

3.3.2. Effect of Pretreatment Parameters on Xylose Yield

Figure 5 illustrates the interactive effects of microwave-assisted pretreatment conditions MP, ET, and SL on xylose yields under MSA pretreatment. Figure 5a points to a strong interaction between MP and ET. It can be observed that the gradual increase in microwave power from 80 to 800 W enhanced the xylose yield from 32 to 43.2 mg/g at low and medium levels of exposure time. A maximum xylose yield (43.2 mg/g) was achieved at 800 W for 10 min. However, increasing the exposure time beyond 10 min caused a decrease in xylose yield, a similar finding to that of Ma et al. [17]. This interactive effect between irradiation time and microwave power level increased biomass digestibility by enhancing hemicellulose removal. However, an extended exposure time with a higher microwave power may lead to a decline in biomass digestibility, as this increases irradiation time and microwave power cause high temperatures within the sample, which could initiate the decomposition of released sugar [31].

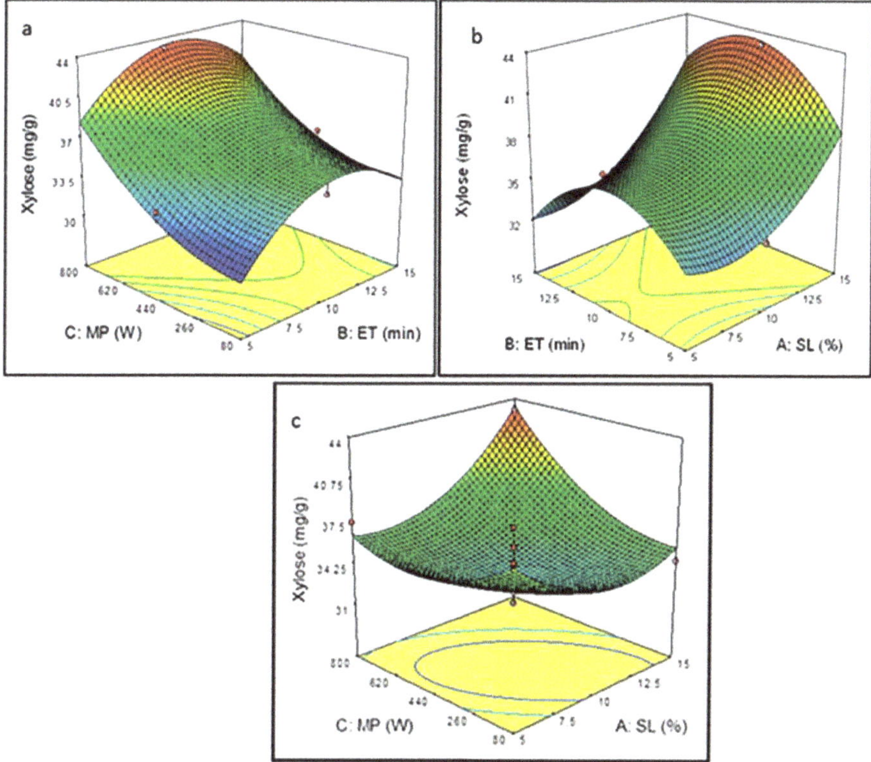

Figure 5. The effect of pretreatment parameters of MSA pretreatment on the xylose yield (**a**) with MP and ET at a constant SL, (**b**) with ET and SL at a constant MP, and (**c**) with MP and SL at a constant ET.

By the same token, as seen in Figure 5b, ET has a stronger influence on xylose yield compared to SL. The xylose yield is less influenced by increases in SL, while the increase in ET results in a higher xylose yield until a specific point. Ma et al. [17] reported that the elliptical nature of the contour plots refers to a prominent interaction between the pretreatment variables, as seen between MP and SL in Figure 5c. An increase in solid loading causes a decrease in biomass digestibility at low and medium levels of MP; an increase in microwave power to 800 W mitigates the negative effect of the increase in solid loading. These results can be attributed to the energy effect for high solid loading samples. By increasing microwave power, the sample is able to receive more energy absorbed by the solvent, which helps to disturb the hemicellulose structure, impacting positively on xylose yield [32].

Similarly, Figure 6a shows the significant interaction between MP and ET at a constant solid loading on xylose yields under MSH pretreatment. The gradual increase in microwave power and exposure time causes a gradual increase in xylose yield. Extending the exposure time and increasing the microwave power, in addition to the high dielectric constant of NaOH (6.8 Debay) [33], can lead to higher heat generation within the sample which, in turn, enhances sugar yield. These results support the view that to facilitate partial hemicellulose fractionation may require a strong pretreatment condition during alkaline pretreatment [34].

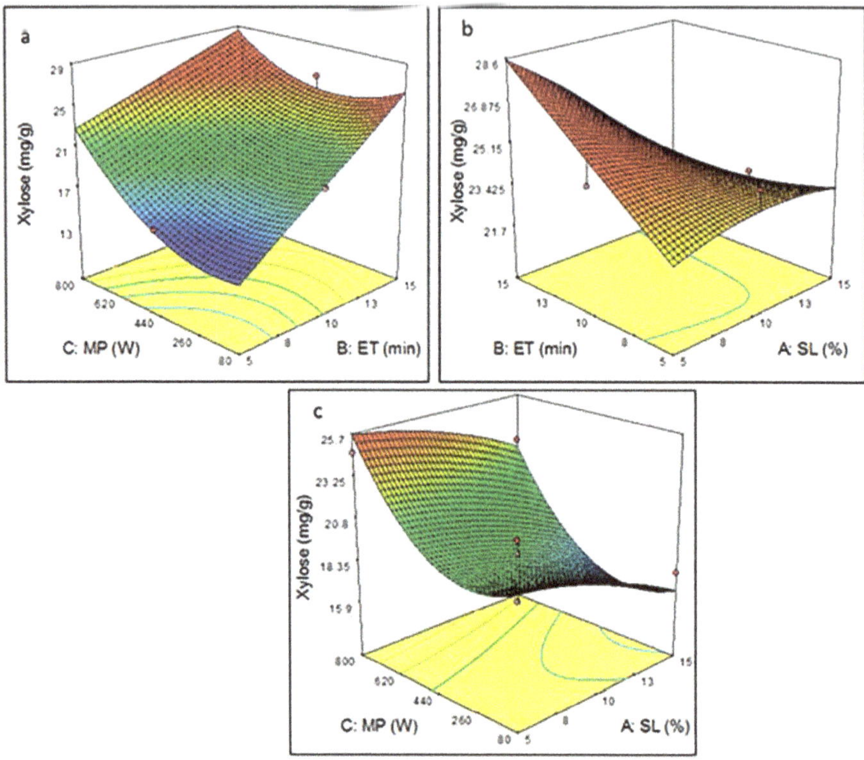

Figure 6. The effect of pretreatment parameters of MSH pretreatment on the xylose yield (**a**) with MP and ET at a constant SL, (**b**) with ET and SL at a constant MP, and (**c**) with MP and SL at a constant ET.

Figure 6b suggests a significant relationship between exposure time and solid loading; a high xylose yield (25.7 mg/g) was found with a longer exposure time (15 min) and a low solid load (5%). Extended exposure time and low solid loadings allow the samples to absorb more energy, which intensifies the generation of heat within the biomass particles. This increases hemicellulose destruction and enhances its hydrolysis, resulting in the swelling of the biomass particles and the improvement of carbohydrate accessibility to enzymes [34]. It has been suggested that alkaline pretreatments may need more time to reach the same level of digestibility offered by other pretreatments [5]. Figure 6c shows that the xylose yield found at a high microwave power and low solid loading decreased when the microwave power was reduced, and solid loading was increased. These results can also be explained by energy absorption, where higher microwave power provided more energy, causing an increase in reaction temperature. In summary, it can now be stated that high microwave power, longer exposure time, and low solid loading during MSH pretreatment can enhance xylose yield during the enzymatic hydrolysis step.

Figure 7a shows that a gradual increase in microwave power and exposure time can increase xylose yield until a tipping point is reached where further increments cause a decrease in xylose yield. This may be related to the formation of carbonic acid in $NaHCO_3$ aqueous solutions which, coupled with heat, cause sodium bicarbonate to act as a raising agent by releasing carbon dioxide [35]. This enhances biomass destruction. However, extended exposure time and microwave power does not enhance the sugar yield; this may be due to loss of carbon dioxide because of the conversion of sodium bicarbonate. The circular contour nature between exposure time and solid loading, as shown Figure 7b, suggests a less prominent or negligible interaction between these two parameters [17]. Figure 7c also illustrates a minor interactive effect between MP and SL, as there is no significant difference in xylose yield (between 5% and 15%) or increase in microwave power (from 80 to 800 W). In other words, the xylose yield is less affected by the interaction of pretreatment variables during MSB pretreatment.

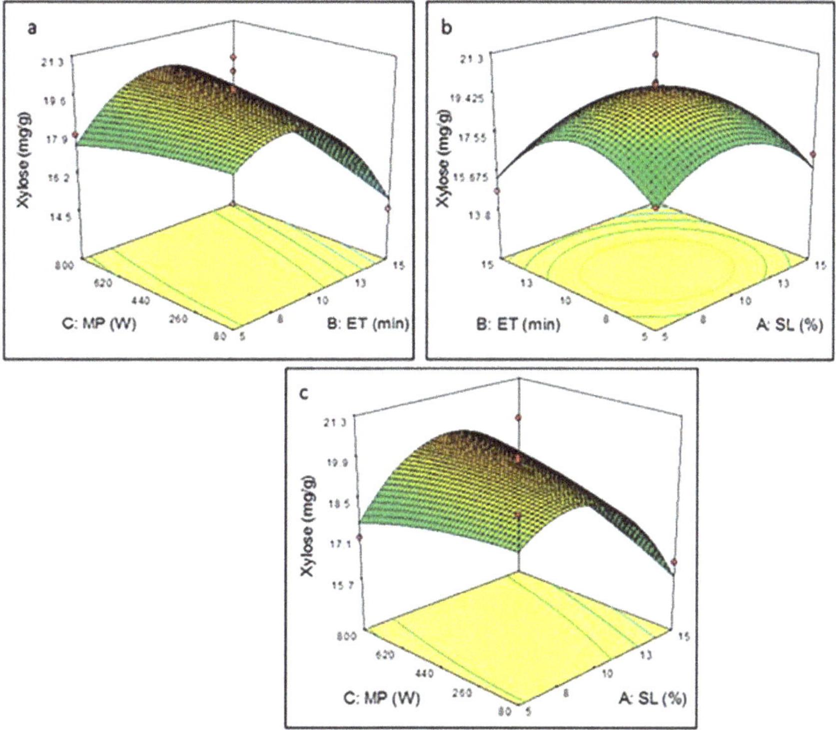

Figure 7. The effect of pretreatment parameters of MSB pretreatment on the xylose yield (**a**) with MP and ET at a constant SL, (**b**) with ET and SL at a constant MP, and (**c**) with MP and SL at a constant ET.

4. Conclusions

The results revealed that MSA pretreatment is the most efficient sago palm bark pretreatment technique, compared to the two other types of microwave-alkali pretreatment (MSH and MSB), for the release of reducing sugar and the yield of xylose. However, MSH pretreatment methods resulted in the higher glucose yield. The analyses of the effects of the pretreatment parameters MP, ET, and SL on the glucose and xylose yield from SPB revealed that microwave-assisted pretreatment parameters showed different patterns of influence on glucose and xylose yield via enzymatic hydrolysis for MSA, MSH, and MSB pretreatment. From this, it can be interfered that variation in microwave-assisted pretreatment parameters and the pretreatment solution play a crucial role on the sugar yield from lignocellulosic biomass. For further studies, the performance of conventional sulphuric acid pretreatments using the same solvent level and same exposure time is recommended. Moreover, the base cost analysis is useful when exploring new technology. Therefore, conducting overall mass-energy balance and economic analyses will help the microwave-assisted pretreatment to be commercialized. Additionally, performing the microwave-assisted pretreatment in a microwave system equipped with a temperature sensor will be useful to develop the kinetic models for sugar yield. Moreover, this will enable the identification of appropriate process parameters that will be useful in the scale-up of microwave-assisted pretreatment processes.

Author Contributions: Conceptualization, S.E.; data curation, S.E. and R.O.; formal analysis, S.E.; R.O.; M.K.S.M. and A.B.D.R.; investigation, S.E.; methodology, S.E.; R.O.; M.K.S.M. and A.B.D.R.; project administration, R.O.; resources, S.E.; R.O.; M.K.S.M. and A.B.D.R.; software, S.E.; M.K.S.M.; supervision, R.O.; M.K.S.M. and A.B.D.R.; writing—original draft, S.E.; writing—review and editing, S.E. and R.O.; funding acquisition, R.O. All authors have read and agreed to the published version of the manuscript.

Funding: This research was funded by IPMS, grant number 9446900.

Acknowledgments: The authors are grateful for the support of the University Putra Malaysia and the University of Thi-Qar for their financial support.

Conflicts of Interest: The authors declare that there is no conflict of interest regarding the publication of this paper.

References

1. Ethaib, S. Solid Waste Situation in Thi-Qar Governorate. In *IOP Conference Series: Material Science Engineering*; IOP Publishing: Bristol, UK, 2019; p. 584012023.
2. Ethaib, S.; Omar, R.; Mazlina, M.; Radiah, A.; Syafiie, S.; Harun, M.Y. Effect of microwave-assisted acid or alkali pretreatment on sugar release from Dragon fruit foliage. *Int. Food Res. J.* **2016**, *23*, S149–S154.
3. Ethaib, S.; Erabee, I.K.; Abdulsahib, A.A. Removal of Methylene Blue Dye from Synthetic Wastewater using Kenaf Core and Activated Carbon. *Int. J. Eng. Technol.* **2018**, *7*, 909–913. [CrossRef]
4. Ethaib, S.; Omar, R.; Mustapa Kamal, S.M.; Awang Biak, D.R.; Syam, S.; Harun, M.Y. Microwave-assisted pretreatment of sago palm bark. *J. Wood Chem. Technol.* **2017**, *37*, 26–42. [CrossRef]
5. Jørgensen, H.; Kristensen, J.B.; Felby, C. Enzymatic conversion of lignocellulose into fermentable sugars: Challenges and opportunities. *Biofuels Bioprod. Biorefin.* **2007**, *1*, 119–134. [CrossRef]
6. Öhgren, K.; Bura, R.; Saddler, J.; Zacchi, G. Effect of hemicellulose and lignin removal on enzymatic hydrolysis of steam pretreated corn stover. *Bioresour. Technol.* **2007**, *98*, 2503–2510. [CrossRef]
7. Blue, D.; Fortela, D.L.; Holmes, W.; LaCour, D.; LeBoeuf, S.; Stelly, C.; Revellame, E.D. Valorization of Industrial Vegetable Waste Using Dilute HCl Pretreatment. *Processes* **2019**, *7*, 853. [CrossRef]
8. Han, L.; Feng, J.; Zhang, S.; Ma, Z.; Wang, Y.; Zhang, X. Alkali pretreated of wheat straw and its enzymatic hydrolysis. *Braz. J. Microbiol.* **2012**, *43*, 53–61. [CrossRef]
9. Min, D.Y.; Xu, R.S.; Hou, Z.; Lv, J.Q.; Huang, C.X.; Jin, Y.C.; Yong, Q. Minimizing inhibitors during pretreatment while maximizing sugar production in enzymatic hydrolysis through a two-stage hydrothermal pretreatment. *Cellulose* **2015**, *22*, 1253–1261. [CrossRef]
10. Ethaib, S.; Omar, R.; Mustapa Kamal, S.M.; Awang Biak, D.R. Comparison of sodium hydroxide and sodium bicarbonate pretreatment methods for characteristic and enzymatic hydrolysis of sago palm bark. *Energy Source Part A Recovery Util. Environ. Eff.* **2020**, 1–11. [CrossRef]
11. Ethaib, S.; Omar, R.; Kamal, S.M.; Biak, D.A. Microwave-assisted pretreatment of lignocellulosic biomass: A review. *J. Eng. Sci. Technol.* **2015**, *2*, 97–109.
12. Mokhtar, N.M.; Ethaib, S.; Omar, R. Effects of microwave absorbers on the products of microwave pyrolysis of oily sludge. *J. Eng. Sci. Technol.* **2018**, *13*, 3313–3330.
13. Rahimi, M.A.; Omar, R.; Ethaib, S.; Mazlina, M.S.; Biak, D.A.; Aisyah, R.N. Microwave-Assisted Extraction of Lipid from Fish Waste. In *IOP Conference Series: Material Science Engineering*; IOP Publishing: Bristol, UK, 2017; Volume 206, p. 012096.
14. Komolwanich, T.; Tatijarern, P.; Prasertwasu, S.; Khumsupan, D.; Chaisuwan, T.; Luengnaruemitchai, A. Comparative potentiality of Kans grass (*Saccharum spontaneum*) and Giant reed (*Arundo donax*) as lignocellulosic feedstocks for the release of monomeric sugars by microwave/chemical pretreatment. *Cellulose* **2014**, *21*, 1327–1340. [CrossRef]
15. Alexandre, A.M.; Matias, A.A.; Bronze, M.R.; Cocero, M.J.; Mato, R. Phenolic Compounds Extraction of Arbutus unedo L.: Process Intensification by Microwave Pretreatment. *Processes* **2020**, *8*, 298. [CrossRef]
16. Ethaib, S.; Omar, R.; Mazlina, M.K.S.; Radiah, A.B.D.; Syafiie, S. Microwave-assisted Dilute Acid Pretreatment and Enzymatic Hydrolysis of Sago Palm Bark. *BioResources* **2016**, *11*, 5687–5702. [CrossRef]
17. Ma, H.; Liu, W.; Chen, X.; Wu, Y.; Yu, Z. Enhanced enzymatic saccharification of rice straw by microwave pretreatment. *Bioresour. Technol.* **2009**, *100*, 1279–1284. [CrossRef]

18. Vani, S.; Binod, P.; Kuttiraja, M.; Sindhu, R.; Sandhya, S.V.; Preeti, V.E.; Pandey, A. Energy requirement for alkali assisted microwave and high pressure reactor pretreatments of cotton plant residue and its hydrolysis for fermentable sugar production for biofuel application. *Bioresour. Technol.* **2012**, *112*, 300–307. [CrossRef]

19. Chan, C.H.; Yusoff, R.; Ngoh, G.C.; Kung, F.W.L. Microwave-assisted extractions of active ingredients from plants. *J. Chromatogr. A* **2011**, *1218*, 6213–6225. [CrossRef]

20. Ethaib, S.; Omar, R.; Mazlina, M.K.S.; Radiah, A.B.D.; Syafiie, S. Development of a hybrid PSO–ANN model for estimating glucose and xylose yields for microwave-assisted pretreatment and the enzymatic hydrolysis of lignocellulosic biomass. *Neural Comput. Appl.* **2018**, *30*, 1111–1121. [CrossRef]

21. Bouras, M.; Chadni, M.; Barba, F.J.; Grimi, N.; Bals, O.; Vorobiev, E. Optimization of microwave-assisted extraction of polyphenols from Quercus bark. *Ind. Crop. Prod.* **2015**, *77*, 590–601. [CrossRef]

22. Chadni, M.; Bals, O.; Ziegler-Devin, I.; Brosse, N.; Grimi, N. Microwave-assisted extraction of high-molecular-weight hemicelluloses from spruce wood. *Comptes Rendus Chim.* **2019**, *22*, 574–584. [CrossRef]

23. Adney, B.; Baker, J. Measurement of Cellulase Activities. In *Laboratory Analytical Procedure*; National Renewable Energy Laboratory (NREL): Golden, CO, USA, 1996; Volume 6.

24. Miller, G.L. Use of dinitrosalicylic acid reagent for determination of reducing sugar. *Anal. Chem.* **1959**, *31*, 426–428. [CrossRef]

25. Jeffries, T.W.; Yang, V.W.; Davis, M.W. Comparative study of xylanase kinetics using dinitrosalicylic, arsenomolybdate, and ion chromatographic assays. *Appl. Biochem. Biotechnol.* **1998**, *70*, 257–265. [CrossRef] [PubMed]

26. Zhang, S.; Wang, W.C.; Li, F.X.; Yu, J.Y. Swelling and dissolution of cellulose in NaOH aqueous solvent systems. *Cellul. Chem. Technol.* **2013**, *47*, 671–679.

27. Box, G.E.; Hunter, J.S. Multi-factor experimental designs for exploring response surfaces. *Ann. Math. Stat.* **1957**, *1*, 195–241. [CrossRef]

28. Manaso, J.; Luengnaruemitchai, A.; Wongkasemjit, S. Optimization of two-stage pretreatment combined with microwave radiation using response surface methodology. World Academy of Science, Engineering and Technology World Academy of Science, Engineering and Technology (WASET). *Int. J. Chem. Mol. Eng.* **2013**, *76*, 599.

29. Mandal, V.; Mohan, Y.; Hemalatha, S. Microwave assisted extraction—An innovative and promising extraction tool for medicinal plant research. *Pharmacogn. Rev.* **2007**, *1*, 7–18.

30. Lu, X.; Xi, B.; Zhang, Y.; Angelidaki, I. Microwave pretreatment of rape straw for bioethanol production: Focus on energy efficiency. *Bioresour. Technol.* **2011**, *102*, 7937–7940. [CrossRef]

31. Kabel, M.A.; Bos, G.; Zeevalking, J.; Voragen, A.G.; Schols, H.A. Effect of pretreatment severity on xylan solubility and enzymatic breakdown of the remaining cellulose from wheat straw. *Bioresour. Technol.* **2007**, *98*, 2034–2042. [CrossRef]

32. Yang, B.; Wyman, C.E. Effect of xylan and lignin removal by batch and flowthrough pretreatment on the enzymatic digestibility of corn stover cellulose. *Biotechnol. Bioeng.* **2004**, *86*, 88–98. [CrossRef]

33. Keshwani, D.R. Microwave Pretreatment of Switchgrass for Bioethanol Production. Ph.D. Thesis, North Carolina State University, Raleigh, NC, USA, 2009.

34. Modenbach, A. Sodium Hydroxide Pretreatment of Corn Stover and Subsequent Enzymatic Hydrolysis: An Investigation of Yields, Kinetic Modeling and Glucose Recovery. Ph.D. Thesis, University of Kentucky, Lexington, KY, USA, 2013.

35. Greenwood, N.N.; Earnshaw, A. *Chemistry of the Elements*, 2nd ed.; Butterworth-Heinemann Publications Elsevier Ltd.: Oxford, UK, 2012.

Article

Evaluation of Napier Grass for Bioethanol Production through a Fermentation Process

Mallika Boonmee Kongkeitkajorn [1,*], **Chanpim Sae-Kuay** [1] **and Alissara Reungsang** [1,2]

[1] Department of Biotechnology, Faculty of Technology, Khon Kaen University, Khon Kaen 40002, Thailand; chanpim0112@gmail.com (C.S.-K.); alissara@kku.ac.th (A.R.)

[2] Research Group for Development of Microbial Hydrogen Production Process, Khon Kaen University, Khon Kaen 40002, Thailand

* Correspondence: mallikab@kku.ac.th

Received: 12 April 2020; Accepted: 7 May 2020; Published: 11 May 2020

Abstract: Ethanol is one of the widely used liquid biofuels in the world. The move from sugar-based production into the second-generation, lignocellulosic-based production has been of interest due to an abundance of these non-edible raw materials. This study interested in the use of Napier grass (*Pennisetum purpureum* Schumach), a common fodder in tropical regions and is considered an energy crop, for ethanol production. In this study, we aim to evaluate the ethanol production potential from the grass and to suggest a production process based on the results obtained from the study. Pretreatments of the grass by alkali, dilute acid, and their combination prepared the grass for further hydrolysis by commercial cellulase (Cellic® CTec2). Separate hydrolysis and fermentation (SHF), and simultaneous saccharification and fermentation (SSF) techniques were investigated in ethanol production using *Saccharomyces cerevisiae* and *Scheffersomyces shehatae*, a xylose-fermenting yeast. Pretreating 15% *w/v* Napier grass with 1.99 M NaOH at 95.7 °C for 116 min was the best condition to prepare the grass for further enzymatic hydrolysis using the enzyme dosage of 40 Filter Paper Unit (FPU)/g for 117 h. Fermentation of enzymatic hydrolysate by *S. cerevisiae* via SHF resulted in the best ethanol production of 187.4 g/kg of Napier grass at 44.7 g/L ethanol concentration. The results indicated that Napier grass is a promising lignocellulosic raw material that could serve a fermentation with high ethanol concentration.

Keywords: Napier grass; bioethanol; biomass fractionation; enzyme hydrolysis; acid pretreatment; alkali pretreatment

1. Introduction

Napier grass (*Pennisetum purpureum* Schumach), known also as elephant grass, is a perennial grass found in tropical regions. Its high yield, easy cultivation, nutrient availability and versatility make it widely popular for use as a fodder crop. Annual production yields of the grass vary depending on cultivars, harvest cycles, fertilization, and climates. The reported Napier grass yield in temperate climates was 20–40 ton/ha [1], while the yield in tropical climates was higher at 50–67 ton/ha [2,3]. Napier grass is classified as a lignocellulosic biomass. Its structural compositions varies depending on weather, variety, and age. Reports on its composition covered 31–41% cellulose and 15–47% hemicellulose [4–7]. Hydrolysis of its structure yields glucose and xylose, which are monomeric sugars that serve as substrates in microbial fermentation to produce various biochemical including biofuels.

Napier grass plantation in Thailand has been for agricultural purposes. As the country has continuously promoted biofuel production, Napier grass is considered as an energy crop. Thailand's 10 Year (2012–2021) Alternative Energy Development Plan has placed the grass as one of the focused energy plants [8]. Napier grass has been successfully used in biogas production [8–10]. Studies on the use of Napier grass for other biofuels production also exist, including for bioethanol production. Pretreatments play an important role in production of liquid fuels. A study showed a higher ethanol yield when using pretreated versus non-pretreated grasses, mainly due to increases in sugar yields obtained in hydrolysis [11].

Pretreatment of Napier grass is an important step to prepare it for further hydrolysis. The process alters physical structure of the biomass so that enzyme has a better access to the cellulose chain of the biomass and improves hydrolysis [12,13]. Studies have been investigating various methods of pretreatment, primarily physical and chemical methods, with the aim to obtain high sugar yields. Some examples of the pretreatments focused on Napier grass included the uses of various alkalis, e.g., aqueous ammonia, $Ca(OH)_2$, NaOH, alkali H_2O_2 [14–16], and ammonia gas [17,18], dilute acid [14], acid-peroxide [19], acid-alkali [20], steam explosion [14,21], hydrothermal treatment [11], and nitrogen explosive decompression [22].

In the current study, we attempted to determine a simple and viable process for ethanol production from Napier grass. The works in each pretreatment and hydrolysis step involved the use of response surface methodology in order to determine the suitable conditions for each pretreatment and hydrolysis routes.

Despite various pretreatment methods available, we chose to investigate the chemical pretreatment of mechanically ground Napier grass due to its operational simplicity. We also chose NaOH and H_2SO_4 as they are common chemicals that are already in use by industries. The suitable conditions for enzymatic hydrolysis of the each pretreatment methods were then determined to obtain the process that resulted in the highest sugar released from the grass. Upon obtaining suitable pretreatment and hydrolysis conditions, ethanol production was investigated using hydrolysate obtained from enzymatic hydrolysis of pretreated grass, through separated hydrolysis and fermentation (SHF), and using the pretreated grass, through simultaneous saccharification and fermentation (SSF). Two yeast strains were used in fermentations, a common yeast and a xylose-fermenting yeast. We evaluated the ethanol production using such combination of substrates and yeasts and suggested the process for ethanol production from Napier grass based on the results of the study.

2. Materials and Methods

2.1. Preparation of Napier Grass

Napier grass (*Pennisetum purpureum* Schumach) Pakchong 1 aged 90–150 days was harvested from the Demonstration field of the Faculty of Agriculture, Khon Kaen University, Thailand. Stems were cut approximately 5 cm above the ground. They were then chopped and oven dried at 70 °C. The dried stems were then milled and sieved through a 10-mesh screen to obtain dried grass used in the study.

2.2. Alkali and Dilute Acid Pretreatments

Pretreatments using alkali and acid followed a similar procedure. In the case of alkali pretreatment, dried Napier grass was soaked in 200 mL sodium hydroxide solution at 15% *w/v* loading in 500 mL high-pressure laboratory bottles and heated in an autoclave. For acid pretreatments, dried Napier grass or alkali-pretreated grass was soaked in sulfuric acid and heated in an autoclave. The concentration, time, and temperature of the pretreatment were varied according to the values shown in Table 1. After pretreatment, solid fractions were collected and washed until the pH was neutral. They were then dried at 70 °C prior to analysis and enzymatic hydrolysis.

Design Expert® (version 7.0 demo, Stat-Ease, Minneapolis, MN, USA, 2005) was used in designing the experimental runs and analysis of results. Box–Benkhen design was applied for all experiments. Treated grasses were analyzed for their susceptibility to cellulase hydrolysis which was used as responses in all pretreatments. Mathematical models obtained from program analysis were used to predict the conditions that resulted in the highest susceptibility for each of the pretreatment methods.

Table 1. Factors and levels used in pretreatment and enzymatic hydrolysis studies.

Treatment Methods	Factors	Low Level (−1)	High Level (+1)
Alkali pretreatment	NaOH, Molarity (M)	0.25	3
	Temperature, °C	50	100
	Time, min	30	180
Acid pretreatment	H_2SO_4, % *v/v*	1	5
	Temperature, °C	80	120
	Time, min	30	120
Acid pretreatment of alkali-treated grass	H_2SO_4, % *v/v*	1	5
	Temperature, °C	50	100
	Time, min	60	180
Enzymatic hydrolysis	Enzyme dosage (Filter Paper Unit (FPU)/g substrate)	10	50
	Incubation time, h	24	120
	Substrate loading, % *w/v*	5	15

2.3. Enzymatic Hydrolysis

Pretreated grasses were hydrolyzed using a cellulase, Cellic® CTec2 (Novozymes, Bagsvaerd, Denmark). Grass was added into 100 mL of 50 mM citrate buffer solution (pH 5.0) in a 500-mL laboratory bottle. The enzyme was then added and mixed, and the mixture was incubated in a water bath at 50 °C with constant mixing. The hydrolysis reaction was stopped by boiling the content for 5 min. The liquid fraction was analyzed for reducing sugar.

Operation factors that were evaluated in enzymatic hydrolysis included enzyme dosage, incubation time and substrate loading (Table 1). The Box–Benkhen design was used in designing experimental runs with the help of Design Expert® software (version 7.0 demo, Stat-Ease, Minneapolis, MN, USA, 2005). The analysis of data was also carried out using the program. Suitable conditions for hydrolysis by the enzyme were determined based on the maximum amount of reducing sugars obtained from model prediction.

2.4. Ethanol Fermentation via SHF and SSF

Saccharomyces cerevisiae TISTR 5339 (Thailand Institute of Scientific and Technological Research, Bangkok, Thailand) and *Scheffersomyces shehatae* ATCC 22984 (The American Type Culture Collection, Manassas, VA, USA) were used in all fermentations. Yeast stocks were maintained in 30% glycerol. They were propagated twice in Yeast Malt (YM) agar. Liquid medium used in fermentation was prepared from enzymatic hydrolysate of alkali-treated grass supplemented with 2 g/L autolyzed yeast powder (FM801, Angel Yeast, Yichang, China) for *S. cerevisiae* or 6 g/L for *S. shehatae*. Fermentations were carried out in 250 mL Erlenmeyer flasks.

In inoculum preparation, a few single colonies were inoculated in diluted hydrolysate medium whose glucose was adjusted to 10 g/L. It was incubated at 30 °C with shaking at 200 rpm for 24 h. Ten percent of the first seed inoculum was transferred to a fresh hydrolysate medium with 20 g/L glucose and incubated at the same conditions for another 24 h.

In separate hydrolysis and fermentation (SHF), 10 mL of inoculum was transferred to 100 mL of hydrolysate medium. For simultaneous saccharification and fermentation (SSF), 100 mL of 15% *w/v* slurry of alkali-treated grass in water, supplemented with autolyzed yeast powder, were used as substrate and medium for fermentation. Ten mL of inoculum was added together with 40 Filter Paper Unit (FPU)/g$_{grass}$ of cellulase, Cellic® CTec2, to start the cultivation. Fermentation conditions were 30 °C with shaking at 100 rpm. Samples were taken at intervals and analyzed for cell growth, sugars, and ethanol concentrations.

2.5. Analysis

Grasses after pretreatment were analyzed by assessing their susceptibility to cellulase hydrolysis ("susceptibility test" in short). One gram of dried grass was hydrolyzed by 30 FPU/g of cellulase (Cellic® CTec2) in 50 mM citrate buffer (pH 5.0) for 72 h at 50 °C. Results from the analysis were reported as susceptibility with the unit of gram of reducing sugar per liter ($g_{RS, suscepibility}$/L).

Structural compositions of grasses before and after pretreatments were analyzed based on extractive-free biomass following the NREL procedure [23] and ASTM standard test E1758-01(2007). Surfaces of grasses were observed under scanning electron microscope (S-3000N, Hitachi, Tokyo, Japan).

Number of cells was reported as cells/mL and determined using cell count on hemocytometer. Reducing sugar was analyzed using dinitrosalicylic acid colorimetric assay. High-performance liquid chromatography (LC-20A, Shimadzu, Kyoto, Japan) was used to determine concentrations of glucose, xylose, and ethanol. It was equipped with an Aminex HPX-87H (Bio-Rad, Hercules, CA, USA) column and a refractive index detector (RID-6A, Shimadzu, Kyoto, Japan) for analysis. Column temperature was set at 40 °C. The mobile phase was 5 mM sulfuric acid, flowing at 0.75 mL/min.

3. Results

3.1. Pretreatments of Napier Grass

Prior to hydrolysis of Napier grass by enzyme or using it in fermentation, pretreatment of the grass is necessary in order to obtain biomass that is more vulnerable to further hydrolysis. Investigations in using alkali, acid and their combinations in pretreatment of Napier grass were carried out to determine for suitable conditions when applying each pretreatment regime.

3.1.1. Alkali Pretreatment

In order to obtain pretreatment conditions using alkali, NaOH concentrations (A in molarity), treatment temperature (B in °C) and time (C in min) were varied within the value ranges indicated in Table 1. By using the Box–Behnken design, experimental runs and their respective results were obtained as shown in Table 2. The data fitted well with this following quadratic equation with R-squared value of 0.9843.

$$y = 7.24 + 19.8\,A + 0.045\,B + 0.168\,C + 0.093\,AB + 0.015\,AC - 0.0014\,BC - 6.46\,A^2 \\ + 0.00081\,B^2 - 0.00029\,C^2 \tag{1}$$

Table 2. Experimental runs and results for determining conditions of alkali pretreatment following the Box–Behnken design.

Run No.	NaOH, A	Temperature, B	Time, C	Susceptibility
	(M)	(°C)	(min)	($g_{RS, susceptibility}$/L)
1	3	75	30	24.3
2	1.625	75	105	31.7
3	0.25	75	180	10.8
4	1.625	100	180	35.8
5	1.625	75	105	32.4
6	1.625	75	105	33.4
7	0.25	100	105	10.0
8	3	100	105	39.1
9	0.25	50	105	9.62
10	1.625	50	180	31.2
11	1.625	75	105	31.7
12	0.25	75	30	9.80
13	1.625	50	30	22.4
14	1.625	100	30	37.7
15	3	75	180	31.4
16	3	50	105	26.0
17	1.625	75	105	26.0

The analysis of variance (ANOVA) for the model (Table 3, NaOH) indicated that all the factors had a significant effect on the susceptibility of the grass to cellulase hydrolysis. In addition, interactions between NaOH concentration and temperature (A, B) and between temperature and time (B, C) also had significant effects as demonstrated in Figure 1. The plots indicate that an increase in the concentration of NaOH resulted in an increase in reducing sugar obtained from the susceptibility test until the concentration was approximately 2 M; then the sugar started to decrease. Furthermore, the time taken in pretreatment has shown its dependence on the temperature used (Figure 1b). At low pretreatment temperature, the grass was more susceptible to cellulase hydrolysis with longer treatment time whereas time had little effect at high treatment temperature.

Table 3. *P*-values obtained from ANOVA (analysis of variance) for response surface quadratic models used in the prediction of suitable conditions for pretreatments of Napier grass.

Source	*p*-Value	Source	*p*-Value	
	NaOH		**H$_2$SO$_4$**	**NaOH + H$_2$SO$_4$**
Model	<0.0001	Model	0.0075	0.0055
A—NaOH	<0.0001	A—H$_2$SO$_4$	0.0022	0.0244
B—Temperature	0.0005	B—Temperature	0.0011	0.0004
C—Time	0.0318	C—Time	0.0979	0.9906
AB	0.0145	AB	0.2714	0.8541
AC	0.1648	AC	0.7242	0.5962
BC	0.0306	BC	0.0766	0.7019
A^2	<0.0001	A^2	0.0923	0.0210
B^2	0.6162	B^2	0.7228	0.0061
C^2	0.1380	C^2	0.1178	0.2965
Lack of Fit	0.1695	Lack of Fit	0.5632	0.0018
R-Squared	0.9843	R-Squared	0.9053	0.9139
Predicted R-Squared	0.8206	Predicted R-Squared	0.3468	−0.3992

Notes: NaOH = pretreatment with NaOH, H$_2$SO$_4$ = pretreatment with H$_2$SO$_4$ and NaOH + H$_2$SO$_4$ = pretreatment with NaOH followed by H$_2$SO$_4$.

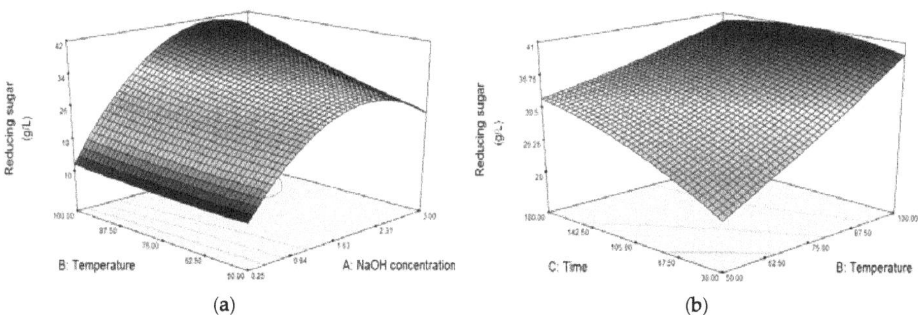

Figure 1. Surface plots showing significant interactions ($\alpha = 0.05$) between (**a**) NaOH concentration and temperature (AB) at 116 min and (**b**) temperature and time (BC) at 1.99 M NaOH in alkali pretreatment on reducing sugar obtained from the susceptibility test.

The equation predicted the pretreatment conditions that resulted in the highest susceptibility (39.3 g$_{RS, \text{susceptibility}}$/L) of 15% dried grass to be 1.99 M NaOH, 95.7 °C and 116 min. A verification test using these conditions in pretreating Napier grass resulted in 38.4 g/L of reducing sugar obtained from the susceptibility test, which was a 2.2% deviation from the model prediction.

3.1.2. Dilute Acid Pretreatment

In the investigation for suitable conditions in pretreatment of Napier grass by dilute acid, three factors of interest were H_2SO_4 concentration (A in % v/v), temperature (B in °C) and time (C in min). Their levels as indicated in Table 1 was applied to the Box–Behnken design and the results of 17 experimental runs are shown in Table 4. The data were fitted with a quadratic equation. The resulting equation gave the best fit with R-squared value of 0.9053:

$$y = -3.40 - 1.06\ A + 0.124\ B + 0.110\ C + 0.008\ AB - 0.0011\ AC - 0.00063\ BC + 0.130\ A^2 \\ - 0.00025\ B^2 - 0.00023\ C^2 \tag{2}$$

Table 4. Experimental runs and results for determining conditions of dilute acid pretreatment following the Box–Behnken design.

Run No.	H_2SO_4, A	Temperature, B	Time, C	Susceptibility
	(%)	(°C)	(min)	($g_{RS,\ susceptibility}$/L)
1	3	120	120	9.57
2	3	100	75	9.17
3	1	100	120	8.36
4	3	100	75	8.02
5	5	100	120	9.77
6	1	80	75	7.76
7	3	100	75	8.62
8	3	100	75	9.37
9	3	80	120	8.32
10	3	80	30	5.95
11	5	120	75	11.5
12	3	120	30	9.47
13	5	80	75	9.13
14	1	120	75	8.86
15	3	100	75	9.33
16	5	100	30	9.73
17	1	100	30	7.92

Analysis of variance for the model (Table 3, H_2SO_4) suggested that the main factors that posed significant influence on the susceptibility of the grass to cellulase hydrolysis at 95% confidence level were H_2SO_4 concentration (A) and temperature (B). Time was also significant but at a higher significance level ($\alpha = 0.1$). In pretreatment with dilute acid, interaction between factors did not significantly affect the susceptibility of the grass. Nonetheless, the interaction between time and temperature (BC) was significant at a 90% confidence level. The surface plot in Figure 2 shows that temperature affected the susceptibility strongly when using a short pretreatment time. Its effect was lesser at longer treatment time. Furthermore, combination of high temperature and long treatment time exerted detrimental effects on the susceptibility of the grass at 5% H_2SO_4.

The quadratic model predicted the maximum reducing sugar from susceptibility test to be 11.6 g/L. The respective conditions were pretreating the grass at 15% loading with 5% v/v H_2SO_4 at 120 °C for 56 min. The verification test resulted in 11.4 ± 0.1 g/L of reducing sugar from susceptibility test. Although the predicted R-squared of this model was low, these results of the conditions were still valid since the conditions were similar to those of run number 11. The small difference in pretreatment time between run number 11 and the predicting value was insignificant, as the pretreatment time did not have a significant effect on the susceptibility test result.

According to the susceptibility results in Table 4, it should be noted that dilute acid pretreatment was generally inferior to alkali pretreatment. Lower reducing sugar released in susceptibility test indicated that the pretreated grass was less susceptible to cellulase hydrolysis.

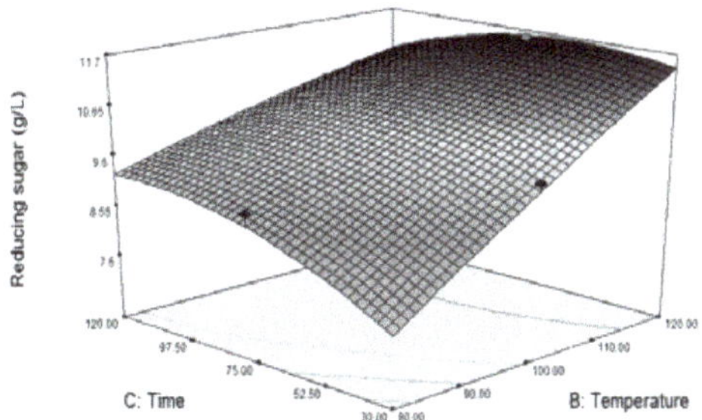

Figure 2. Surface plot showing interactions ($\alpha = 0.10$) between temperature and time (*BC*) at optimal H_2SO_4 concentration (5% *v/v*) in dilute acid pretreatment on reducing sugar obtained from the susceptibility test.

3.1.3. Alkali Followed by Acid Pretreatment

In this pretreatment, Napier grass was firstly pretreated with NaOH under the best conditions. The NaOH-treated grass was subjected to the treatment using the same factors as in the dilute acid pretreatment but the levels were adjusted as shown in Table 1. The results from all 17 experimental runs showed that the difference between the lowest and highest values of reducing sugars from susceptibility test was quite narrow (Table 5). This circumstance indicated the more profound effect of alkali pretreatment on the biomass such that further treatment by acid resulted in small changes in susceptibility values.

Table 5. Experimental runs and results for determining conditions of alkali followed by dilute acid pretreatments following the Box–Behnken design.

Run No.	H_2SO_4, A	Temperature, B	Time, C	Susceptibility
	(%)	(°C)	(min)	($g_{RS, susceptibility}$/L)
1	5	75	180	29.5
2	1	75	180	33.9
3	5	100	120	23.7
4	1	50	120	30.7
5	5	75	60	28.1
6	3	50	60	34.3
7	1	100	120	25.8
8	5	50	120	29.2
9	1	75	60	31.1
10	3	75	120	32.0
11	3	50	180	32.7
12	3	100	180	25.2
13	3	75	120	32.3
14	3	75	120	31.8
15	3	75	120	31.5
16	3	75	120	31.6
17	3	100	60	27.9

Analysis of variance (Table 3, NaOH + H_2SO_4) suggested that acid concentration and temperature were the two factors that influenced the susceptibility of the grass to cellulase hydrolysis at 97.6% and more than 99.9% confidence levels, respectively. The ANOVA also indicated that there was no interaction between factors.

The experimental data was fitted to a quadratic model with the resulting equation:

$$y = 15.2 + 2.82\,A + 0.522\,B - 0.027\,C - 0.0026\,AB - 0.0031\,AC - 0.00018\,BC - 0.490\,A^2$$
$$- 0.0041\,B^2 + 0.00021\,C^2 \tag{3}$$

This equation fitted well with the experimental data with R-squared value of 0.9139. The model predicted the best conditions that yielded the highest susceptibility of 33.5 $g_{RS,susceptibility}$/L to be the use of 2.53% v/v H_2SO_4 at 62 °C for 60 min at 15% grass loading. Although the lack of fit test was shown as significant (p-value = 0.0018) and the predicted R-squared was negative, the result of the independent susceptibility test using the predicted conditions (33.0 ± 0.5 $g_{RS,susceptibility}$/L) was very close to the predicted value. This situation was possible regardless of the significant lack of fit and due to the following supporting reason. Since the time was not a significant factor and there was no interaction between factors, it would not affect the response. Considering runs with the same temperature, the optimum values of acid concentration (2.53% v/v) laid between run 9 and 10 (Table 5) where the temperature was 75 °C. The responses of the two runs were very similar at 31.6 and 31.9 $g_{RS,susceptibility}$/L. Following the same principle, optimum temperature (62 °C) laid between values in run 10 and 11 (Table 5) where acid concentration was the same. The responses of the 2 runs were also similar at 31.9 and 33.3 $g_{RS,susceptibility}$/L. From these explanations, it was evident that changes in acid concentration and temperature between those ranges barely affected the responses. Therefore, the confirmation test using the predicted conditions was valid. However, low predicted R-squared values indicated that this equation is not suitable for predicting the response of the conditions that were not at the experimental points.

3.2. Physical Structure and Composition of Napier Grass after Pretreatments

Napier grass obtained after the three pretreatment regimes were analyzed for their physical structures and structural compositions. The results in Table 6 demonstrated that all pretreatments are able to reduce the amount of lignin in Napier grass, although at different capability. Reduction in lignin content was more pronounced when treating with NaOH than using the acid. The acid directly affects the hemicellulose part of the grass as reflected in the more than 50% reduction in hemicellulose in the pretreated samples. NaOH also dissolves some hemicellulose causing a drop in hemicellulose content, although not to the same extent caused by the dilute acid. As the results of pretreatment, an increase in the cellulose fraction was evident.

Table 6. Structural composition of native and pretreated Napier grasses.

Composition	Native Grass	Treated Grass with NaOH	Treated Grass with H_2SO_4	Treated Grass with (NaOH + H_2SO_4)
% Lignin	19.4 ± 0.1 [a]	9.90 ± 0.4 [c]	15.0 ± 0.6 [b]	4.27 ± 0.34 [d]
% Cellulose	35.2 ± 1.2 [b]	46.5 ± 0.6 [a]	46.6 ± 0.0 [a]	48.3 ± 1.7 [a]
% Hemicellulose	18.1 ± 0.3 [a]	14.0 ± 0.2 [b]	7.72 ± 0.0 [c]	13.9 ± 0.2 [b]
% Ash	1.73 ± 0.18 [a]	0.58 ± 0.03 [c]	0.98 ± 0.08 [b]	0.25 ± 0.07 [d]
Susceptibility ($g_{RS,\ susceptibility}$/L)	n/a	38.4 ± 0.14 [a]	11.4 ± 0.1 [c]	33.0 ± 0.5 [b]

Notes: (1) The statistical comparisons were between the same compositions of the different treatments. The same letter indicated that the values were not significantly different at 95% confidence level. (2) All samples were extractive-free. n/a = not available.

When matching the susceptibility data to the structural composition of the pretreated grasses, it is evident that the presence of lignin relates to a much lesser vulnerability of the biomass to hydrolysis by cellulase. Although the pretreated grasses have similar polysaccharide contents, they do not result in the same level of hydrolysis as indicated by the resulting reducing sugar from the susceptibility test.

Surfaces of native grass observed under SEM shows smooth and packed-layer surface (Figure 3a). The layering surface still exists but with loose, wrinkled, and swelling appearance after the pretreatment with NaOH, both with and without acid pretreatment (Figure 3b,d). A different effect when using acid pretreatment is observed such that the grass appeared fibrous with signs of degradation but the appearance is still packed (Figure 3c).

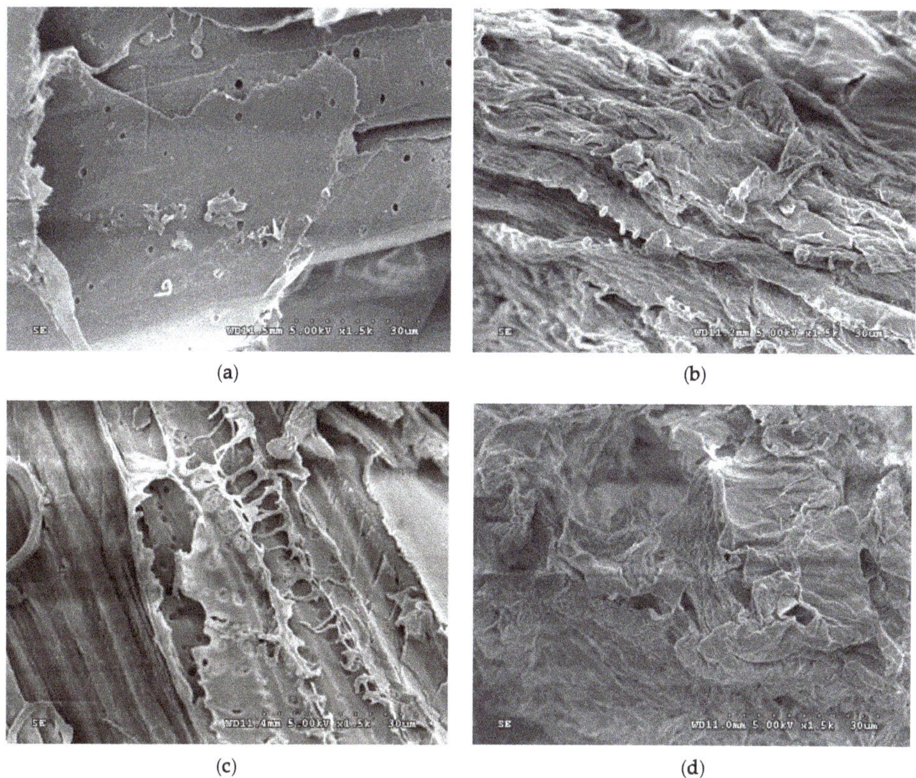

(a)

(b)

(c)

(d)

Figure 3. Physical appearances of Napier grass (**a**) in native form, (**b**) when pretreated with NaOH, (**c**) when pretreated with H_2SO_4 and (**d**) pretreated with NaOH followed by H_2SO_4 under a scanning electron microscope at 1500× magnification.

According to the results on susceptibility to cellulase hydrolysis, the structural compositions and the physical structures of the grass, we conclude that the pretreatment of 15% dried grass by 1.99 M NaOH at 95.7 °C for 116 min is suitable as it makes the grass most susceptible to further enzymatic hydrolysis. The treatment with dilute acid following the NaOH pretreatment is not necessary, as the susceptibility did not improve with the extra treatment.

3.3. Enzymatic Hydrolysis of Pretreated Grass

Although the previous results on pretreatment methods have suggested the use of NaOH alone in pretreating the Napier grass, we decided to investigate the enzymatic hydrolysis of the grass from all three pretreatment methods. Pretreated grasses were dried prior to the hydrolysis. The hydrolysis conditions followed those shown in Table 1. The reducing sugar concentration in the liquid hydrolysate was the response used for the analysis of the results and the prediction of the suitable hydrolysis conditions.

The experimental results of enzymatic hydrolysis of grass pretreated with the three methods were demonstrated in Table 7. In general, both pretreatments involving NaOH resulted in similar reducing sugar concentrations and the values were significantly higher than that treated with dilute acid. The ANOVA of the results in Table 8 show that all factors (enzyme dosage, hydrolysis time, and substrate loading) affect the amount of sugar obtained in the liquid hydrolysate, except for the grass pretreated with H_2SO_4 where enzyme dosage does not affect the reducing sugar obtained.

Table 7. Experimental runs based on the Box–Behnken design and results for determining conditions of enzymatic hydrolysis of pretreated Napier grass.

Run	Enzyme, A	Time, B	Substrate Loading, C	Reducing Sugars (g/L)		
	(FPU/g)	(h)	(%)	NaOH	H_2SO_4	NaOH + H_2SO_4
1	30	120	5	65.5	19.3	57.1
2	30	72	10	83.6	33.4	99.9
3	50	120	10	98.9	36.1	112.5
4	10	120	10	71.6	32.5	89.5
5	30	24	5	53.2	11.6	50.9
6	10	72	5	43.3	9.85	46.6
7	30	72	10	83.0	32.3	98.2
8	10	72	15	56.9	52.1	75.9
9	10	24	10	50.3	18.8	82.9
10	30	72	10	84.5	32.8	98.1
11	30	120	15	128.6	54.5	125.7
12	50	72	5	28.7	23.2	73.1
13	30	72	10	84.2	33.7	98.2
14	50	24	10	57.5	15.3	74.9
15	30	72	10	83.2	34.2	99.9
16	50	72	15	110.3	38.1	107.3
17	30	24	15	81.6	31.0	93.6

Table 8. ANOVA for response surface quadratic models used in prediction of suitable conditions for enzyme hydrolysis of pretreated Napier grass.

Source	p-Value		
	NaOH	H_2SO_4	NaOH + H_2SO_4
Model	<0.0001	<0.0001	<0.0001
A—Enzyme dosage	<0.0001	0.7787	<0.0001
B—Time	<0.0001	<0.0001	<0.0001
C—Substrate loading	<0.0001	<0.0001	<0.0001
AB	<0.0001	0.0009	<0.0001
AC	<0.0001	<0.0001	<0.0001
BC	<0.0001	<0.0001	<0.0001
A^2	<0.0001	<0.0001	0.0100
B^2	<0.0001	<0.0001	<0.0001
C^2	<0.0001	0.0789	<0.0001
Lack of Fit	0.0904	0.7141	0.2125
R-Squared	0.9993	0.9989	0.9987
Predicted R-Squared	0.9910	0.9939	0.9865

Notes: NaOH = grass pretreated with NaOH, H_2SO_4 = grass pretreated with H_2SO_4 and NaOH + H_2SO_4 = grass pretreated with NaOH followed by H_2SO_4.

Quadratic equations for predicting the suitable hydrolysis conditions for the pretreated grass using different pretreatment regimes are as follows:

Grass pretreated with NaOH:

$$y = 34.32 + 1.12\,A - 0.459\,B + 1.47\,C + 0.0052\,AB + 0.170\,AC + 0.036\,BC - 0.046\,A^2 + 0.0018\,B^2 - 0.224\,C^2 \tag{4}$$

Grass pretreated with H_2SO_4:

$$y = -25.60 + 1.01\ A + 0.221\ B + 3.07\ C + 0.0018\ AB - 0.068\ AC + 0.018\ BC - 0.0077\ A^2 \\ + 0.0019\ B^2 - 0.026\ C^2 \tag{5}$$

Grass pretreated with H_2SO_4 followed by NaOH:

$$y = -29.77 + 0.879\ A - 0.131\ B + 14.59\ C + 0.0081\ AB - 0.096\ AC + 0.027\ BC - 0.0052\ A^2 \\ - 0.0030\ B^2 - 0.409\ C^2 \tag{6}$$

where y = reducing sugar (g/L), A = enzyme dosage (FPU/g), B = hydrolysis time (h) and C = substrate loading (% *w/v*). All model equations have high R-squared values, which indicate that the models fit well with the experimental data. In addition, the high values of predicted R-squared suggest that the model equations can be used to predict the reducing sugar obtained from the hydrolysis.

The surface plots in Figure 4 show the reducing sugar concentrations obtained from varying enzyme dosages and times when hydrolyzing grasses from different pretreatments. Hydrolysis time greatly affects the sugar obtained, such that longer time results in higher sugar concentration. By observing the surface shapes, the pretreatment involving the dilute acid causes a different response pattern of sugar obtained from hydrolysis when compared with the pretreatment with NaOH alone. Pretreating the grass with dilute acid seems to gain some benefit from the use of lower enzyme dosage with a long hydrolysis time, as shown in Figure 4b,c. While removing lignin by NaOH could loosen the grass structure and allow for more cellulase accessibility to the cellulose, dissolving hemicellulose by dilute acid could also increase the cellulose accessibility [24]. Since the cellulose accessibility is the major factor on cellulase hydrolysis [12,13], the extra increase in this accessibility allows better penetration of the enzyme during hydrolysis even at a lower enzyme dosage.

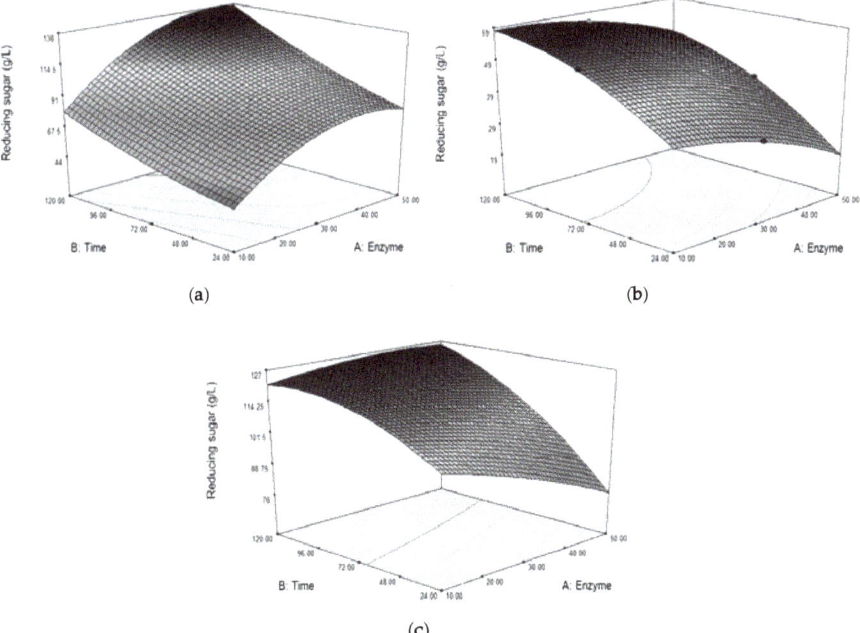

Figure 4. Surface plots showing significant interactions ($\alpha = 0.05$) between enzyme dosage and time (AB) for the enzymatic hydrolysis of Napier grass (**a**) pretreated with NaOH, (**b**) pretreated with H_2SO_4, and (**c**) pretreated with NaOH followed by H_2SO_4, each at their optimal conditions.

The hydrolysis conditions obtained from the model equations for pretreated grasses of different pretreatment regimes are shown in Table 9, together with the predicted and actual reducing sugar concentrations obtained from the hydrolysis. Pretreatment solely with NaOH resulted in the highest reducing sugar concentrations when the pretreated grass was hydrolyzed with cellulase. Confirmation tests using the predicted optimal conditions were carried out for all pretreated grasses and the reducing sugar concentrations were very close to the predicting values.

Table 9. Optimal values of operating conditions for enzymatic hydrolysis and reducing sugar obtained.

	NaOH		H_2SO_4		NaOH + H_2SO_4	
	Optimal Values	Sugar (g/L)	Optimal Values	Sugar (g/L)	Optimal Values	Sugar (g/L)
Enzyme dosage (FPU/g)	40		11		33.6	
Time (h)	117	140.5 (139.7 ± 0.8)	105	57.3 (57.2 ± 1.0)	119.2	125.8 (124.7 ± 0.7)
Substrate loading (% *w/v*)	15		14.8		14.9	

Note: Sugar concentrations are reducing sugar. Values are the predicted concentration from model equations. Values in parentheses are from confirmation tests.

According to the results, cellulase hydrolysis of Napier grass pretreated with NaOH and with NaOH followed by H_2SO_4 resulted in similar reducing sugar concentrations, with the slightly higher concentration in the pretreatment with NaOH. The sugar concentrations obtained from the NaOH-pretreated grasses were high and in accordance with the polysaccharide contents in the pretreated grasses. However, the amount of sugar obtained from the acid-pretreated grass was much lower, regardless of the similar polysaccharide contents. These results emphasized the importance of cellulase susceptibility of the biomass on the hydrolysis rather than the polysaccharide contents of the biomass.

Both regimes involving the use of NaOH in pretreatment could be used for preparing liquid hydrolysates for fermentations due to the high sugar concentrations obtained. However, based on the number of steps involved and the higher sugar concentration obtained, we chose to pretreat the Napier grass in a single step with NaOH followed by cellulase hydrolysis to prepare the hydrolysate as a substrate for ethanol fermentation. To pretreat the Napier grass, 15% dried grass is to be soaked in 1.99 M NaOH and subjects to wet heat at 95.7 °C for 116 min. Dried pretreated grass is hydrolyzed using cellulase (Cellic® CTec2) at 40 FPU/g for 117 h and the grass loading of 15%.

3.4. Ethanol Production Potentials of Napier Grass

In evaluating the ethanol production potential, we employed two yeast strains, which are *S. cerevisiae* TISTR 5339 and *S. shehatae* ATCC 22984, and two fermentation techniques viz. SHF (separate hydrolysis and fermentation) and SSF (simultaneous saccharification and fermentation). The SHF technique utilizes liquid hydrolysate obtained from enzymatic hydrolysis of NaOH-pretreated Napier grass, while the SSF technique uses the NaOH-pretreated grass in the fermentation.

3.4.1. Ethanol Production by *S. cerevisiae* TISTR 5339

Ethanol production by *S. cerevisiae* was carried out using two cultivation techniques, SHF and SSF. The liquid hydrolysate used in SHF contained 134.9 g/L reducing sugar, of which 90.8 g/L was glucose and 18.8 g/L was xylose. In SHF (Figure 5a), the maximum ethanol of 44.7 g/L was obtained after 24 h cultivation. Glucose utilization was in accord with increasing ethanol concentration. Yeast growth also related to glucose utilization with the maximum specific growth rate of 2.79 ± 0.09 1/d. The growth ceased when glucose consumption ended. In addition, xylose decreased throughout the fermentation with the total xylose uptake of 6.71 g/L, which was 32.9% of the total amount. Slight increase of xylitol was observed with the final concentration of 2.47 g/L. The fermentation profiles suggested that ethanol was mainly produced from glucose with the yield of 0.49 g/g.

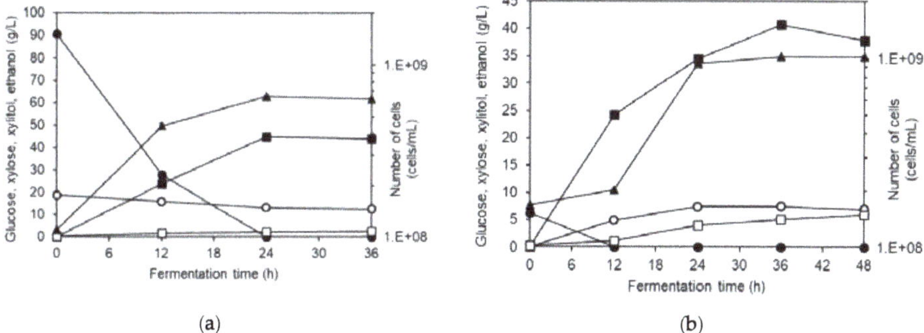

(a)   (b)

Figure 5. Fermentation profiles of *S. cerevisiae* TISTR 5339 when cultivated in (**a**) liquid hydrolysate of NaOH-pretreated Napier grass using SHF and (**b**) NaOH-pretreated grass using SSF. Symbols: ●—glucose; ○—xylose; ■—ethanol; □—xylitol and ▲—cell number.

In ethanol production using SSF by *S. cerevisiae* TISTR 5339 (Figure 5b), glucose remained at a low level throughout the fermentation. Glucose was gradually released and taken up by the yeast to produce ethanol and 40.8 g/L of ethanol was produced after 36 h. Yeast growth appeared to be very slow at the beginning of the fermentation. The use of high solid loading in the fermentation could result in this initial apparent slow growth where clear separation of the solid and liquid phases was evident. The growing yeast could attach to the surface of the grass and undetected in the liquid phase. After 12 h, the physical appearance of the fermentation became slurry and the cells could detach from the grass into the liquid resulting in an increase in cell count. The maximum specific growth rate of 3.08 ± 0.03 1/d was slightly higher than that of SHF and the final cell count was slightly higher than that in SHF. Xylose accumulated during the first 24 h together with xylitol accumulation. The final xylitol concentration was 5.96 g/L. The simultaneous increase of xylose and xylitol indicated that xylitol was produced along with the release of xylose from the grass.

3.4.2. Ethanol Production by *S. shehatae* ATTC 22984

Both SHF and SSF were also employed in the study of ethanol production by *S. shehatae* ATTC 22984 (Figure 6a,b). The sugar concentrations in the Napier grass hydrolysate was similar to those used in the SHF study using *S. cerevisiae*. In SHF, production of ethanol directly related to glucose utilization with the highest concentration of 31.3 g/L, which was significantly lower than that obtained using *S. cerevisiae*. Xylose utilization observed during the fermentation was as expected as the yeast is a xylose-fermenting yeast. However, profile analysis suggested that xylose did not contribute to ethanol production but to xylitol production with the final xylitol concentration of 2.5 g/L. The growth profile in SHF showed the continuous growth during the first 48 h of fermentation. The uncoupling ethanol production after 48 h suggested a possibility of product inhibition effect on growth of *S. shehatae*.

When cultivating *S. shehatae* using SSF (Figure 6b), the growth profile in relation to ethanol profile was similar to that occurred in SHF, suggesting also a possibility of growth inhibition by ethanol. The glucose profile showed a slight increase in concentration during the first 24 h before dropped and remained at low level throughout the fermentation. The small accumulation of glucose suggested a lower growth rate of the yeast as compared to the rate of sugar released from the grass by the action of enzyme. Accumulation of xylose confirmed the ineffective xylose utilization, especially at high ethanol concentration. The final ethanol concentration obtained from SSF was 34 g/L, which was significantly higher than that obtained from SHF. Along with ethanol production, xylitol was also produced and the final concentration was 3.33 g/L.

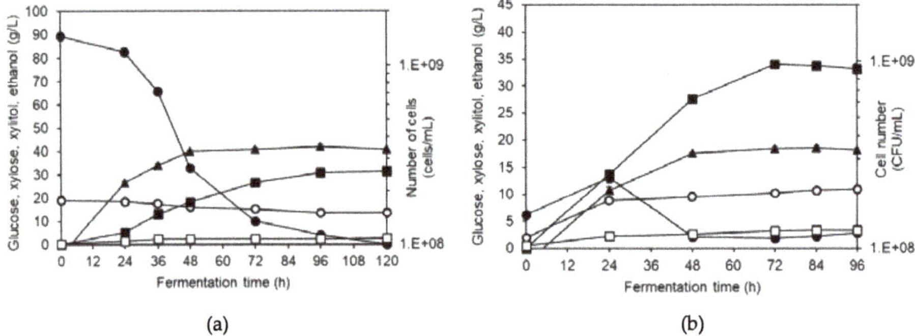

Figure 6. Fermentation profiles of *S. shehatae* when cultivated in (**a**) liquid hydrolysate of NaOH-pretreated Napier grass using SHF and (**b**) NaOH-pretreated grass using SSF. Symbols: ●—glucose; ○—xylose; ■—ethanol; □—xylitol and ▲—cell number.

The summary in Table 10 shows that *S. cerevisiae* TISTR 5339 is able to deliver a superior ethanol production than *S. shehatae* ATCC 22984, in terms of concentration, productivity, and ethanol yield. *S. shehatae*, a xylose-fermenting yeast, did not deliver the expected xylose-fermenting ability when fermenting in hydrolysate that contained high glucose concentration. The overall ethanol production yield from Napier grass suggested that the grass could be used for ethanol production with the best yield of 187.4 g/kg of dried Napier grass obtained with SHF fermentation by *S. cerevisiae*. The results of this study are in the range, or above the average, of other studies both in terms of concentration obtained during fermentation and in terms of ethanol yield based on the amount of biomass used (Table 11).

Table 10. Summary on ethanol production from pretreated Napier grass using SSF and hydrolysate of the pretreated grass using SHF by *S. cerevisiae* and *S. shehatae*.

Parameters	*S. cerevisiae* TISTR 5339		*S. shehatae* ATCC 22984	
	SHF	**SSF**	**SHF**	**SSF**
μ_{max} (l/day)	2.79 ± 0.09 [b]	3.08 ± 0.03 [a]	0.94 ± 0.05 [a]	0.91 ± 0.03 [a]
Ethanol (g/L)	44.7 ± 0.4 [a]	40.8 ± 0.8 [b]	31.3 ± 0.6 [b]	34.0 ± 0.6 [a]
$Q_{p,\,fermentation}$ (g/L.h)	1.87 ± 0.02 [a]	1.14 ± 0.02 [b]	0.26 ± 0.01 [b]	0.47 ± 0.01 [a]
$Q_{p,\,overall\,process}$ (g/L.h)	0.31 ± 0.00 [b]	1.14 ± 0.02 [a]	0.13 ± 0.00 [b]	0.47 ± 0.01 [a]
Y_{ps} (mg/g treated grass)	298.0 ± 0.0 [a]	272.0 ± 0.0 [b]	208.7 ± 0.0 [b]	227.3 ± 0.0 [a]
Y_{ps} (mg/g grass)	187.4 ± 0.0 [a]	171.1 ± 0.0 [b]	131.2 ± 0.0 [b]	142.6 ± 0.0 [a]

Notes: (1) $Q_{p,\,fermentation}$ calculation is based only on fermentation time. (2) $Q_{p,\,overall\,process}$ calculation includes hydrolysate preparation time (5 days) and fermentation time. (3) Comparisons are between different fermentation methods and yeast strains within the same parameter. The same letter indicated that the values were not significantly different at 95% confidence level. (4) Y_{ps} (mg/g treated grass) = ethanol yield based on pretreated grass (5) Y_{ps} (mg/g grass) = ethanol yield based on non-pretreated grass.

4. Discussion

Regarding the use of dilute acid or NaOH for pretreatment of Napier or elephant grass, the results varied among several studies but they were all towards the same trend. A study showed that dilute acid reduced both lignin and hemicellulose contents of the grass, while NaOH treatment only reduced lignin content but not hemicellulose [14]. Our results on dilute acid pretreatment followed a similar reduction trend of both lignin and hemicellulose. However, we found that NaOH solubilized not only lignin, but also the hemicellulose fraction (Table 6). These results agreed with some other studies where hemicellulose reduction was reported in alkali treatment but to a smaller extent when compared to acid pretreatment [20,24]. Both lignin and hemicellulose removals by alkali pretreatment were also evident in other biomass, such as sweet sorghum straw [25] and sugarcane bagasse [26].

Table 11. Comparison of ethanol production from Napier grass in various studies.

Microorganism (Mode of Fermentation)	Pretreatment	Ethanol, g/L (Yield)	References
S. cerevisiae NBRC 2044 + *E. coli* KO11 (SSF+PF *)	Low-moisture anhydrous ammonia	34.2 g/L ** (241 mg/g [#])	[18]
S. cerevisiae NBRC 2044 + *E. coli* KO11 (SSCF)	Low-moisture anhydrous ammonia	19.4 g/L (204 mg/g)	[17]
Ethanol Red (SSF)	NaOH	26.05 g/L (-)	[14]
S. cerevisiae (SSF)	H_2SO_4 + NaOH	24 g/L (127.9 mg [%]/g)	[20]
S. cerevisiae CAT-1 (SHF)	Steam explosion	~4.5 g/L (87.2 mg/g)	[21]
S. cerevisiae CAT-1 (SHF)	Milling + enzyme	6.1 g/L (-)	[27]
Z. mobilis + fungi (SSCF)	None	0.51 g/L (30 mg/g)	[28]
Ethanol Red (SSF)	NaOH	30.2 g/L (143 mg/g)	[29]
Scheffersomyces stipitis (Evolved strain) (SHF)	Low-moisture ammonia hydroxide	22.7 g/L (247 mg [%]/g)	[1]
S. cerevisiae TISTR 5339 (SHF)	NaOH	44.7 g/L (187.4 mg/g) (298 mg/g [#])	This study

Notes: (1) SSF = simultaneous saccharification and fermentation, SSCF = simultaneous saccharification and co-fermentation, SHF = separate hydrolysis and fermentation. * PF = pentose fermentation, where ethanol from SSF was removed prior to PF, ** ethanol concentration from glucose fraction via SSF by *S. cerevisiae*, [#] pretreated biomass and [%] converted from volume using ethanol density at 20 °C (789.2 kg/m^3).

Treatment with NaOH alters Napier grass structure by causing swelling of the structures, while the acid-treated grass maintained a smooth surface, as evident in Figure 3b,c. The swelling is the result of a saponification reaction at the ester bonds between 4-O-methylglucuronic acid and xylan by NaOH [30]. The swelled structure allows enzymes to penetrate, resulting in a more efficient hydrolysis [31]. This explanation supports the results shown in Tables 2, 4 and 5 where susceptibility to cellulase hydrolysis of the grass was higher when treated with NaOH as compared to dilute acid. Severity of pretreatment has shown effects on the hydrolysis step. The use of mild pretreatments such as hot water and dilute NaOH resulted in low ethanol concentrations obtained even with further enzymatic hydrolysis of pretreated biomass [28,32]. However, it should be noted that a suitable strength of pretreatment is also important. A study on the pretreatment of switch grass showed that a high level of delignification reduces the cellulose accessibility to cellulase due to the change in cellulose structure that led to the reduction in available surface area for enzyme adsorption [13]. The result from that study explains our results in Figure 1a where the susceptibility appeared to decrease when NaOH exceeded 2 M.

In addition, extra lignin removal in the grass pretreated with NaOH followed by H_2SO_4 (Table 6) did not enhance the susceptibility of the grass to cellulose hydrolysis. The susceptibility indicates an ability of cellulase to access and hydrolyse a structure. The results agreed with those reported earlier that the key parameter for hydrolysis of lignocellulosic materials was cellulose accessibility, rather than lignin content or crystallinity of the biomass [12,13,16].

The effect of alkali and dilute acid on the susceptibility of the grass during pretreatment process reflects on the results of hydrolysis studies. Since the dilute acid pretreatment resulted in the pretreated grass with the structure that was less susceptible to hydrolysis by cellulase, as indicated by its low susceptibility value, the enzyme dosage required for its hydrolysis was much lower than that for the alkali-treated grasses (Table 9). Similar to the case of excessive delignification, a lesser area accessible by the enzyme resulted in lower sugars obtained from the hydrolysis of the grass pretreated by dilute acid. In addition, the lesser accessible area could cause the limit in hydrolysis even at low enzyme dosage, such that increase in enzyme dosage did not result in a further increase in reducing sugars as evident in Figure 4b. ANOVA of the acid pretreated grass in Table 8 also confirmed this result as the enzyme dosage did not exert a significant effect on the hydrolysis.

Regarding the fermentation for ethanol production, decrease in xylose during *S. cerevisiae* fermentation was not expected, as the yeast is not reported as a xylose-fermenting yeast. The decrease may be the result of xylose transport into the cells. Xylose intake through glucose transport system was formerly reported in *S. cerevisiae* [33]. In addition, *S. cerevisiae* also inherits xylose reductase and xylitol dehydrogenase activities. Xylose reductase is induced by xylose and xylitol produced accumulates without further conversion to ethanol [33,34].

When comparing the cultivations of *S. cerevisiae* via SHF and SSF (Table 10), it is evident that SHF yields higher ethanol concentration and productivity. The hydrolysis of Napier grass at its optimal temperature yields high sugar concentration, whereas the hydrolysis in SSF occurs at 30 °C at which the enzyme has lower activity. A 24-h hydrolysis test of pretreated Napier grass at 30 °C resulted in 119.1 g/L reducing sugar in which contained 82.1 g/L glucose and 14.1 g/L xylose. These values were slightly lower than the hydrolysate used in SHF. The maximum specific growth rate in SSF was, in contrary, slightly higher than that in SHF. Substrate inhibition on growth could explain the result obtained, as glucose at the concentration higher than 80 g/L showed an inhibition effect on *S. cerevisiae* [35]. Despite the higher ethanol obtained from SHF, the process involved the hydrolysis period, which resulted in a significantly lower overall ethanol productivity for the process. Reducing the hydrolysis time in SHF would improve this overall productivity value.

In SHF of *S. shehatae*, which is a xylose-fermenting yeast, the yeast utilized only 33% of the xylose. The issue should not relate to the presence of inhibitors normally found in acid hydrolysate, as we used enzymatic hydrolysis. This result contradicts our initial assumption that the yeast could effectively utilize xylose. We based the assumption on our previous study that showed the yeast's ability to ferment both glucose and xylose to ethanol [36]. However, the current work involved the use of higher concentration of sugars (in hydrolysate). It could be possible that the higher ethanol concentration at the time glucose depleted in this study could have a more negative effect to xylose transport than in our previous work.

The use of SHF and SSF in ethanol production by *S. shehatae* resulted differently from the cultivations of *S. cerevisiae*. In this case, the SSF by *S. shehatae* resulted in higher ethanol concentration regardless of a similar specific growth rate of the yeast between both fermentation techniques (Table 10). By comparing the ethanol production profiles during the first 24 h of fermentation (Figure 6), it is evident that the ethanol produced in SSF was higher than that in SHF, resulting also in a higher ethanol concentration obtained in SSF at the end of the fermentation. The faster ethanol production could be the result of lower sugar concentration during the fermentation that alleviated the substrate inhibition effect on product formation. The conversion yield of sugar in the hydrolysate appears more effective in SSF for *S. shehatae* as the ethanol was higher than that obtained from SHF regardless of the lower sugar obtained from hydrolysis at 30 °C.

5. Conclusions

Napier grass has delivered satisfactory results as a second-generation feedstock for ethanol production. Pretreatment of the grass by NaOH is an important step for further hydrolysis by cellulase, both in SHF and SSF for ethanol production. The process for pretreatment involved the use of 1.99 M NaOH at 95.7 °C for 116 min with the grass loading of 15% *w/v*. After washing off the alkali, the grass could be used in ethanol production by *S. cerevisiae* via SHF or SSF, with SHF yielded a slightly higher ethanol concentration. The enzyme (Cellic® CTec2) at the dosage of 40 FPU/g could be applied directly in SSF. The hydrolysate for SHF could be prepared using the same enzyme dosage to hydrolyse 15% *w/v* of the pretreated grass at 50 °C for 117 h to obtain fermentable sugars. The overall process could yield 171–187 g of ethanol/kg of Napier grass at the ethanol concentrations of 41–45 g/L.

Author Contributions: Conceptualization: M.B.K.; data curation: M.B.K.; formal analysis: M.B.K. and C.S.-K.; funding acquisition: M.B.K. and A.R.; investigation: C.S.-K.; methodology: M.B.K. and C.S.-K.; project administration: M.B.K. and C.S.-K.; resources: M.B.K. and A.R.; supervision: M.B.K. and A.R.; validation: M.B.K. and C.S.-K.; visualization: M.B.K.; writing—original draft: M.B.K.; writing—review and editing: M.B.K. and A.R. All authors have read and agreed to the published version of the manuscript.

Funding: This research was funded by (1) Thailand Research University Network under Bio-conversion group (2) Energy Conservation and Promotion Fund, Energy Policy and Planning Office (EPPO), Ministry of Energy Thailand (3) National Research Council of Thailand (NRCT), the Higher Education Research Promotion, and National Research University Project of Thailand through Biofuels Research Cluster of Khon Kaen University, project number NRU543030 and (4) TRF Senior Research Scholar, grant number RTA6280001.

Acknowledgments: The authors appreciate the donation of Napier grass from Department of Agronomy, Faculty of Agriculture, Khon Kaen University. C.S.-K. is grateful for her partial financial support from Center for Alternative Energy Research and Development. We would also like to thank Prawphan Yuvadetkun for her valuable assistance during the course of the project operation.

References

1. Dien, B.S.; Anderson, W.F.; Cheng, M.H.; Knoll, J.E.; Lamb, M.; O'Bryan, P.J.; Singh, V.; Sorensen, R.B.; Strickland, T.C.; Slininger, P.J. Field productivities of Napier grass for production of sugars and ethanol. *ACS Sustain. Chem. Eng.* **2020**, *8*, 2052–2060. [CrossRef]

2. De Favare, H.G.; de Abreu, J.G.; de Barros, L.V.; da Silva, F.G.; Ferreira, L.M.M.; Barelli, M.A.A.; Neto, I.M.d.S.; Cabral, C.E.A.; Peixoto, W.M.; Campos, F.I.d.S.; et al. Effect of Elephant grass Genotypes to Bioenergy production. *J. Exp. Agric. Int.* **2019**, *38*, 1–11. [CrossRef]

3. Wijitphan, S.; Lowilai, P. Effects of cutting interval on yields and nutritive values of King Napier grass (*Pennisetum purpureum* cv. King grass) under Irrigation Supply. *KKU Res. J.* **2011**, *16*, 215–224.

4. Cardona, E.; Rios, J.; Pena, J.; Penuela, M.; Luis Rios, L. King Grass: A very promising material for the production of second generation ethanol in tropical countries. *Biomass Bioenergy* **2016**, *95*, 206–213. [CrossRef]

5. Kamarullah, S.H.; Marlina Mohd Mydin, M.M.; Omar, W.S.A.W.; Harith, S.S.; Noor, B.H.M.; Alias, N.Z.A.; Manap, S.; Mohamad, R. Surface morphology and chemical composition of Napier grass fibers. *Malays. J. Anal. Sci.* **2015**, *19*, 889–895.

6. Li, Y.; Zhang, Y.; Zheng, H.; Du, J.; Zhang, H.; Wu, J.; Huang, H. Preliminary evaluation of five elephant grass cultivars harvested at different time for sugar production. *Chin. J. Chem. Eng.* **2015**, *23*, 1188–1193. [CrossRef]

7. Mohammed, I.Y.; Abakr, Y.A.; Kazi, F.K.; Yusup, S.; Alshareef, I.; Soh, A.; Chin, S.A. Comprehensive characterization of Napier grass as a feedstock for thermochemical conversion. *Energies* **2015**, *8*, 3403–3417. [CrossRef]

8. Napier Grass. Available online: http://weben.dede.go.th/webmax/content/napier-grass (accessed on 6 April 2020).

9. Narinthorn, R.; Choorit, W.; Chisti, Y. Alkaline and fungal pretreatments for improving methane potential of Napier grass. *Biomass Bioenergy* **2019**, *127*, 105262. [CrossRef]

10. Suaisom, P.; Patiroop Pholchana, P.; Aggarangsi, P. Holistic determination of suitable conditions for biogas production from *Pennisetum purpureum* x *Pennisetum americanum* liquor in anaerobic baffled reactor. *J. Environ. Manag.* **2019**, *247*, 730–737. [CrossRef]

11. Bensah, E.C.; Kádár, Z.; Mensah, M.Y. Ethanol production from hydrothermally-treated biomass from West Africa. *BioResources* **2015**, *10*, 6522–6537. [CrossRef]

12. Arantes, V.; Saddler, J.N. Cellulose accessibility limits the effectiveness of minimum cellulase loading on the efficient hydrolysis of pretreated lignocellulosic substrates. *Biotechnol. Biofuels* **2011**, *4*, 3. [CrossRef] [PubMed]

13. Rollin, J.A.; Zhu, Z.; Sathitsuksanoh, N.; Zhang, Y.-H.P. Increasing cellulose accessibility is more important than removing lignin: A comparison of cellulose solvent-based lignocellulose fractionation and soaking in aqueous ammonia. *Biotechnol. Bioeng.* **2011**, *108*, 22–30. [CrossRef] [PubMed]

14. Eliana, C.; Jorge, R.; Juan, P.; Luis, R. Effects of the pretreatment method on enzymatic hydrolysis and ethanol fermentability of the cellulosic fraction from elephant grass. *Fuel* **2014**, *118*, 41–47. [CrossRef]

15. Pensri, B.; Aggarangsi, P.; Chaiyaso, T.; Chandet, N. Potential of fermentable sugar production from Napier cv. Pakchong 1 grass residue as a substrate to produce bioethanol. *Energy Procedia* **2016**, *89*, 428–436. [CrossRef]

16. Phitsuwan, P.; Sakka, K.; Ratanakhanokchai, K. Structural changes and enzymatic response of Napier grass (*Pennisetum purpureum*) stem induced by alkaline pretreatment. *Bioresour Technol.* **2016**, *218*, 247–256. [CrossRef]

17. Yasuda, M.; Nagai, H.; Takeo, K.; Ishii, Y.; Ohta, K. Bio-ethanol production through simultaneous saccharification and co-fermentation (SSCF) of a low-moisture anhydrous ammonia (LMAA)-pretreated napiegrass (*Pennisetum purpureum* Schumach). *Springerplus* **2014**, *3*, 333. [CrossRef]

18. Yasuda, M.; Takeo, K.; Nagai, H.; Uto, T.; Yui, T.; Matsumoto, T.; Ishii, Y.; Ohta, K. Enhancement of ethanol production from Napiergrass (*Pennisetum purpureum* Schumach) by a low-moisture anhydrous ammonia pretreatment. *J. Sustain. Bioenergy Syst.* **2013**, *3*, 179–185. [CrossRef]

19. Bohórquez, C.; González, E.A.; Reina, M.R. Effect of pretreatment dilute acid—Peroxide on napier grass (*Pennisetum purpureum* schumach) to enhance reducing sugar yield by enzymatic hydrolysis. *AVANCES Investigación en Ingeniería* **2014**, *11*, 48–55. [CrossRef]

20. Camesasca, L.; Ramírez, M.B.; Guigou, M.; Ferrari, M.D.; Lareo, C. Evaluation of dilute acid and alkaline pretreatments, enzymatic hydrolysis and fermentation of napiergrass for fuel ethanol production. *Biomass Bioenergy* **2015**, *74*, 193–201. [CrossRef]

21. Scholl, A.L.; Menegol, D.; Pitarelo, A.P.; Fontana, R.C.; Filho, A.Z.; Ramos, L.P.; Dillon, A.J.P.; Camassola, M. Ethanol production from sugars obtained during enzymatic hydrolysis of elephant grass (*Pennisetum purpureum*, Schum.) pretreated by steam explosion. *Bioresour Technol.* **2015**, *192*, 228–237. [CrossRef]

22. Meneses, L.R.; Otor, O.F.; Bonturi, N.; Orupõld, K.; Kikas, T. Bioenergy yields from sequential bioethanol and biomethane production: An optimized process flow. *Sustainability* **2020**, *12*, 272. [CrossRef]

23. Sluiter, A.; Hames, B.; Hyman, D.; Payne, C.; Ruiz, R.; Scarlata, C.; Sluiter, J.; Templeton, D.; Wolfe, J. *Determination of Total Solids in Biomass and Total Dissolved Solids in Liquid Process Samples*; Technical Report NREL/TP-510-42618; National Renewable Energy Laboratory: Golden, CO, USA, 2011.

24. Sun, S.; Sun, S.; Cao, X.; Sun, R. The role of pretreatment in improving the enzymatic hydrolysis of lignocellulosic materials. *Bioresour. Technol.* **2016**, *199*, 49–58. [CrossRef] [PubMed]

25. Dong, M.; Wang, S.; Xu, F.; Wang, J.; Yang, N.; Li, Q.; Chen, J.; Li, W. Pretreatment of sweet sorghum straw and its enzymatic digestion: Insight into the structural changes and visualization of hydrolysis process. *Biotechnol. Biofuels* **2019**, *12*, 276. [CrossRef] [PubMed]

26. Zhang, H.; Wei, W.; Zhang, J.; Huang, S.; Xie, J. Enhancing enzymatic saccharification of sugarcane bagasse by combinatorial pretreatment and Tween 80. *Biotechnol. Biofuels* **2018**, *11*, 309. [CrossRef]

27. Menegol, D.; Fontana, R.C.; Dillon, A.J.P.; Camassola, M. Second-generation ethanol production from elephant grass at high total solids. *Bioresour. Technol.* **2016**, *211*, 280–290. [CrossRef]

28. Liu, Y.K.; Chen, W.C.; Huang, Y.C.; Chang, Y.K.; Chu, I.M.; Tsai, S.L.; Wei, Y.H. Production of bioethanol from Napier grass via simultaneous saccharification and co-fermentation in a modified bioreactor. *J. Biosci. Bioeng.* **2017**, *124*, 184–188. [CrossRef]

29. Tsai, M.H.; Lee, W.C.; Kuan, W.C.; Sirisansaneeyakul, S.; Savarajara, A.A. Evaluation of different pretreatments of Napier grass for enzymatic saccharification and ethanol production. *Energy Sci. Eng.* **2018**, *6*, 683–692. [CrossRef]

30. Tarkow, H.; Feist, W.C. A mechanism for improving the digestibility of lignocellulosic materials with dilute alkali and liquid ammonia. In *ACS Advances in Chemistry Series*; Hajny, G.J., Reese, E.T., Eds.; American Chemical Society: Washington, DC, USA, 1969; Volume 95, pp. 197–218.

31. Hendriks, A.T.W.M.; Zeeman, G. Review Pretreatments to enhance the digestibility of lignocellulosic biomass. *Bioresour. Technol.* **2009**, *100*, 10–18. [CrossRef]

32. Soares, I.B.; Marques, O.M.; Benachour, M.; Abreu, C.A.M.d. Ethanol production by enzymatic hydrolysis of elephant grass. *J. Life Sci.* **2011**, *5*, 157–161.

33. Van Zyl, C.; Prior, B.A.; Kilian, S.G.; Kock, J.L.F. D-xylose utilization by *Saccharomyces cerevisiae*. *J. Gen. Microbiol.* **1989**, *135*, 2791–2798. [CrossRef]

34. Batt, C.A.; Carvallo, S.; Eassond, D.D.; Akedom, M.; Sinskeya, A.J. Direct evidence for a xylose metabolic pathway in *Saccharomyces cerevisiae*. *Biotechnol. Bioeng.* **1986**, *28*, 549–553. [CrossRef] [PubMed]

35. Zhang, Q.; Wu, D.; Lin, Y.; Wang, X.; Kong, H.; Tanaka, S. Substrate and product inhibition on yeast performance in ethanol fermentation. *Energy Fuels* **2015**, *29*, 1019–1027. [CrossRef]

36. Yuvadetkun, P.; Boonmee, M. Ethanol production capability of *Candida shehatae* in mixed sugars and rice straw hydrolysate. *Sains Malays.* **2016**, *45*, 581–587.

 *processes*

Article

Optimizing Yield and Quality of Bio-Oil: A Comparative Study of *Acacia tortilis* and Pine Dust

Gratitude Charis [1,*], **Gwiranai Danha** [1] **and Edison Muzenda** [1,2]

[1] Department of Chemical, Materials and Metallurgical Engineering, Botswana International University of Science and Technology, Palapye P Bag 016, Botswana; danhag@biust.ac.bw (G.D.); muzendae@biust.ac.bw (E.M.)

[2] Department of Chemical Engineering Technology, School of Mining, Metallurgy and Chemical Engineering, University of Johannesburg, P O Box, 17011, Doornfontein, Johannesburg 2094, South Africa

* Correspondence: gratitude.charis@studentmail.biust.ac.bw; Tel.: +267-72-483-242

Received: 9 January 2020; Accepted: 27 February 2020; Published: 8 May 2020

Abstract: We collected pine dust and *Acacia tortilis* samples from Zimbabwe and Botswana, respectively. We then pyrolyzed them in a bench-scale plant under varying conditions. This investigation aimed to determine an optimum temperature that will give result to maximum yield and quality of the bio-oil fraction. Our experimental results show that we obtain the maximum yield of the oil fraction at a pyrolysis temperature of 550 °C for the acacia and at 500 °C for the pine dust. Our results also show that we obtain an oil fraction with a heating value (HHV) of 36.807 MJ/kg using acacia as the feed material subject to a primary condenser temperature of 140 °C. Under the same pyrolysis temperature, we obtain an HHV value of 15.78 MJ/kg using pine dust as the raw material at a primary condenser temperature of 110 °C. The bio-oil fraction we obtain from *Acacia tortilis* at these condensation temperatures has an average pH value of 3.42 compared to that of 2.50 from pine dust. The specific gravity of the oil from *Acacia tortilis* is 1.09 compared to that of 1.00 from pine dust. We elucidated that pine dust has a higher bio-oil yield of 46.1% compared to 41.9% obtained for acacia. Although the heavy oils at condenser temperatures above 100 °C had good HHVs, the yields were low, ranging from 2.8% to 4.9% for acacia and 0.2% to 12.7% for pine dust. Our future work will entail efforts to improve the yield of the heavy oil fraction and scale up our results for trials on plant scale capacity.

Keywords: *Acacia tortilis*; biofuel; biomass; pine dust; pyrolysis

---

1. Introduction

The advanced biofuels market has been forecasted to expand at a compounded annual growth rate (CAGR) of 44% between the years 2017 and 2021. This trend is driven by higher demand for greener and cleaner energy, regional economic growth and more favorable trade balances through substitution of fossil fuel imports [1,2]. These aspirations have resulted in a spate of supportive policies from influential fuel traders like the United States and European Union, especially for advanced second-generation (2G) or third-generation (3G) biofuels that do not use edible feedstocks. Bio-oil is one such biofuel obtained from biomass pyrolysis, which has received much attention in research. The overarching goal is to find routes and processes that give higher yields and better quality products at lower costs or economic downstream upgrading methods. The option that is closer to the market is to obtain a moderately upgraded product that can substitute fuel oil for industrial applications, power generation and marine fuelling [3].

Lignocellulosic feedstocks are attractive for the production of such biofuels, due to their abundance and the possibility for higher energy recoveries since a larger fraction of the biomass is utilized compared to the 1G counterparts, where only choice parts such as the grains are useful [4,5]. The most preferred lignocellulosic feedstocks are forestry/agricultural residues, or dedicated energy crops grown in marginal lands with no competition for food production. Charis et al. [6] reviewed the research and development (R&D) metrics of 2G biofuels compared to 1G for the period 2012–2017. They reported that 2G biofuels had 15% more researches and 23% more collaborations than 1G during the period, showing an increased interest in the former field.

1.1. Socio-Economic Background

Authors such as [4,7,8], have highlighted the need for Southern Africa to step up its participation in 2G biofuel research and to migrate from 1G biofuels that have limited feedstocks. Furthermore, using 1G feedstocks compromises food security directly, or indirectly through land-use change, which also results in increased greenhouse gas emissions. Southern Africa, in particular, has a wealth of lignocellulosic feedstocks in the form of agricultural waste, forestry residues and invasive species. However, the primary energy uses for biomass have been, mostly, the traditional open fire cooking methods. Since women and children are the most involved in such cooking activities, they are exposed to respiratory problems and endure arduous tasks such as gathering and chopping firewood [9,10]. Southern African nations are susceptible to energy poverty, especially in rural and remote areas, despite the high potential for renewable energy sources like solar, wind and biomass. The cases identified in two Southern African nations are examples of areas off the main electricity grid with a good inventory of biomass. The biomass identified in this research is encroacher bushes in Botswana and pine sawmill residues from Zimbabwe. In such cases, mini-grid power generation through the use of bio-oil in a stationary fuel oil generator could be an alternative to solar systems, especially when the pyrolysis system can be self-sufficient in terms of energy or can use a cheap, renewable energy source. Such a venture would be feasible if the bio-oil obtained does not need rigorous upgrading to match the counterpart fuel oils in terms of physical and fuel properties. Power generation is a priority application for energy impoverished Africa compared to the highly considered alternative use of bio-oil as a maritime fuel. Blending options with up to 25% pine bio-oil have proved to be miscible with heavy fuel oil (HFO) giving improved fuel flow properties while maintaining good engine properties [11]. If bio-oils with better fuel properties in terms of heating value can be obtained, blending ratios could be increased in favor of the bio-oil, and there could be a possibility of substituting the HFO entirely.

In this research, we assessed the conversion of the two types of forestry waste biomass to a 2G crude biofuel through pyrolysis and the subsequent potential application. Pyrolysis is a promising future source of fuels and chemicals that is resource-efficient, potentially auto-thermal, technically simpler, less capital intensive and operable at smaller scales compared to other thermochemical conversions. Pyrolysis can also be a good contributor to greening a brown economy by re-directing biomass feedstock that could otherwise have been burnt openly to be processed in a system largely regarded as carbon neutral. Moreover, its major products (bio-oil and char) both have high-value potential uses and are easier to handle than gas. Previous research by the same authors in [12,13] focused on the sources of lignocellulosic wastes, socio-economic and environmental drivers for their valorization.

1.2. Technical Background

Pyrolysis is a thermochemical conversion method in which a substance is heated in an oxygen-starved atmosphere to obtain solid, liquid and gaseous products. The yield of the various products depends on the process conditions, equipment used, feedstock types and pre-treatment regimes [14]. The liquid product (bio-oil), which is the primary target phase for this study, can be used as a fuel for heating, power and transportation after various levels of upgrading. It can also be a source of building blocks for the manufacture of various chemicals through polymerization of the monomer substrates. Bio-oil from pyrolysis also has its challenges that impede its uptake as a fuel including the high water content, high viscosity, low pH, instability, presence of solids, high oxygen content and low calorific value [15–17]. These properties explain the associated corrosive tendencies; ageing and phase separation; immiscibility with hydrocarbon fuels; bad engine performance (due to injector blockages by solids and poor atomization) and high pumping costs [15].

1.3. The State of Research

In light of the challenges stated in Section 1.2, we review the progress in research on optimizing the quantity and quality of bio-oil obtained from pyrolysis in this section. We cover the intermediate to fast pyrolysis (IFP) range, which comprises moderate to high liquid yields (40–80 wt %) and hot vapor residence of 1–20 s. Most research within this scope has focused on optimizing pyrolysis process conditions and reactor designs for higher recoveries of bio-oil [15,18–21], while there has been limited research on conditions focusing on bio-oil quality improvement [14]. Temperature and heating rate have been identified as the most influential variables on bio-oil yields for a given biomass feedstock, among other factors like particle size and residence time. Consequently, the highest recoveries of bio-oil have been achieved with reactors that allow maximum heat transfer rate such as fluidized bed and free fall reactors, with temperatures and inert gas flow rates that facilitate fast pyrolysis [20]. Table 1 summarizes the operation philosophy of some of the most promising pyrolysis technologies and the status of their research, development or commercialization.

Table 1. Trending pyrolysis technologies. Adapted from [22,23].

Pyrolysis Reactor	Description and Operation Philosophy	Operation Complexity and Max Oil Yield	Scale Up	Inert Gas Flow Rate	Particle Size	R&D Highest Status
Fixed bed	Biomass is placed immobile, above the inert gas distributor plate. Char remains in the reactor while oil and gas are collected downstream	Medium; up to 75 wt % oil	Hard	Low	Large	Few at pilot scale; Multiple lab scale
Bubbling fluidized bed (BFB)	Comprises reactor section with a continuous feed of biomass and high flow of inert gas to fluidize the particles. Char and sand are collected using cyclones.	Medium; up to 75 wt % oil	Easy	High	Small	Multiple demo and lab-scale plants
Circulating fluidized bed (CFB)	Similar to BFB, but collected char and sand are recycled through a combustor, which supplies hot sand to the fluidized bed.	High; up to 75 wt % oil	Hard	High	Medium	Multiple pilot and lab-scale plants
Ablative	Heat transfer to the biomass is direct from the walls of the reactor; no fluidizing gas. Biomass melts and vaporizes rapidly to form pyrolysis vapors.	High; up to 75 wt % oil	Hard	Low	Large	Few at pilot scale
Rotating cone	Heat transferred by reactor wall and hot sand, introduced into the rotating cone along with the biomass. The hot pyrolysis vapor is recovered from the bottom of the cone.	Medium; up to 70 wt % oil	Medium	Low	Medium	Demo/industrial scale
Screw/auger	Heat is mainly transferred by the wall surfaces. The biomass is moved along a heated cylindrical reaction zone by a screw	Low; up to 70 wt % oil	Easy	Low	Medium	Multiple pilot and lab-scale

Amongst the in-process techniques used to improve the bio-oil quantity and quality, catalysis has enjoyed more attention than methods such as fractional condensation of condensable vapors and hot vapor filtration. Hot vapor filtration, which targets only the removal of ash, has been a big challenge with minimal research on long-term operations. This is due to complex requirements in terms of material specifications for the filter assembly and regeneration or cleaning mechanisms. Moreover, the trapped char catalytically cracks the pyrolysis vapors, reducing bio-oil yields by up to 20%. Simple in-process catalysis has been employed successfully using catalysts such as selected metals and zeolites to improve bio-oil yields and quality [15].

Condensers have traditionally been used to recover the condensable off-gases from the pyrolyzed matter. However, it is only the fractionated condenser systems that enable the recovery of targeted products of defined qualities. Fractional condensation has always employed multiple condensers, set at different temperatures, to selectively optimize the recovery of high calorific value oil and fractions with a high composition of speciality chemicals like methanol. Since the first condenser is typically set above 80 °C, water and light carboxylic acids are not recovered here. Therefore, the collected oil has higher stability, lower water content and about 2%–3% of acids, compared to 10%–15% in unfractionated bio-oil [3,24]. Since this is done within the same pyrolysis step, using the same process heat, the method is less energy-intensive than trying to distil the bio-oil product. Molecular distillation would result in the thermal decomposition of the largely unstable compounds in the bio-oil unless done carefully, with slower ramp-up rates at some stages such as coking. Such slow ramp-up rates would, however, lead to a very high overall energy consumption [24].

Several researchers have reported on various pyrolysis operation scales with multistage fractional condensation systems to obtain liquid bio-products of targeted physical or thermal properties. Gooty in [3] modeled the effect of both pyrolysis and condenser temperatures for a three-staged condenser system from the fast pyrolysis of birch bark and Kraft lignin. Gooty's system was composed of two cyclonic condensers and a condenser cum electrostatic precipitator (ESP) at outlet vapor temperatures of 105 °C, 0 °C and 38–56 °C respectively. For birch bark, the author obtained a yield of 35% of the dry oil (<1% water) with a higher heating value (HHV) of 31 MJ/kg. Chen et al. in [25], utilized four condensers in series at coolant temperatures of 32–44 °C, 25–27 °C, 22–25 °C and 22–25 °C, respectively, followed by an ESP. Their drier oils were obtained in condensers 2–4 with a water composition of 7.45%–7.82%, HHV of 22.6–23.5 MJ/kg and total yield of 14.3%. Oasmas et al. [26] pyrolyzed red oak wood in a fluidized bed with five stages of bio-oil recovery at vapor temperatures of 102 °C, 129 °C (ESP 1), 65 °C, 77 °C (ESP 2) and 18 °C. Most of the researches done on this subject have covered the continuous, fast pyrolysis regime using reactors such as fluidized bed and auger, with a few around the intermediate regime and reactors such as the fixed bed. According to [24], very little research on fractional condensation has been conducted at the lab scale.

1.4. The Purpose and Significance of This Research

In this research, we mainly sought to generate a profile for *Acacia tortilis* as a pyrolysis feedstock as, to our best knowledge, no work has been reported on this species. We conducted a comparative study with pine dust, a well-researched species. The pine dust residues in that geographical context have, however, not been experimented on before using pyrolysis. We aimed to find the optimum temperature for the maximum overall oil yield of both feedstocks, then the optimum primary (first) condenser temperature that gave the best fuel quality of bio-oil in a bespoke three-stage condensation line. The properties we considered for the bio-oil were HHV, viscosity, pH and specific gravity (SG), which we compared to fuel oil properties to assess the bio-oil's potential to substitute fuel oil in power generation engines. We also evaluated the chemical composition of the bio-oils using scanning calorimetry to investigate the presence of significant concentrations of extractable speciality chemicals. The research provides essential elementary information about the fuel and yield potential of acacia in comparison with pine when process and condenser temperatures are manipulated using a basic fixed bed reactor.

2. Materials and Methods

2.1. Methods of Feedstock Preparation and Characterization

Adequate knowledge of the supply capacity and the suitability of the biomass properties for thermochemical conversions is imperative for bioenergy projects to be sustainable [12]. To this end, we conducted proximate, ultimate and calorimetric analyses to determine the biomass quality and suitability for conversion through pyrolysis. We firstly milled dry pine dust of <10% moisture and *A. tortilis* shrubs using a JF 2-D chopper cum hammer mill (Staalmeester, Hartbeesfontein, South Africa) with a 0.8 mm sieve. We then sieved the mill product to obtain particles below 250 μm for the characterization tests. We used the American Society of Testing Materials (ASTM) standards for the characterization tests. Detailed test procedures were provided by the authors [27] and the properties are reported in the results section.

We carried out proximate analysis using a Leco 701 thermogravimetric analyzer (Leco TGA, St Joseph, MI, USA), which was intrinsically set according to ASTM standard E 1131–03 for compositional analysis using thermogravimetry. We then determined the moisture content (MC), volatile matter (VM) and ash and fixed carbon (FC) compositions using the mass loss plot from the TGA.

We also carried out a CNHSO composition analysis using a Flash 2000 CHNS analyzer (ThermoFisher Scientific, Waltham, MA, USA). Oxygen (O) was found by the difference.

We determined the higher heating value (HHV) of the two lignocellulosics using a CAL2K-2 bomb calorimeter (Digital Data Systems, Randburg, South Africa) employing the ASTM D5468–02 standard. The calorimeter was calibrated with benzoic acid. The HHV was determined as the average of two runs.

2.2. Pyrolysis

Of the many variables tested for optimum yields of bio-oil, temperature has been identified as the most fundamental and influential parameter on bio-oil yields [20]. It was our initial focus on these previously unstudied biomasses, from both a geographical (pine residues) and species (*A. tortilis*) viewpoint. We also optimized the quality of bio-oil by varying the primary condenser temperatures in the pyrolysis system while maintaining constant values for the secondary condensers.

We firstly comminuted the feedstocks using a hammer-chopper mill with a 5 mm sieve. We then screened the biomass to obtain a size range between 1.70 and 5.00 mm for the pyrolysis runs. We placed a mass of 200–220 g of the biomass inside a bench-scale fixed bed reactor (Figure 1). The reactor shell comprised an outside cylinder 55 cm high with an external diameter of 25.9 cm and inside diameter of 10.3 cm. The reaction took place in a removable cuboid-shaped holder of dimensions L—7.2 cm, W—7.2 cm and H—35 cm. The space between the inside and outside diameter had a compartment for the heating coils, next to the inner wall, and another one for cladding to insulate the reactor. We weighed the container weight and mass of biomass before each new run, while the mass of char was determined after the run. We also measured the weights of the bio-oil receivers before each run so that the net masses of the oils and the char could be determined by subtracting the mass of container from the mass of the container + oil or char.

We used the heat traced standpipe as the primary condenser (ES01) and Liebig shell and tube type for the secondary condensers (ES02 and ES03). We then adjusted the set point for the primary condenser was to 125 °C, while the secondary condensers were both set at 25 °C.

Subsequently, we purged the whole system for 2–3 min using inert nitrogen gas. We switched on the heating system at the SCADA graphical user interface (GUI) and ramped it at 66 °C per minute until the 450 °C setpoint, then maintained this temperature until the end of the run. The SCADA system captured the volumetric oil yields, and gas yields while showing the real-time temperature variations at different points of the bench-scale plant system. We ran each experiment until there were no more bubbles in the scrubber. We then weighed the bio-oil and chars and calculated the overall solid and liquid yields from the initial mass of feed. We repeated the pyrolysis runs at 500 °C, 550 °C and 600 °C. The optimum temperature was regarded as the one at which the highest total bio-oil wt % yield was obtained. We selected the temperature range based on literature, which catalogues optimum pyrolysis ranges of 480–520 °C [15] and 400–600 °C [20] for various lignocellulosics such as herbaceous grasses, filter cakes and woody biomass. Bridgwater [15] comments that woody lignocellulosics occupy the upper end of these ranges due to the higher composition of temperature recalcitrant lignin, explaining why the authors selected a range of 450–600 °C.

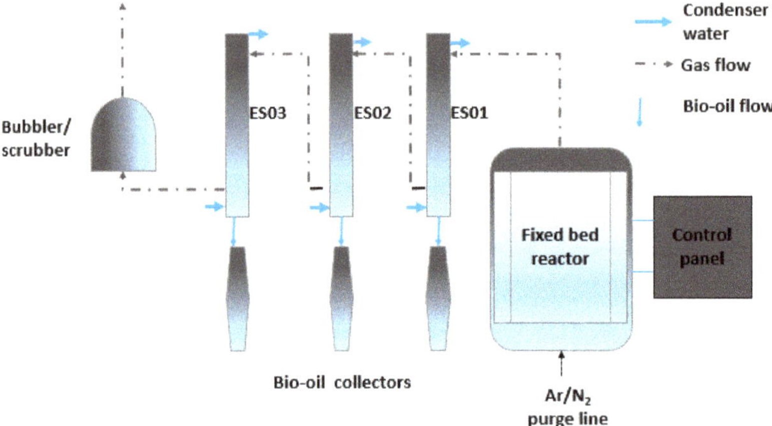

Figure 1. Schematic for the fixed bed pyrolysis unit.

To obtain the optimum primary condenser temperature, we ran the bench-scale plant at a pyrolysis temperature of 450 °C, while varying the primary condenser temperatures from 90 °C, 100 °C and then 110 °C, 125 °C and 140 °C [28]. Both secondary condensers were kept at 25 °C, which is below the dew point of water (50 °C) and some light acids so that these molecules would be collected at these points. We varied the primary condensation temperature above the 80 °C threshold recommended by [24] in such a way that only compounds of higher boiling and dew points with greater molecular weights would be retained successively at higher condenser temperatures. The highest condenser temperature of the exit hot vapor used, according to [24]'s review of multi-stage condensation systems was 135 °C, therefore, the authors went up to 140 °C. We initially regarded the optimum condenser temperature as the one that gave bio-oil of the highest specific gravity (SG), implying the least aqueous phase. However, the calorimetric tests carried out later on the oils proved to be a more decisive metric.

2.3. Characterization of Products

The product of interest was the bio-oil as a fuel for power generation and potentially, for vehicular applications. It was necessary to compare the properties of the bio-oil to diesel and fuel oils, hence the need for characterization. We also characterized both the primary and secondary condensates by gas chromatography (GC) Agilent 5975C (Agilent Technologies, Waldbronn, Germany) and a mass spectrometer (MS) (Agilent Technologies, Waldbronn, Germany) to assess if there were any useful extractable substances of significant concentrations. Finally, we characterized the residual biochar to assess its calorific fuel value, if it could be used to make charcoal briquettes.

The important physicochemical properties that we measured for comparison purposes include the SG, pH and viscosity. We obtained the SG by expressing the density of the bio-oil (mass/volume) as a ratio to the density of water and used it as an indicator of the water composition in the absence of a Karl Fischer titrator. We measured the pH was using a JENCO pH 6810 meter (Jenco instruments, San Diego, CA, USA) calibrated with pH 4.00, 6.86 and 9.18 solutions, while the viscosity was determined using a Thermo scientific Haake viscotester E (Waltham, MA, USA). We then measured the bio-oil thermal and fuel properties (represented by the calorific value) using an IKA C1 bomb calorimeter (IKA, Staufen,

Germany). Only the primary condenser samples could be tested as the secondary condenser bio-oils had a high aqueous composition that made it difficult for any combustion, even using at least 80 w/w% of a combustion aid. We poured each bio-oil sample was into a metal crucible then connected ignition wire and fuse combination before closing the bombshell. We then ran the enclosed system according to the DIN 51,900 standards and the calorific value was read off the panel. Where there was a difficulty in obtaining complete combustion, the test was repeated using a proportion of combustion aid of between 40% and 80%, whose calorific value was predetermined. We also tested the char calorific values using the same procedure. The actual HHVs of the char or bio-oil was then calculated using Equation (1).

$$HHV_{oil\ or\ char} = \frac{Total\ HHV - m.f\ of\ CA \times HHV_{CA}}{m.f\ of\ bio - oil} \tag{1}$$

where *CA* denotes the combustion aid, which was either paraffin or diesel; and m.f denotes mass fraction.

We assessed the chemical composition of the bio-oils using the GC–MS Agilent 5975C (Agilent Technologies, Waldbronn, Germany) equipped with a library. We initially filtered the samples using a syringe fitted with a 1 μm filter and a needle to eliminate small solid particles. We then dissolved them with 99.9% HPLC grade dimethyl ether, which had proved to be a better solvent compared to acetone. There are no standard tests and parameters yet for bio-oils set for engine applications, including procedures for GC–MS [29]. Therefore, we adopted this GC-MS test to resemble tests conducted on other bio-oils as well for fatty acid methyl ester (FAME) diesel [30–33]. We ran the samples in a GC–MS 5975C (Agilent Technologies, Waldbronn, Germany) using a DB-1HT 30 m × 250 μm diameter column with a 0.1 μm film. The flow rate of helium was set at 3 mL/min. We initially set the column temperature at 60 °C for 5 min, then, ramped at 5 °C/min to 235 °C and sustained this for 10 min. We then ramped it to 290 °C at 10 °C/min.

3. Results

This research brings together the results of overall pyrolysis optimization tests and selective optimization tests for the pine dust compared to the acacia. We had presented some characterizations of bio-oils from the selective optimization tests, namely the SG, pH and HHV, in [28]. However, we conducted further tests on the viscosity of the oils and their chemical composition and comprehensive comparative analysis of all results in this research to establish the feasible design and operation schemes for these residues.

We compared the properties of the bio-oil reported in this work to the properties of conventional types of diesel and fuel oils. The physicochemical properties of the bio-oil determine if the fuel could be used sustainably in unmodified form without affecting the engine through erosion or corrosion; or if they could be blended successfully. We also measured the HHV values of the chars to build a more holistic perspective on the socio-economic relevance of valorizing the wastes through pyrolysis.

3.1. Characterization of Feedstocks

Table 2 shows the results of the ultimate, proximate and calorimetric tests on the acacia and pine dust in comparison to other results for similar or related feedstocks.

Table 2. Characterization results: ultimate, proximate and higher heating value (HHV).

Biomass	Ultimate Composition (%)					Proximate Composition (Dry Basis) (%)				HHV (MJ/kg)
	C	H	N	S	O *	Ash	* FC	VM	MC	
A. tortilis	41.47	5.15	1.23	nd	52.15	3.90	19.59	76.51	3.72	17.27
Pine dust	45.76	5.54	0.039	nd	48.66	0.83	20.00	79.16	6.50	17.57

nd—Not Detected; * by difference.

We could not find any literature on the characterization of *A. tortilis*; therefore, comparisons were made using other acacia species. We concluded that the results are comparable to those obtained in other literature, notably [34,35] where characterizations of *Acacia holosericea*, *Acacia mangium* and *Acacia auriculiformis* were discussed; and [36,37] where pine dust from two other nations was characterized. Both acacia and pine dust qualify as cleaner thermochemical feedstocks due to the good HHV, low N&S, acceptable ash content (below critical 6% for slagging and fouling) and high VM, which specifically favors higher pyrolysis oil yields. We had initially found the 'as received' pine dust to have an MC of 65.17%, which is higher than the recommended '<10%' for pyrolysis. Consequently, we sun-dried the biomass for at least 3 weeks to achieve an MC of 6.50%, below the '<10%' requirement. Due to the higher C and H content in pine and lower ash compared to acacia, one would expect pine to have a considerably higher HHVs; however, the acacia had a low MC, half that of pine. This could explain why the pine's HHV was not as high as expected. The pine dust from Zimbabwe had relatively lower N&S content compared the other pine forms, and the N was considerably lower compared to the acacia, showing that it was a cleaner thermochemical feedstock. Further discussions on these results will be made jointly with the results of pyrolysis and product characterization in Section 4.

3.2. Optimum Pyrolysis Conditions

3.2.1. Optimum Pyrolysis Temperature

Figures 2 and 3 show the variation of pyrolysis yields with temperature for both biomass forms. The maximum bio-oil yield (41.9 wt %) was obtained at 550 °C for the acacia, while pine dust had its maximum yield (46.1 wt %) at 500 °C. The optimum pyrolysis temperature for pine dust (500 °C) corresponded to that obtained for Indian pine and Canadian mixed sawdust where maximum yields of 39.37% and 45% were obtained in a fixed bed reactor [38,39]. Bridgwater [15] claims in his review of fast pyrolysis that optimum bio-oil yields from lignocellulosic biomass should be obtained between 480 and 520 °C, with grasses occupying the lower end and woody biomass the higher end.

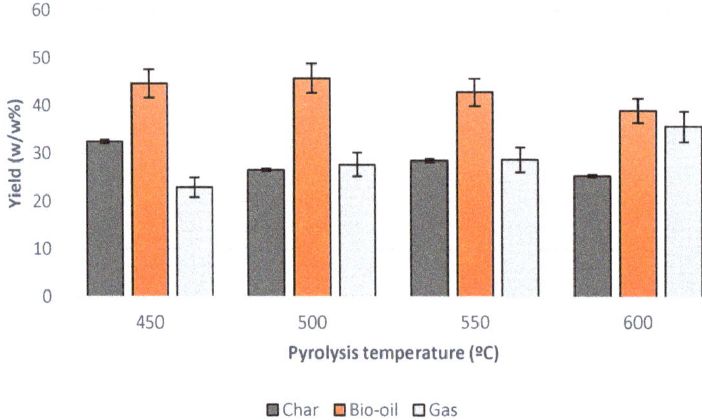

Figure 2. Variation of product yields with pyrolysis temperature for pine dust.

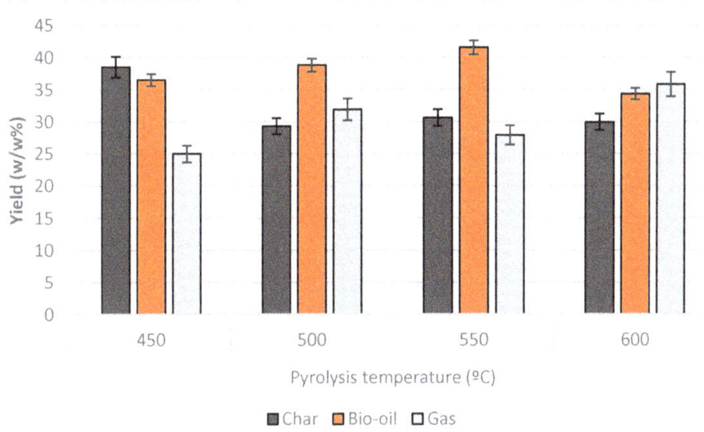

Figure 3. Variation of product yields with pyrolysis temperature for acacia.

With no research reported on *Acacia tortilis*, we could only compare the pyrolysis results from this research to the results obtained on other acacia species. Ahmed et al. [40] compared the bio-oil yields from the intermediate pyrolysis of various parts of *Acacia cincinnata* and *Acacia holosericea* in a fixed bed reactor at a temperature (500 °C. They obtained yields of 45.31–52.95 wt % for the *A. cincinnata* and 41.24–46.92 wt % with the *A. holosericea*, which are comparable to the maximum *Acacia tortilis* yield of 41.9 wt %. Reza et al. [34] used a fixed bed reactor to pyrolyze *A. holosericea* and experimented at two temperatures, 500 °C and 600 °C, obtaining a maximum bio-oil yield of 37.61 wt % at 600 °C. The optimum pyrolysis temperature of 550 °C for *Acacia tortilis* was therefore credible and higher than that of pine due to the higher temperature recalcitrance of the hardwood (acacia) compared to the softwood (pine).

As stated in Section 1.3, temperature and heating rate have been the most experimented variables deemed as the most influential to pyrolysis oil yields and the product distribution [20]. In this case, we only manipulated the temperature since it is difficult to maintain a constant local heating rate for large particle sizes and where several reactions occur. The overall pyrolysis temperature was, therefore, the distinguishing variable affecting the oil yield when equal ramp rates and biomass particle sizes were used. Raising the pyrolysis temperatures until the optimum point increases the heat transfer rates, which favor oil yields to higher localized heating rates. In the fluidized bed, conical spouted bed and circulating bed reactors the small particle sizes and fluidization effect cause even higher heating rates and higher oil yields.

On the other hand, ablative and rotating cone reactors use the high rate of heat transfer from the walls of the reactor, which are typically at elevated temperatures. These higher heating rates, according to Uzun et al. [41], accentuate the reduction of the water content in bio-oil explaining such fast pyrolysis conditions would give higher quality bio-oils compared to the intermediate regime in this research. Typically, such high heating rates should be coupled with moderately high inert gas flow rates to reduce the hot vapor residence time; otherwise, they crack or repolymerize, leading to low oil yields [20]. Such a continuous purge stream was another missing aspect of achieving a fast pyrolysis mode in the case of the fixed bed reactor where the inert gas was used to purge out *the* air at the initial stages of the experiment only.

The ensuing sections focus on the results we obtained from the attempt to optimize the quality of bio-oil by separating the heavier oil from the aqueous light phase using various condenser temperatures.

3.2.2. The Effect of Fractionation on pH

Figure 4 shows the pH of the separated fractions, the heavier oil or primary condensate and the lighter oil or secondary condensate.

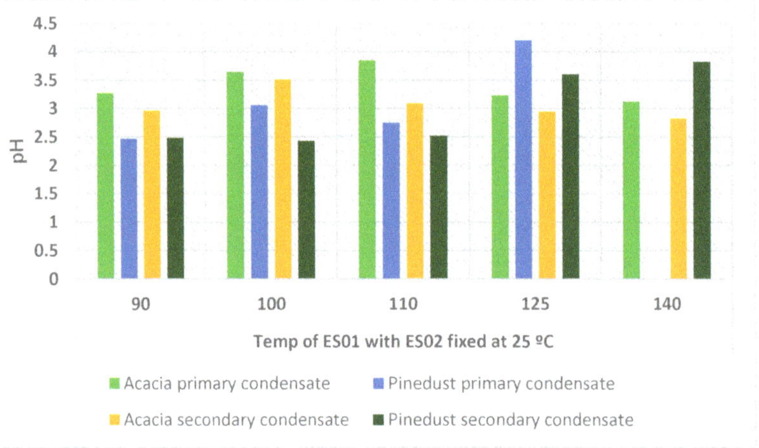

Figure 4. pH of various bio-oil fractions at different primary condenser temperatures for a sample.

The results indicate that the heavier oil fraction obtained in the primary condenser (ES01) has a higher pH, suggesting that most acids preferentially reported to the more aqueous secondary condensates. The *A. tortilis* primary condensates had a higher average pH (3.42) compared to pine dust (average 3.12), although the highest pH was obtained for the pine (4.20). These averages are however, higher and consequently better than the average of 2.5 estimated in the literature for woody bio-oils in general [15]. High acidity in the bio-oil makes it corrosive, especially with metallic vessels or pipes.

We also observed that there was a drastic reduction in the heavier oil yield for the pine at the primary condenser temperature of 140 °C. This could suggest that most of the molecular compounds in pine pyrolysis vapors have low dew points, vaporizing at higher condensation temperatures such as 140 °C and resulting in very low heavy oil yields. We could not conduct the pH, SG and HHV characterizations on the pine bio-oil obtained at this temperature due to the low yield of 0.2% (Table 3). This explains why there are gaps in Figures 4 and 5.

Table 3. Actual yield of primary condensate at various primary condenser temperatures.

Temp of ES01		90 °C	100 °C	110 °C	125 °C	140 °C
Acacia tortilis	Heavy oil yield	6.0% ± 0.2%	5.4% ± 0.1%	4.9% ± 0.1%	2.8% ± 0.1%	3.9% ± 0.1%
	Total oil yield	38.8% ± 1.0%	41.9% ± 1.0%	40.8% ± 1.0%	36.5% ± 0.9%	37.5% ± 1.0%
Pine dust	Heavy oil yield	15.4% ± 0.4%	10.7% ± 0.3%	12.7% ± 0.3%	7.2% ± 0.2%	0.2%
	Total oil yield	34.7% ± 0.9%	46.1% ± 1.2%	44.8% ± 1.1%	44.4% ± 1.1%	33.5% ± 0.9%

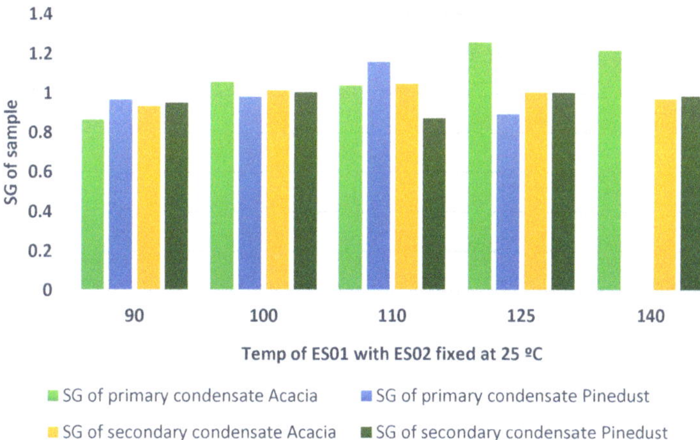

Figure 5. Specific gravity (SG) of various bio-oil fractions at different primary condenser temperatures.

3.2.3. The Effect of Fractionation Temperature on SG, Heavy Oil Yields and Quality

Figure 5 shows the variation of SG with primary condenser temperature, which was used as an indicator of the aqueous composition in the oil.

The higher the SG, the lower the aqueous content and the higher the perceived quality of the product. This is because pure pyrolysis oil has a higher SG than water; therefore, as the purity of oil increases, the SG increases too. Using this metric, the optimum quality bio-oil would be at 125 °C for acacia (SG of 1.257) and at 110 °C for pine (SG of 1.159). At primary condenser temperatures above the boiling point of water (100 °C), the quality of the oil increased, as expected, since more of the water reported to the secondary receivers. However, for pine, beyond the 110 °C we observed a reduction in the yield of the heavier oil (see Table 3) and the least total yield of bio-oil was obtained at 140 °C. One of the pine dust pyrolysis runs at 125 °C also had a zero yield, although it was not used in the calculation of the average. We suspect that the liquid that accumulated in the other two runs could be the result of back mixing of the vapors at that temperature, which caused the accumulation of an oil water mixture of a low SG. This could have been due to the design or orientation of the tubes at the condensers inlets and outlets. The smaller yield of the dry, heavier oil at higher temperatures could be due to lower dew points of molecular compounds in the pine bio-oil [24]. The 140 °C condenser temperature did not seem to have a major effect however on the acacia, most probably because of its higher temperature recalcitrance as a hardwood, therefore the total and heavier oil yields were not significantly reduced.

3.2.4. Properties of the Bio-Oils Obtained at Various Condenser temperatures

Physico-Chemical Properties of the Heavier (Choice) Bio-Oil

Since we target to use the bio-oil in a moderately upgraded form as a fuel for power generation, it is necessary to compare its properties to those of conventional fuels such as diesel, heavy and light fuel oil whose engines could be compatible with the bio-oil. These average HHV values are shown in Table 4, along with the measured values for the viscosity. We used the heavier bio-oil fraction properties in the comparison since they were more comparable with conventional fuels, especially the HHVs. We could not test the secondary condenser oils for HHV due to the high water composition that prevented combustion, even with combustive aids.

Table 4. Various bio-oil physicochemical properties compared to conventional fuels.

		Viscosity (mPa·s) at 25 °C		HHV (MJ/kg)		SG	
	90 °C	*A. tortilis* oil	Pine dust oil	*A. tortilis* oil	Pine dust oil	*A. tortilis* oil	Pine dust oil
		2217.6	3043	4.310	5.227	0.867	0.968
Primary condenser (ES01) Temp °C	100 °C	4804.8	13,951	21.412	9.235	1.057	0.982
	110 °C	Too little	10,151	23.610	15.780	1.040	1.159
	125 °C	Too little	2142	26.191	0.6338	1.257	0.894
	140 °C	Too little	Very little	36.809	Very little	1.217	Very little
Conventional diesel		2178		43–45		0.844	
Light fuel oil (LFO)		-		42–44		0.85–0.910	
Heavy fuel oil (HFO)		>17,800 at 50 °C		40		0.940–0.989	

The viscosities of the two bio-oils are on the higher end of the range reported in the literature for woody bio-oils, of 30–12,000 mPa·s or more [15]. For the best quality of pine bio-oil obtained at 110 °C, we obtained a viscosity of 10,151 mPa·s, which is four times the viscosity of diesel (2178 mPa·s) and almost half that of HFO (17,800 mPa·s) at 50 °C. The bio-oils can, therefore, readily substitute HFO in blends and even lower energy requirements for heating and pumping HFO. The blending ratio with pine bio-oil would, however, be limited given its low HHV compared to the acacia bio-oil [11]. Unfortunately, we could not measure the viscosities of acacia bio-oil obtained at higher temperatures due to their small volumes; however, if the trend against the pine dust values continued as expected, they would still be way below the HFO values. Meanwhile, the acacia bio-oil's HHV values also compare well to HFO, though the SG is about 1.2 times more, tallying with the average value of bio-oil density (1200 kg/m^3) provided by [42]. This means that the acacia heavy bio-oil at ES01 of 140 °C has 92% of the energy content in HFO on a weight basis, but has 113% on a volumetric basis. This bio-oil could, therefore, be a good substitute provided that it is obtained in reasonably large quantities to make the process economically justifiable. The 15.780 MJ/kg obtained for the pine dust bio-oil is slightly lower than the 17 MJ/kg reported in the literature and much less than the HHVs of diesel, LFO or HFO [15].

The Chemical Composition of the Bio-Oils

In this section, we have only presented the GC–MS results of best quality primary condensates and their corresponding secondary condensates as representative samples. We managed to identify 95 compounds from the primary condensate and 76 from the secondary condensates for the *A. tortilis* bio-oil at the primary condenser temperature of 140 °C. For the pine bio-oil at a temperature of 110 °C, 107 compounds were detected in the primary condensate, while only 87 were identified in the secondary condensate. Bio-oil is typically composed of more than 300 oxygenated compounds [17]. However, not all of the compounds can be easily identified by the conventional GC–MS due to their complex matrix in the bio-oil; therefore we recommend a GC × GC, with a higher resolution in 3D, to identify most of the compounds. Tables 5 and 6 show the eight compounds with the largest peak areas for the sampled primary and secondary condensates.

Table 5. GC–MS analysis for *Acacia tortilis* primary and secondary condensates obtained at optimal condition (primary condenser temp of 140 °C).

	Compound	Compound Class	Molecular Weight (g/mol)	Area %	Retention Time (min)
Primary condensate	Cresol	Methylphenol	138.16	7.759	8.23
	Phenol, 4-ethyl-2-methoxy-	Phenol	152.19	5.325	10.31
	Phenol, 2,6-dimethoxy-	Phenol	154.16	5.120	11.83
	Phenol, 2-methoxy-	Phenol	164.20	4.847	5.89
	Benzene, 1,3-bis(1,1-dimethylethyl)-	Hydrocarbon	190.33	4.390	10.09
	Phenol, 2,4-bis(1,1-dimethylethyl)-	Phenol	278.50	4.254	16.28
	5-tert-Butylpyrogallol	Phenol	182.22	3.302	16.16
	Heneicosane	Alkane hydrocarbon	296.583	3.3148	18.57
	Others				
	Benzoic acid	Carboxylic acid	122.12	0,6791	8.48
	Cyclopenten-1-one, 2-hydroxy-3-methyl-	Ketone	68.12	2,1704	4.73
	Octane, 3-ethyl-	Alkyl hydrocarbon	142.28	1.073	5.54
	Furaldehyde phenylhydrazone OR Furfuraldehyde	Aldehyde	96.0841	0.1729	19.53
Secondary condensate	Phenol, 2-methoxy-	Phenol	164.20	15.638	6.06
	Phenol, 2,6-dimethoxy-	Phenol	154.16	12.377	12.38
	2-Cyclopenten-1-one, 2-hydroxy-3-methyl-	Ketone	112.13	7.533	5.01
	Hydroquinone mono-trimethylsilyl ether	Phenol	110.03	6.052	16.29
	4-Methoxy-2-methyl-1-(methylthio)benzene	Hydrocarbon (phenylpropanes)	168.26	5.103	14.40
	Benzene, 1,3-bis(1,1-dimethylethyl)-	Hydrocarbon	190.32	3.960	10.12
	2-Cyclopenten-1-one, 3-ethyl-2-hydroxy-	Ketone	126.15	3.764	6.83
	Cyclohexanol, 2,2-dichloro-1-methyl-	Alcohol	183.03	2.138	4.49
	Others				
	Trans 2-(2-Pentenyl)furan	Furan	136.19	0.2457	7.36
	3,4-dimethylcyclohexanol	Alcohol	128.21	0.3821	4.66

Table 6. GC–MS analysis for pine dust primary and secondary condensates obtained at optimal condition (primary condenser temp of 110 °C).

	Compound	Compound Class	Molecular Weight (g/mol)	Area %	Retention Time (min)
Primary condensate	Phenol, 2-methoxy-	Phenol	164.20	11.066	6.01
	Cresol	Methylphenol	138.16	7.154	8.26
	Benzene, 1,3-bis(1,1-dimethylethyl)-	Hydrocarbons	190.33	4.935	10.10
	1,2-Cyclopentanedione, 3-methyl-	Ketone	183.07	4.851	4.77
	Hexadecane	Hydrocarbon		4.1684	18.80
	Phenol, 2,4-bis(1,1-dimethylethyl)-	Phenol	278.5	4.047	16.27
	Homovanillyl alcohol	Alcohol	168.19	3.8349	16.04
	Phenol, 4-ethyl-2-methoxy-		152.19	3.6806	10.32
	Others				
	trans-Isoeugenol	Phenol	164.20	2.544	13.37
	Tridecane, 7-hexyl-	Hydrocarbon	268.5	1.6788	22.42
	Propanal, 2-propenylhydrazone	Aromatic aldehyde	126.15		
	Methoxyacetic acid, nonyl ester	Monocarboxylic acid and ether	216.32	0.2737	7.23
Secondary condensate	Phenol, 2-methoxy-	Phenol	164.20	14.526	5.99
	Cresol	Methylphenol	138.16	14.093	8.38
	Phenol, 4-ethyl-2-methoxy-	Phenol	152.19	9.343	10.42
	2-Cyclopenten-1-one, 2-hydroxy-3-methyl-	Ketone	112.13	7.509	4.93
	Phenol, 2-methoxy-4-(1-propenyl)-, (Z)-	Phenol	164.20	4.767	14.44
	Phenol, 2-methoxy-4-propyl-	Phenol	166.22	2.871	12.53
	Benzene, 1,3-bis(1,1-dimethylethyl)-	Hydrocarbon	190.32	2.281	10.10
	Eugenol	Guaiacol (phenol)	164.20	2.276	12.23
	Others				
	Propanoic acid, 2-methyl-, 2,2-dimethyl-1-(2-hydroxy-1-methylethyl)propyl ester	Carboxylic acid and ester	244.37	0.7445	5.76
	Maltol	Pyranones (ketones)	126.11	1.47	6.47

For *A. tortilis*, we observed that cresol had the highest composition detected in the primary condensate, while it was next to Phenol, 2-methoxy- in the pine dust primary condensate. In the secondary condensates, Phenol, 2-methoxy and Phenol, 2, 6-dimethoxy were the most concentrated in the acacia, while Phenol, 2-methoxy- and cresol were the highest in the pine. The secondary condensates had characteristically high concentrations of specific compounds in both biomass forms and could be used as a source of chemicals since they are not useful for fuel. The phenols in the secondary condensers were of significant concentrations on a dry basis and could be extracted for use as food additives—mostly as flavorings, spices, seasonings and colorings—while cresol is an active ingredient in disinfectants [43]. We also identified other compounds in both bio-oils including acids like acetic acid, benzoic acid and homovanillic acid; esters such as succinic acid and the nonyl tetrahydrofurfuryl ester;

food flavorants and their precursors such as maltol and guaiacol (also used as a pesticide) and ketones such as acetone and other types of phenols. We did not manage to detect Levoglucosan as expected in such lignocellulosic biomass [44] probably because it was present in tiny quantities. However, we detected another type of anhydrous sugar compounds, 1,4:3,6-Dianhydro-.alpha.-d-glucopyranose, in the dry pine bio-oil with a peak area of 1.22%. Lyu et al. [45] also observed a shallow composition of anhydrosugars (<1 wt %) at a pyrolysis temperature of 480 °C, which increased at 500 °C and 520 °C, attaining a maximum of 4.1 wt % at 550 °C.

The GC–MS results that we obtained are comparable to those obtained by Ahmed et al. [40] for the trunk of *Acacia cincinnata* and *Acacia holosericea*, who reported that phenolic compounds and their derivatives were the major compounds detected in the bio-oil samples. For instance, the *Acacia holosericea* trunk was found to have the following sampled areas: phenol, 2,6-dimethoxy- (5.36%), cresol (3.39%), (E)-2,6-dimethoxy-4- phenol (3.35%), phenol, 2-methoxy- (3.23%), benzene, 1,2,3-trimethoxy-5-methyl- (3.22%), 2- cyclopenten-1-one, 2-hydroxy-3-methyl- (2.04%), 2-furancarboxaldehyde (2.99%), isoeugenol 2 (2.08%), guaiacol, 4-ethyl- (2.06%) and 2-methoxy-4-vinylphenol (1.72%) [40]. The high presence of phenolics in both the acacia and pine primary oils is typical as reported by Muda et al. [46], who recorded a high composition of phenolics in bio-oil obtained from the subcritical water treatment of oil palm trunk. The compound with the highest composition in their bio-oil was phenol with a peak area of 17.11%, while 2,6-dimethoxy-phenol and 4-ethyl-2-methoxy-phenol had 7.10% and 1.78% respectively. Lyu et al. [45] have also identified phenol peak areas of up to 20% for the pyrolysis of wood and rice straw, while Kim et al. [47] mention a composition of 0.1–3.8 wt %. In terms of peak areas, the findings in our research are, therefore in agreement with the literature cited. Kim et al. [47] also mentioned that the pyrolysis of softwood lignins yielded mainly guaiacols, including the eugenols, vanillin and homovanillin, which had a significant presence in the pine bio-oil. The composition of the acacia bio-oil that we observed was also consistent with the observation that hardwood lignins give both guaiacols and syringols, including phenols, methoxy phenols and cresol. The high composition of phenols that we observed could also be due to the low pyrolysis temperature of 450 °C used for the fractional condensation experiments, corroborating with research done by Lyu et al. [45]. In their research, the phenolic content in bio-oil increased from 3.9 to 5.5 wt % when the temperature was decreased from 550 to 480 °C.

The reported compositions of bio-oil differ widely in the literature depending on the biomass source and also on process conditions used [20]. At a higher condensation temperature for instance, in the case of acacia, the product was a tar-like oil of high HHV, rich in complex phenolics (48% peak area) with ~30% hydrocarbons and the rest of the area occupied by other compounds—on a dry basis. The dominance of complex phenolics and aromatics in such tar-like oil corroborates with literature such as Muda et al. [46] and these authors also obtained an exceptionally high HHV value for that bio-oil (33.2 MJ/kg). The characteristically high HHV of this tar-like oil, which is similar to that obtained in this research, seems to contradict with the general trend that the bio-oil with a higher percentage of phenolics could be unuseful as a fuel [20]. The high HHV could be due to the complex type of phenolics comprising methyl and methoxy groups and the significant amount of long-chain carboxylic acids, hydrocarbons and ketones (total 38% peak area) in this tar-like oil. Coupled with the low aqueous composition indicated by the high SG, it is logical that the primary condensate should have such a high calorific value. On the contrary, the secondary condensates had ~26% peak area occupied by hydrocarbons, ketones and carboxylic acids and a higher aqueous composition demonstrated by the low SG.

3.2.5. Uncertainty Analysis

We assessed the margin of error that could occur in the pyrolysis temperature range for the bio-oil, char and gas separately by obtaining four readings at a single temperature then calculating the relative standard error (%). This value of the error was then applied across the various temperatures. For instance, from Figure 3, the chars were liable to an error of 4.24% while the bio-oil and gas had

7. Gasparatos, A.; von Maltitz, G.P.; Johnson, F.X.; Lee, L.; Mathai, M.; De Oliveira, J.P.; Willis, K.J. Biofuels in sub-Sahara Africa: Drivers, impacts and priority policy areas. *Renew. Sustain. Energy Rev.* **2015**, *45*, 879–901. [CrossRef]

8. Amigun, B.; Kaviti, J.; Stafford, W. Biofuels and sustainability in Africa. *Renew. Sustain. Energy Rev.* **2011**, *15*, 1360–1372. [CrossRef]

9. Von Maltitz, G.P.; Setzkorn, K.A. A typology of Southern African biofuel feedstock production projects. *Biomass Bioenergy* **2013**, *59*, 33–49. [CrossRef]

10. Blimpo, M.P.; Cosgrove-Davies, M. *Electricity Access in Sub-Saharan Africa: Uptake, Reliability, and Complementary Factors for Economic Impact*; The World Bank: Washington, DC, USA, 2019.

11. Kass, M. *Evaluation of Bio-oils for Use in Marine Engines*; The Bioenergy Technology Office of the U.S. Department of Energy: Oak Ridge, TN, USA, 2019.

12. Charis, G.; Danha, G.; Muzenda, E. Waste valorisation opportunities for bush encroacher biomass in savannah ecosystems: A comparative case analysis of Botswana and Namibia. *Procedia Manuf.* **2019**, *35*, 974–979. [CrossRef]

13. Charis, G.; Danha, G.; Muzenda, E. A review of timber waste utilization: Challenges and opportunities in Zimbabwe. *Procedia Manuf.* **2019**, *35*, 419–429. [CrossRef]

14. Montoya, J.I.; Valdés, C.; Chejne, F.; Gómez, C.A.; Blanco, A.; Marrugo, G.; Osorio, J.; Castillo, E.; Aristóbulo, J.; Acero, J. Bio-oil production from Colombian bagasse by fast pyrolysis in a fluidized bed: An experimental study. *J. Anal. Appl. Pyrolysis* **2015**, *112*, 379–387. [CrossRef]

15. Bridgwater, A.V. Review of fast pyrolysis of biomass and product upgrading. *Biomass Bioenergy* **2011**, *38*, 68–94. [CrossRef]

16. Pratap, A.; Chouhan, S. Critical Analysis of Process Parameters for Bio-oil Production via Pyrolysis of Biomass: A Review. *Recent Patents Eng.* **2013**. [CrossRef]

17. Badger, P.; Badger, S.; Puettmann, M.; Steele, P.; Cooper, J. Techno-Economic Analysis: Preliminary Assessment of Pyrolysis Oil Production Costs and Material Energy Balance Associated with a Transportable Fast Pyrolysis System. *Bioresources* **2011**, *6*, 34–47.

18. Fonts, I.; Juan, A.; Gea, G.; Murillo, M.B.; Sa, L. Sewage Sludge Pyrolysis in Fluidized Bed, 1: Influence of Operational Conditions on the Product Distribution. *Ind. Eng. Chem. Res.* **2008**, *47*, 5376–5385. [CrossRef]

19. Sirijanusorn, S.; Sriprateep, K.; Pattiya, A. Bioresource Technology Pyrolysis of cassava rhizome in a counter-rotating twin screw reactor unit. *Bioresour. Technol.* **2013**, *139*, 343–348. [CrossRef] [PubMed]

20. Guedes, R.E.; Luna, A.S.; Torres, A.R. Operating parameters for bio-oil production in biomass pyrolysis: A review. *J. Anal. Appl. Pyrolysis* **2018**, *129*, 134–149. [CrossRef]

21. Ellens, C.J.; Brown, R.C. Optimization of a free-fall reactor for the production of fast pyrolysis bio-oil. *Bioresour. Technol.* **2012**, *103*, 374–380. [CrossRef]

22. Bridgwater, A.V.; Meier, D.; Radlein, D. An overview of fast pyrolysis of biomass. *Org. Geochem.* **1999**, *30*, 1479–1493. [CrossRef]

23. Uddin, M.N.; Techato, K.; Taweekun, J.; Rahman, M.M.; Rasul, M.G.; Mahlia, T.M.I.; Ashrafur, S.M. An overview of recent developments in biomass pyrolysis technologies. *Energies* **2018**, *11*, 3115. [CrossRef]

24. Papari, S.; Hawboldt, K. A review on the condensing systems for biomass pyrolysis process. *Fuel Process. Technol.* **2018**. [CrossRef]

25. Chen, T.; Deng, C.; Liu, R. Effect of Selective Condensation on the Characterization of Bio-oil from Pine Sawdust Fast Pyrolysis Using a Fluidized-Bed Reactor. *Energy Fuels* **2010**, *24*. [CrossRef]

26. Oasmaa, A.; Sipilä, K.; Solantausta, Y.; Kuoppala, E. Quality Improvement of Pyrolysis Liquid: Effect of Light Volatiles on the Stability of Pyrolysis Liquids. *Energy Fuels* **2005**, *19*, 2556–2561. [CrossRef]

27. Charis, G.; Danha, G.; Muzenda, E. Thermal and chemical characterization of lignocellulosic wastes for energy uses. In *Wastes: Solutions, Treatments and Opportunities, Proceedings of the 5th International Conference Wastes 2019, Lisbon, Portugal, 4–6 September 2019*; CRC Press: Boca Raton, FL, USA, 2018; pp. 2–7.

28. Charis, G.; Danha, G.; Muzenda, E. Selective bio-oil fraction optimization and characterization for waste lignocellulosic feedstocks. **2019**. Unpublished.

29. Olarte, M.V.; Olarte, M.V.; Christensen, E.D.; Padmaperuma, A.B.; Connatser, R.M.; Stankovikj, F.; Meier, D.; Paasikallio, V. Standardization of chemical analytical techniques for pyrolysis. *Biofuels Bioprod. Biorefin.* **2016**, *10*, 496–507.

30. Ngo, T.; Kim, J.; Kim, S. Fast pyrolysis of palm kernel cake using a fluidized bed reactor: Design of experiment and characteristics of bio-oil. *J. Ind. Eng. Chem.* **2013**, *19*, 137–143. [CrossRef]

31. Lira, C.S.; Berruti, F.M.; Palmisano, P.; Berruti, F.; Briens, C.; Pécora, A.A.B. Fast pyrolysis of Amazon tucumã (*Astrocaryum aculeatum*) seeds in a bubbling fluidized bed reactor. *J. Anal. Appl. Pyrolysis* **2013**, *99*, 23–31. [CrossRef]

32. Pauls, R.E. A Review of Chromatographic Characterization Techniques for Biodiesel and Biodiesel Blends. *J. Chromatogr. Sci.* **2011**, *49*, 384–396. [CrossRef]

33. Mccormick, R.L.; Ratcliff, M.; Moens, L.; Lawrence, R. Several factors affecting the stability of biodiesel in standard accelerated tests. *Fuel Process. Technol.* **2007**, 651–657. [CrossRef]

34. Reza, S.; Ahmed, A.; Caesarendra, W.; Abu Bakar, M.S.; Shams, S.; Saidur, R.; Aslfattahi, N.; Azad, A.K. Acacia Holosericea: An Invasive Species for Bio-char, Bio-oil, and Biogas Production. *Bioengineering* **2019**, *6*, 33. [CrossRef]

35. Ahmed, A.; Bakar, M.S.A.; Azad, A.K.; Sukri, R.S.; Mahlia, T.M.I. Potential thermochemical conversion of bioenergy from Acacia species in Brunei Darussalam: A review. *Renew. Sustain. Energy Rev.* **2018**, *82*, 3060–3076. [CrossRef]

36. Chaula, Z.; Said, M.; John, G. Thermal Characterization of Pine Sawdust as Energy Source Feedstock. *J. Energy Technol. Policy* **2014**, *4*, 57–65.

37. Olatunji, O.; Akinlabi, S.; Oluseyi, A.; Peter, M.; Madushele, N. Experimental investigation of thermal properties of Lignocellulosic biomass: A review. *IOP Conf. Ser. Mater. Sci. Eng.* **2018**, *413*. [CrossRef]

38. Mishra, R.K.; Mohanty, K. Thermal and catalytic pyrolysis of pine sawdust (*Pinus ponderosa*) and Gulmohar seed (*Delonix regia*) towards production of fuel and chemicals. *Mater. Sci. Energy Technol.* **2019**, *2*, 139–149. [CrossRef]

39. Salehi, E.; Abedi, J.; Harding, T. Bio-oil from sawdust: Pyrolysis of sawdust in a fixed-bed system. *Energy Fuels* **2009**, *23*, 3767–3772. [CrossRef]

40. Ahmed, A.; Bakar, M.S.A.; Azad, A.K.; Sukri, R.S.; Phusunti, N. Intermediate pyrolysis of Acacia cincinnata and Acacia holosericea species for bio-oil and biochar production. *Energy Convers. Manag.* **2018**, *176*, 393–408. [CrossRef]

41. Uzun, B.B.; Pütün, A.E.; Pütün, E. Fast pyrolysis of soybean cake: Product yields and compositions. *Bioresour. Technol.* **2006**, *97*, 569–576. [CrossRef]

42. Bridgwater, T. Challenges and Opportunities in Fast Pyrolysis of Biomass. *Johns. Matthey Technol. Rev.* **2018**, *62*, 118–130. [CrossRef]

43. Gauthier-maradei, P.; Manrique, M.; Gauthier, P.; Mantilla, S.V. Methodology for extraction of phenolic compounds of bio-oil from agricultural biomass wastes. *Waste Biomass Valorization* **2015**, *6*, 371–383. [CrossRef]

44. Balat, M. Bio-oil production from pyrolysis of black locust (*Robinia pseudoacacia*) wood. *Energy Explor. Exploit.* **2010**, *28*, 173–186. [CrossRef]

45. Lyu, G.; Wu, S.; Zhang, H. Estimation and comparison of bio-oil components from different pyrolysis conditions. *Front. Energy Res.* **2015**, *3*, 28. [CrossRef]

46. Muda, N.A.; Yoshida, H.; Ishak, H.; Ismail, M.H.S.; Izhar, S. Conversion of oil palm trunk into bio-oil via treatment with subcritical water. *J. Wood Chem. Technol.* **2019**, *39*, 255–269. [CrossRef]

47. Kim, J. Bioresource Technology Production, separation and applications of phenolic-rich bio-oil—A review. *Bioresour. Technol.* **2015**, *178*, 90–98. [CrossRef]

48. Fahmy, T.Y.A.; Fahmy, Y.; Mobarak, F.; El-Sakhawy, M.; Abou-Zeid, R.E. Biomass pyrolysis: Past, present, and future. *Environ. Dev. Sustain.* **2018**, 1–16. [CrossRef]

 *processes*

Article

Electrical Resistivity of Carbonaceous Bed Material at High Temperature

Gerrit Ralf Surup [1,*], Tommy Andre Pedersen [1], Annah Chaldien [2], Johan Paul Beukes [2] and Merete Tangstad [1]

[1] Department of Materials Science and Engineering, Norwegian University of Science and Technology, 7491 Trondheim, Norway; tommyap@stud.ntnu.no (T.A.P.); merete.tangstad@ntnu.no (M.T.)
[2] Chemical Resource Beneficiation, North-West University, Potchefstroom 2520, South Africa; 23032898@student.g.nwu.ac.za (A.C.); Paul.Beukes@nwu.ac.za (J.P.B.)
* Correspondence: gerrit.r.surup@ntnu.no; Tel.: +47-934-73-184

Received: 1 July 2020; Accepted: 23 July 2020; Published: 3 August 2020

Abstract: This study reports the effect of high-temperature treatment on the electrical properties of charcoal, coal, and coke. The electrical resistivity of industrial charcoal samples used as a reducing agent in electric arc furnaces was investigated as a renewable carbon source. A set-up to measure the electrical resistivity of bulk material at heat treatment temperatures up to 1700 °C was developed. Results were also evaluated at room temperature by a four-point probe set-up with adjustable load. It is shown that the electrical resistivity of charcoal decreases with increasing heat treatment temperature and approaches the resistivity of fossil carbon materials at temperatures greater than 1400 °C. The heat treatment temperature of carbon material is the main influencing parameter, whereas the measurement temperature and residence time showed only a minor effect on electrical resistivity. Bulk density of the carbon material and load on the burden have a large impact on the electrical resistivity of each material, while the effect of particle size can be neglected at high heat treatment temperature or compacting pressure. The mechanical durability of charcoal slightly increased after heat treatment and decreased for coal and semi-coke samples. The results indicate that charcoal can be used as an efficient carbon source for electric arc furnaces.

Keywords: charcoal; electrical resistivity; coal; coke; high-temperature treatment; pyrolysis

---

1. Introduction

Biomass and its derivatives (e.g., charcoal) are considered as a possible feedstock to reduce anthropogenic CO_2 emissions produced in industry. Besides its common usage as an energy carrier in power production, charcoal can be used as base material in fuel cells [1], batteries [2], soil amendment [3,4], and as a carbon source in metallurgical industry [5–10]. In the latter, fossil fuels such as anthracite, coal, coal char, semi-coke, petroleum coke, and metallurgical coke are the main reducing agents. These reductants are used in blast furnaces, electric arc furnaces (EAF), and submerged arc furnaces (SAF) to reduce metal oxides to their metallized form. SAF are particularly used in silicon, ferrosilicon, ferrochrome, and ferromanganese production. The consumption of fossil fuels in EAF and SAF generates approximately 1.83 kg CO_2 per kg of steel [11], 1.04 to 1.15 kg CO_2 per kg ferromanganese [12], 1.4 to 6.9 kg CO_2 per kg of silicomanganese, and 2.5 to 4.8 kg CO_2 per kg ferrosilicon [13–15]. The total CO_2 equivalent of manganese alloys is stated as 6.0 kg per kg of alloy when electricity is produced by coal combustion [15]. Biomass and charcoal have the potential to reduce these emissions, e.g., by up to 12% in EAF, or 58% in integrated routes of steel production [16]. A better understanding of charcoal properties will promote the increased use of renewable resources in EAFs and SAFs.

Charcoal has a higher porosity, lower mechanical stability, and higher reactivity than fossil reducing agents [3,8]. The high CO_2 reactivity of charcoal can increase its consumption in the burden by the Boudouard reaction [17]. Approximately 500,000 tonnes of CO_2 emissions generated by ferromanganese and silicomanganese production are due to the Boudouard reaction [18], corresponding to ~30% of the annual emissions. The lower mechanical stability of charcoal can result in the formation of fines, negatively affecting gas permeability of the charge and resulting in an increased risk of bridging and slag boiling [19]. Previous studies have shown that charcoal properties can be improved by increasing the heat treatment temperature. For example, CO_2 reactivity of charcoal approaches that of fossil reducing agents at a heat treatment temperature higher than 1600 °C [20,21]. Thus, it is crucial to understand the properties of charcoal at high temperatures to provide a stable operation of the furnace.

To fully, or significantly, replace fossil fuel resources in SAF, a renewable resource with specific chemical, mechanical, and electrical properties is required. Fossil fuel bed material provides a low electrical resistivity and high mechanical stability, in which the latter is compulsory for a stable operation of the furnace [22,23]. A high mechanical stability is required to maintain its structure for a good gas permeability, whereas power input and temperature profile are affected by the electrical properties of the bulk material. While the chemical and mechanical properties of charcoal as reducing agents have been reviewed in the last decade [24–26], knowledge of its electrical resistivity at high temperature is scarce.

The main parameters that influence the electrical resistivity of carbon bed material are heat treatment temperature [3,5,27,28], compression pressure [3,28–31], particle size [32–34], and volume fraction of the carbonaceous material [27,34]. Previous studies focused on the measurement of the electrical resistivity of powder materials [28], dry coke beds [19,22,32,35], or coke metal-oxide blends [27] of fossil fuels, whereas knowledge on the charcoal beds is limited.

In this study, the impact of heat treatment temperature, particle size, and compaction pressure was investigated for charcoal, in comparison with other fossil fuel materials as references. The specific objectives of this study were to investigate the (1) electrical resistivity of charcoal and fossil fuels samples at high heat-treatment temperature, (2) influence of particle size and compaction pressure, and (3) importance of present temperature on electrical resistivity.

2. Materials and Methods

2.1. Sample Materials

Three industrial charcoals, one coal, one semi-coke, and one metallurgical coke were chosen as sample materials for this study. When this study was initiated, the charcoal, coal, and semi-coke were used in silicon and ferrosilicon production, whereas the metallurgical coke was used for silicomanganese production. The aforementioned lumpy reductants were crushed by a jaw crusher (Retsch, Haan, Germany) and sieved to three particle sizes: fines (d < 2 mm), small particles (2 ≤ d < 4.75 mm), and large particles (4.75 ≤ d < 9.5 mm). Prior to an experiment, the sample was dried overnight in air (at 106 °C) to drive off moisture.

2.2. Resistivity at High-Temperature

Electrical resistivity measurements at elevated temperature were performed using a four-probe point measurement system (SINTEF, Trondheim, Norway) installed in an induction furnace (Inductotherm Europe, Droitwich Spa, UK), as presented in Figure 1. An alumina tube (Alsint®, length: 300 mm; inner diameter: 80 mm; wall thickness: 10 mm) was placed in a graphite crucible (height: 400 mm; diameter: 150 mm; wall thickness: 18 mm). Two molybdenum (Mo) wires (diameter: 1 mm) were installed at a distance of 50 and 150 mm from the bottom of the alumina tube. The sample was distributed homogeneously inside the alumina tube, which was filled to ~25 cm height and compacted to constant volume by dynamic sample compaction with an installed electrode (weight: 1.8 kg), in which drop height was set to 10 cm. Bulk height was measured by a caliper (readability: 0.1 mm) before the crucible was installed into the furnace. A constant load of ~6 kN·m^{-2} was set on the material by the graphite electrode and an additional weight on top of the electrode. Direct current was provided by a welding power source IPM15 (Kempower, Lahti, Finland) between the graphite electrode and the graphite crucible bottom. An upper limit of 100 A was set. The voltage between the Mo-wires in the sample was measured by a NI TB-9214 module (National Instruments, Debrecen, Hungary) connected to the Mo-wires. Current and voltage were measured every 25 °C by a pulse train initiated by the Labview program (National Instruments, Austin, TX, USA). The associated temperature (T_C) was measured in the center of the bulk. The pulse train consists of 8 pulses with increasing voltage adjustment (2× 17%, 33%, 60%, and 90% of the desired voltage). The electrical resistivity was calculated according to Equation (1):

$$\rho = \frac{V \cdot A}{I \cdot l} \tag{1}$$

where ρ is the electrical resistivity, V the measured voltage between the wires, A is the cross-sectional area, I the measured current, and l the distance between the Mo-wires.

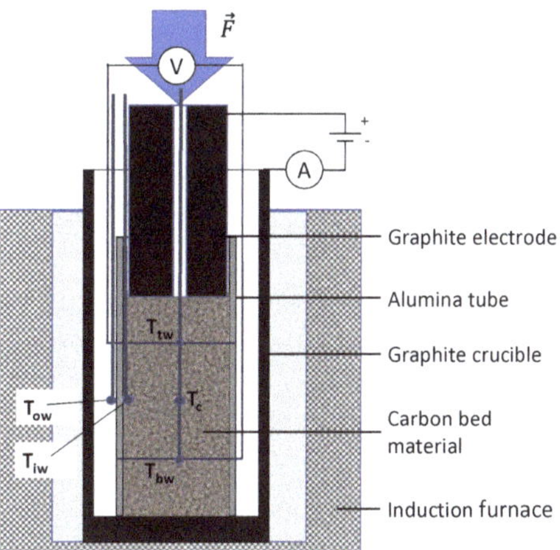

Figure 1. Schematic of the electrical measurement set-up at NTNU/SINTEF.

The sample was heated at 3 °C·min^{-1} up to 250 °C and kept at that temperature for 10 min to minimize thermal stresses in the alumina tube. Subsequently, the sample was further heated at 10 °C·min^{-1} to an outer wall temperature (T_{OW}) of 1600 °C and kept at that temperature for 60 min to ensure a constant temperature profile in the crucible. After the heating program was finished, the height of the bulk was measured, the furnace was turned off, and the sample was cooled overnight. After cooling, voltage and current were measured at room temperature before the sample was removed and stored in sealed sample bags.

2.3. Resistivity Under Load

The electrical resistivity under load was investigated by a four-point probe set-up at room temperature (called the 4-probe set-up), which was installed in a materials testing machine Z2.5 test Control II (Zwick/Roell, Ulm, Germany). Measurements were carried out for the particle sizes: fines (d < 2 mm), small particles were split into (2 < d < 3.35 mm), and (3.35 < d < 4.75 mm), as well as large particles (4.75 < d < 9.5 mm) in an alumina tube (inner diameter: 50 mm, height: 150 mm). The sample was filled and compacted to a constant volume by shaking. Current was provided by a laboratory DC power supply GPR-3030 (GW instek, New Taipei City, Taiwan) and controlled at 2 A current for each measurement. A top graphite and a bottom copper electrode were used to pass the current through the bulk. Basic load on the material by the top electrode was $2.5\,kN\cdot m^{-2}$. The voltage was measured over a distance of 60 mm (30 to 90 mm from the bottom) by a multimeter Model 2000 (Keithley Instruments, Beaverton, OR, USA).

Additional load was introduced and measured by the materials testing machine. The pressure was increased to 500 N in 25 N steps, in which vertical compaction was noted with a readability of 1 µm. After a maximum force of 500 N was reached, the compacting force was released in 25 N steps and voltage and vertical compaction of the bulk were measured.

2.4. Solid Characterization

Proximate analysis

The volatile matter content and yield of crucible coke of the heat-treated material were measured according to DIN 51720. Ash content of coal, coke, and its derivatives was measured according to DIN 51719, whereas ash content of charcoal samples was determined according to DIN EN 14775. Fixed carbon content was calculated by difference (Fixed carbon = 100% − ash content − volatile matter content).

Elemental analysis

Elemental analysis of the feedstocks and solid residues was performed on an Elemental Analyser 2400 CHNS/O Series II (Perkin Elmer, Waltham, MA, USA) by Analytik-Service Gesellschaft mbH. Acetanilide was used as a reference standard. Sulfur content was investigated for coal, semi-coke, and metallurgical coke according to ASTM D 4239:2017. Oxygen content was calculated by difference, in which ash content from the proximate analysis was used.

Scanning electron microscopy

SEM analysis of the char residue was conducted on a high-resolution microscope ULTRA 55 (Zeiss, Oberkochen, Germany) under high vacuum in order to understand structural properties of the heat-treated material.

Compression strength

Compressive strength of the feedstock and heat treated particles was investigated using the materials testing machine Z2.5 test Control II (Zwick/Roell, Ulm, Germany). Compression was carried out with a constant feed of $1\,mm\cdot min^{-1}$. The compression force and compaction distance were collected by testXpert II software (Zwick/Roell, Germany). The first major drop in compression force was allocated to particle breakage. Results were reported as an average of 40 particles.

Mechanical durability

The mechanical durability (also called abrasions strength) of large particles was investigated in a drum test (diameter: 200 mm, 4 lifters) as described elsewhere [35]. Samples were subjected to 1200 revolutions at 40 rpm. Particles less than 3.15 mm in size were separated by sieving, for which mechanical durability was calculated as the weight percentage of the fraction remaining on the sieve in relation to the initial sample weight. Prior to the experiment, particles less than 5 mm in size were removed by sieving.

3. Results

3.1. Feed Material Composition

The proximate and ultimate analyses of the untreated coal, coke, and charcoal samples are shown in Table 1. The volatile matter content of the charcoal samples were in the range 13.1 to 15.9% that revealed a primary heat treatment temperature between 500 and 700 °C [36]. A fixed carbon content greater than 85% is referred to a as a metallurgical-grade charcoal [37]. Oxygen content was 7 to 9 wt.% for the charcoals, as well as 13.8 wt.% for the coal, 2.6 wt.% for the semi-coke, and <0.1 wt.% for the metallurgical coke.

Table 1. Proximate and ultimate analyses of feedstocks.

Feedstock	Charcoal A	Charcoal B	Charcoal C	Coal	Semi-Coke	Metallurgical Coke
Proximate analysis (wt.%)						
Moisture (ar)	27.5	12.1	5.02	12.2	8.8	<0.1
Volatile matter (db)	15.9	13.8	13.1	38.8	5.4	0.9
Fixed carbon (db)	82.8	84.8	85.2	59.3	88.6	87.9
Ash (db)	3.97	1.89	1.75	1.25	5.87	10.90
Ultimate analysis (wt.%, dry basis)						
C	85.5	83.5	88.1	78.3	88.5	87.7
H	1.95	2.28	2.51	4.72	1.24	0.21
N	<0.50	<0.50	<0.50	1.61	1.44	1.75
O	7.40	8.68	7.14	13.77	2.64	<0.10
S	<0.01	<0.01	<0.01	0.35	0.31	0.57

3.2. Product Yield

A summary of the product yield after the high-temperature resistivity treatment with respect to proximate analysis is shown in Figure 2, while the detailed product yield and composition are presented in the supplemental material (Table A1). As is evident from these results, the mass loss was between 2 to 5% larger than the volatile matter of the feedstock. Volatile matter content in the solid residue after the heat treatment at 1600 °C was reduced from 13.1, 13.8, and 15.9 wt.% to 0.3 wt.% for charcoal samples, which is similar to charcoal samples which were heat treated under the same operating conditions [36]. Volatile matter content of fossil fuel reductants reduced from 0.9, 5.4, and 38.8 wt.% to 0.1, 0.1, and 0.3 wt.% for metallurgical coke, semi-coke, and coal, respectively. The additional mass loss occurs due to higher heat treatment temperature and the release of ash elements such as alkali metals [38]. The bulk density of large particles slightly decreased from approximately 240 to 230, from 275 to 270 and 265 to 260 kg·m^{-3} for charcoal A, charcoal B, and charcoal C, respectively, and from 700 to 535 and 640 to 615 kg·m^{-3} for coal and metallurgical coke, respectively, whereas bulk density of semi-coke slightly increased from 420 to 425 kg·m^{-3}. Only minor differences were observed for the fraction of small particles compared to fraction of large particles, whereas bulk density of fines was up to 50% higher than that of the fraction of large particle, as summarized in the supplemental material (Table A2).

Hydrogen and oxygen content of heat-treated material decreased to less than 0.1 wt.%. Sulfur content of fossil fuels remained nearly unchanged at 0.3 to 0.5 wt.%. Thus, heat treated coal, coke, and charcoal exhibited a similar organic composition at 1600 °C on dry ash free basis. The molar H/C and O/C ratios decreased by the thermal decomposition and release of volatile oxygenates, such as organic acids, alcohols, and phenols as well as hydrogen, methane, carbon monoxide, and carbon dioxide [39,40]. Dehydrogenation reaction at high temperature can form small aromatic cluster in the carbon matrix [20,21], in which the increased number of clusters result in an enhanced aromaticity of the solid residue. Ultimate analysis verified results from proximate analysis that solid residue is composed of carbon and inorganic matter.

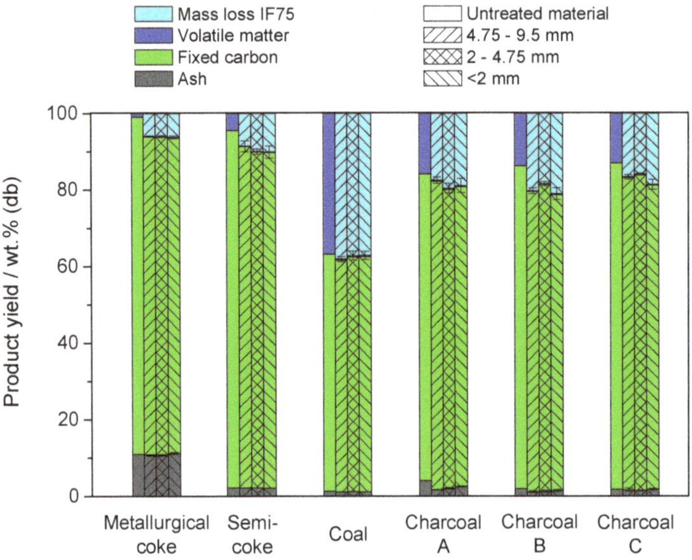

Figure 2. Product yields of char for fines, small particles, and large particles after high temperature measurement. The char yield is separated in ash, fixed carbon, and volatile matter by its proximate analysis. The error bars characterize the standard deviations between the experiments.

3.3. Mechanical Properties

Table 2 summarizes the mechanical properties for the fraction of large particles (4.75–9.5 mm) of the feedstock and solid residue produced in high-temperature set-up. Thermal shrinkage (TS) describes the decrease in particle size due to thermal decomposition, shrinkage and generation of fines by compaction. It was determined by the weight ratio of particles less than 4.75 mm to product yield and measured in the range of 16 to 20% for charcoal, 7 to 8% for semi-coke, 4 to 5% for coal, and ≤2% for metallurgical coke. TS was in good agreement with bed shrinkage during heating and fines generation during compacting, where up to 7% of fine material was formed for charcoal samples. The low TS of coal is related to its caking, whereas only minimal mass loss and shrinkage for metallurgical coke was observed.

Prior to tumbling, particles less than 5 mm in size were removed by sieving. The mechanical durability increased by 2 to 4% points to ~96% for heat-treated charcoal samples, whereas it decreased by approximately 5% points for coal, 1.5% points for semi-coke, and 0.2% points for metallurgical coke. The higher carbonization degree can enhance the mechanical stability by cross-linking and larger aromatic structures, whereas a lower bulk density is related to a higher porosity. Fines and small particles are harder and less porous than larger particles, providing a smaller contact area between particles [34]. Inner surface area decreased at temperature greater than 1000 °C by collapsing of small pores [20]. Thus, decreasing bulk density is related to increasing void fraction and large pores, which resulted in a decreasing mechanical durability. In addition, storage may have affected tumbling resistance for carbonaceous feedstock [41,42].

The compression strength was investigated at random particle orientation. Untreated charcoal A particles started to break at a force of 20 to 60 N, and 30 to 60 N for charcoal B and charcoal C, whereas heat-treated charcoal particles withstood a compression force of 40 to 80 N. Small amounts of fines were generated at lower compacting force by breaking small supporting edges to stabilize particle position in the set-up. Charcoal particles were generally crushed to fines and compacted at compaction forces greater 200 N. The large difference in charcoal samples is correlated to a 3 to 4 times greater compressive strength in direction to wood fibers compared to the perpendicular direction [8]. Coal particles withstood a compaction force of ~50 N before breaking, whereas crushing of broken particles occurred between 150 and 200 N. Semi-coke and metallurgical coke particles broke after a minimum force of 80 N and 100 N, and crushing resistance was measured between 200 and 300 N. The lower compression strength of charcoal was related to its lower envelope density and larger porosity.

Table 2. Mechanical properties of charcoal, coal, and semi-coke before and after heat treatment.

Sample	TS/% Heat-Treated	Mechanical Durability/%		Compression Strength/N	
		Raw	Heat-Treated	Raw	Heat-Treated
Charcoal A	17.6 ± 1.2	93.6	95.6 ± 1.3	20–60	40–80
Charcoal B	16.6 ± 2.4	93.5	96.1 ± 0.4	30–60	40–80
Charcoal C	19.4 ± 3.1	91.8	96.0 ± 0.1	30–60	40–80
Coal	4.5 ± 0.6	99.4	94.3 ± 1.1	50	70
Semi-coke	7.6 ± 0.6	99.8	98.4 ± 0.6	80	100
Metallurgical coke	1.9 ± 0.1	99.8	99.6 ± 0.1	100	100

3.4. Electrical Resistivity at Elevated Temperature

Figure 3 shows the electrical resistivity of charcoal, coal, semi-coke, and metallurgical coke in the temperature range from 950 to 1650 °C. Only semi-coke and metallurgical coke exhibited a detectable electrical conductivity below 950 °C, as shown in the supplemental material (see Figure A1). For both coke samples, electrical resistivity slightly increased from room temperature to 300 °C and subsequently decreased with increasing heat treatment temperature. Electrical resistivity of charcoal decreases from approximately 10^7 Ω·m at 600 °C to 100 Ω·m at 800 °C [3], respectively from $9 \cdot 10^3$ Ω·m to 200 Ω·m for coal in the same temperature range [5]. At elevated temperature, electrical resistivity of large particles further decreased from 70 mΩ·m at 1000 °C to approximately 12 mΩ·m at 1650 °C for semi-coke, from approximately 11.5 to 5.5 mΩ·m for metallurgical coke, and from 20 to 6 mΩ·m for coal. Similar values were reported for coal and coke in the literature [22,27,34,43]. Electrical resistivity of charcoal particles decreased from higher than 100 mΩ·m at 1000 °C to a range of 14 to 18 mΩ·m at 1600 °C.

Electrical resistivity continued to decrease slightly after final temperature was reached, as shown in the supplemental material (see Figure A2). This drop was related to a decreasing temperature gradient in the measuring zone, especially between the bulk center and the top Mo-wire, as shown in the supplemental material (see Figure A3). No further reduction in electrical resistivity was observed after a stationary temperature was reached. The electrical resistivity at final temperature varied between 8.2 to 15 mΩ·m for charcoal, 3.5 to 6.5 mΩ·m for metallurgical coke, 8.5 to 12 mΩ·m for semi-coke, and 4.5 to 6.5 mΩ·m for coal. The decrease in electrical resistivity was attributed to the removal of oxygen-containing functional groups and the restructuring of the carbon matrix [44–48]. Differences in electrical resistivity by feedstock were related to the bulk density and final temperature of the samples, summarized in Figure 4.

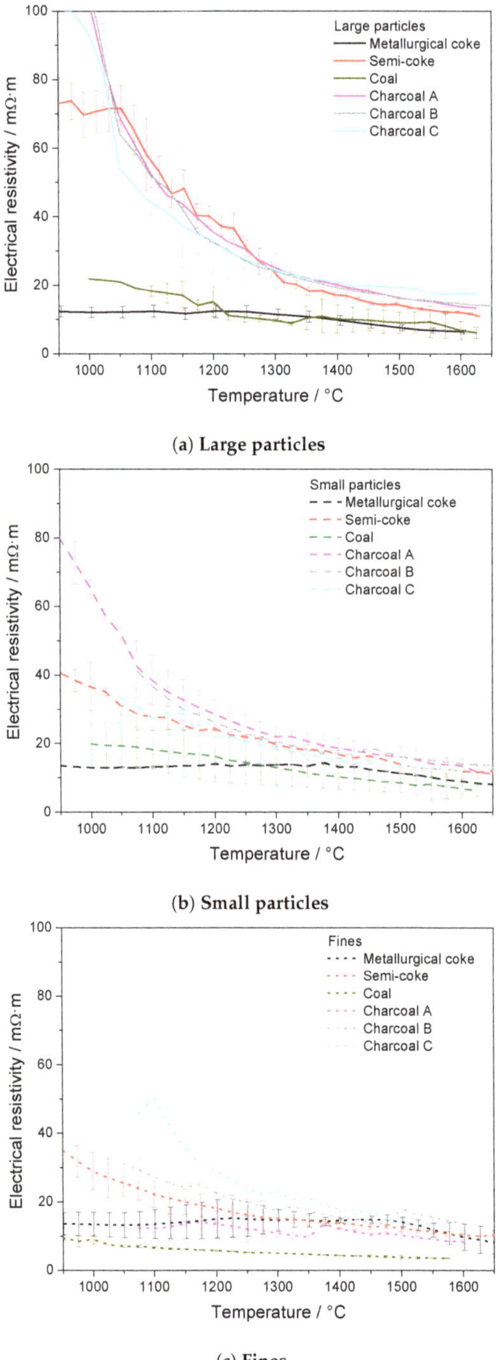

(a) **Large particles**

(b) **Small particles**

(c) **Fines**

Figure 3. Temperature dependence of electrical resistivity of different grain sizes: (**a**) large particles (4.75–9.5 mm), (**b**) small particles (2.00–4.75 mm), and (**c**) fines (<2 mm).

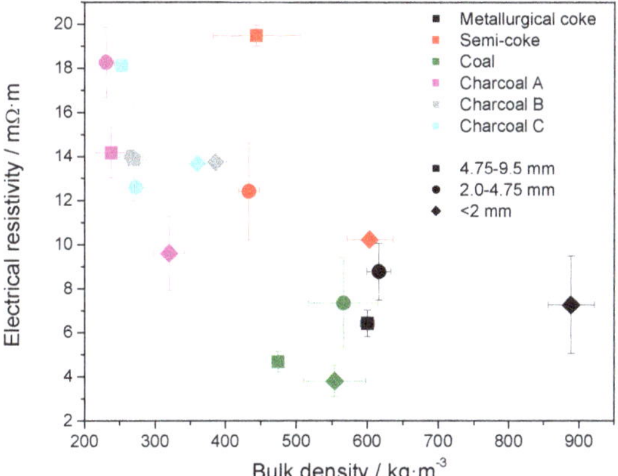

Figure 4. Electrical resistivity at 1600 °C after 1 h residence time versus bulk density.

After the samples were cooled overnight, electrical resistivity was evaluated at room temperature in same set-up. The last measurement at elevated temperature versus electrical resistivity measurement at room temperature is shown in Figure 5. The electrical resistivity at room temperature was scattered around measurements at elevated temperature for small and large particles, which resulted in an overestimation or underestimation at room temperature by an average of 25%. Differences were related to cooling and removing the crucible from furnace, which resulted in a thermal shrinkage of 0.5 to 1% and micromovement of bulk material. Coefficient of determination for small and large particles was calculated as 56%. However, a large difference was observed for charcoal fines, in which electrical conductivity was approximately twofold lower at room temperature compared to hot state.

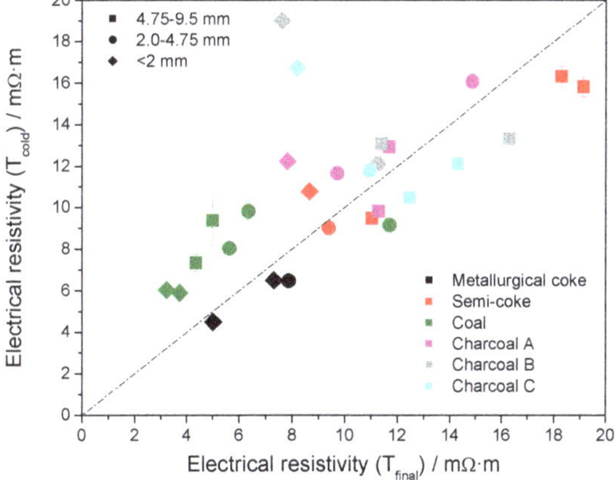

Figure 5. Comparison of electrical resistivity measurement at elevated temperature to measurement at room temperature after cooling.

3.5. Electrical Resistivity under Load

The electrical resistivity under load of heat treated material is shown in Figure 6 for the releasing compaction force. Increasing compression force on bulk material resulted in a compaction and increased bulk density of the material, as shown in the supplemental material (see Figures A4 and

A5. The main compaction occurred at a force less than 100 N (50 kPa), in which the increasing force resulted in the compaction of bed material and decrease in electrical resistivity, from approximately 15 to 4 mΩ·m for metallurgical coke, 35 to 2.4 mΩ·m for semi-coke, 37 to 2.9 mΩ·m for coal, and 33 to 3 mΩ·m for charcoal. Bulk density increased by approximately 2–6% by filling the void fraction between the particles. The resistivity under load was less affected by particle size than by its initial compaction. Electrical resistivity under load of large and small particles was in an acceptable similarity to the measurements at high temperature, as shown in the supplemental material (see Figure A6).

Minor amounts of fines were formed by compaction to 250 kPa, confirming the results from the compression strength. Based on the compression strength of single particles, a pressure greater than 400 kPa can be introduced before breaking charcoal particles, and greater than 1000 kPa for metallurgical coke particles. Thus, only limited amounts of generated particles fill the void fraction of bulk material. Otherwise, increasing compaction force improves the contact between neighboring particles by increasing the number of contact points [22,32] and decreasing the air gap between particles [28]. Contact resistance can be two to ten times larger than electrical resistance from solid material [22,32,34], in which a decreasing particle size results in an increasing electrical resistivity. However, this effect is balanced by an increased number of contacts and conductive paths through the bulk.

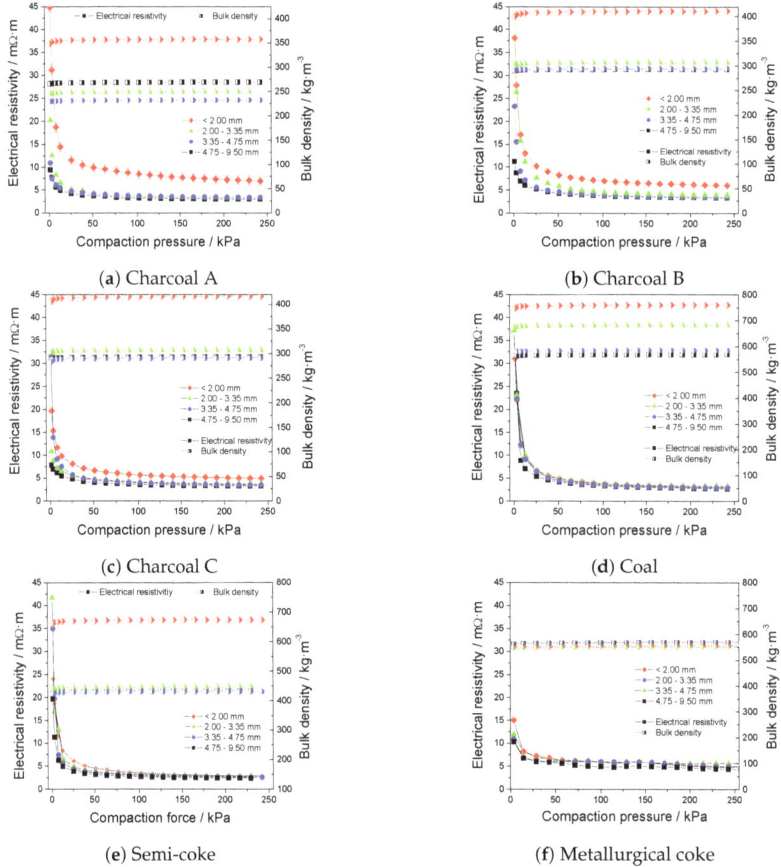

Figure 6. Electrical resistivity and bulk density versus compaction pressure for (**a**) charcoal A, (**b**) charcoal B, (**c**) charcoal C, (**d**) coal, (**e**) semi-coke, and (**f**) metallurgical coke.

4. Surface Morphology

The surface of heat-treated material was investigated by scanning electron microscopy. Charcoal samples heat treated at 1600 °C showed similar structure to its feedstock. Charcoal A exhibited mostly longitudinal tracheids cells and resin canals, schematically shown in Figure 7a, whereas a large number of vessels was observed for charcoals B and C, as shown in Figure 7b,c. The results reveal that charcoal A is a softwood, whereas charcoals B and C are hardwoods [49].

Heat-treated coal exhibited a molten surface as shown in Figure 7d. The molten surface confirmed secondary tar reactions of the metaplast [50], leading to caking of the coal particles, as shown in the supplemental material (see Figure A7). Cracks were observed in the outer surface area with a low amount of pores. Semi-coke particles on the other hand exhibited a porous outer surface, schematically shown in Figure 7e. Metallurgical coke contained molten areas with a pore structure as shown in Figure 7f.

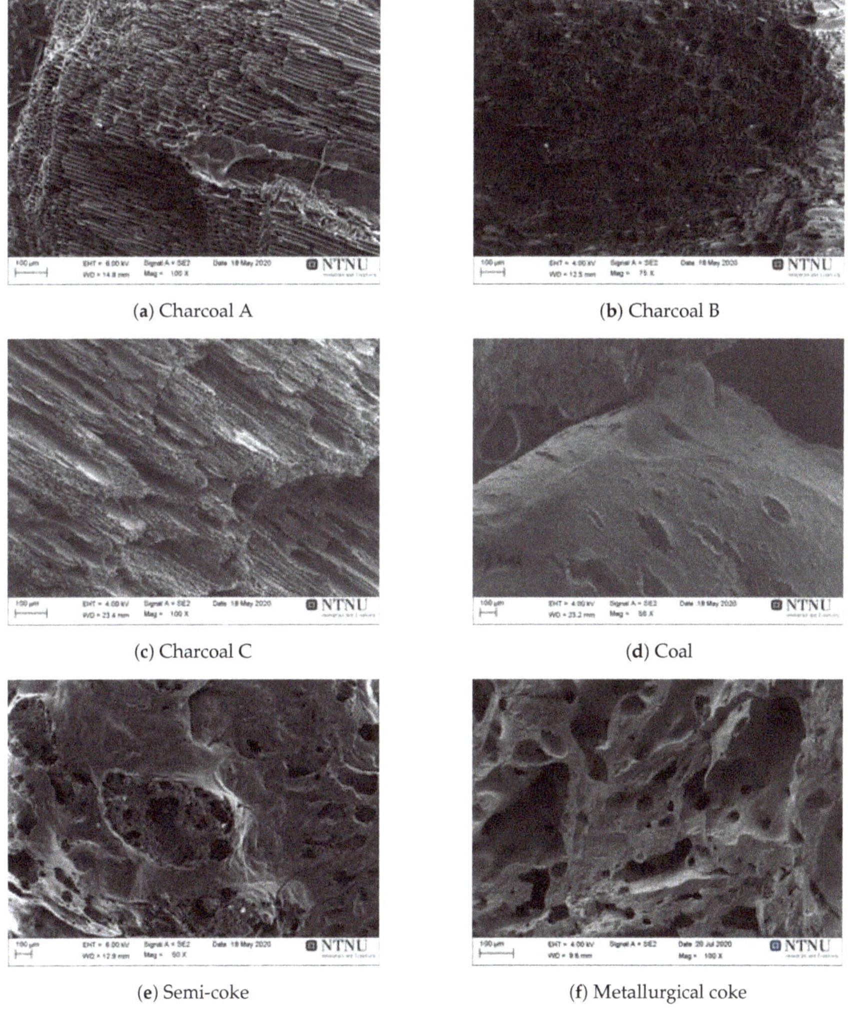

(**a**) Charcoal A

(**b**) Charcoal B

(**c**) Charcoal C

(**d**) Coal

(**e**) Semi-coke

(**f**) Metallurgical coke

Figure 7. SEM images of (**a**) heat-treated charcoal A, (**b**) heat-treated charcoal B, (**c**) heat-treated charcoal C, (**d**) heat-treated coal, (**e**) heat-treated semi-coke, and (**f**) heat-treated metallurgical coke.

5. Discussion

The electrical resistivity of carbonaceous reducing agents decreased with increasing heat treatment temperature [8,22,27,28,32,35], increasing bulk density and increasing load. At temperatures less 950 °C, high oxygen content and disordered carbon structure act as an insulator in biomass, charcoal, and coal [3,44,45,47]. Hydrogen and oxygen content decrease with increasing heat treatment temperature by thermal decomposition and release of volatile matter, such as oxygenates CO, CO_2, CH_4, and H_2 [51], which result in a decrease in electrical resistivity by 5 orders of magnitude [1,3]. Semi-coke and metallurgical coke on the other hand have been produced at ~900 °C and between 1100 to 1400 °C in their production processes [52], and thus already provide a reduced electrical resistivity in low temperature range.

Simulation have shown that 5 to 15% of electrical current is conducted through the burden of SAF by using metallurgical coke [22]. The electrical properties of charcoal and coal at low heat treatment temperature inhibit a conduction of current in the higher regions of the SAF [1,43]. With traveling carbonaceous material to the high temperature zone of SAF, the electrical properties of charcoal approach the properties of fossil reductants such as coal and semi-coke, but remain higher than for metallurgical coke. A higher electrical resistivity enables lower tip position in submerged arc furnaces, and thus improves the heat generation in the lower part of the furnace [19].

The electrical resistivity of metallurgical coke decreased from 13 mΩ·m at 1000 °C to approximately 6.5 mΩ·m at 1600 °C for a particle size of 4.75–9.5 mm, similar to that described in the literature for larger particles (10–20 mm) [35]. It was reported that electrical resistivity decreased by approximately 50% by increasing particle size from 5–10 mm to 15–20 mm [19]. Charcoal particles larger than 2 mm were measured to approximately 50 mΩ·m at 1100 °C and 17 mΩ·m at 1600 °C, similar to larger grain sizes (5–35 mm) reported in the literature [35]. The current results indicate that the particle size affects electrical conductivity less than bulk density. The minor effect of the particle size can be explained by the increased number of contact points in carbon bed, in which particle size decreases similar to the increased number of contact points [31]. However, in industrial furnaces the contact between particles of the carbon bed can be affected by the presence of slag, which can result in an increased particle-to-particle resistance [19]. However, softwood, with the lowest density, provided an electrical conductivity similar to both hardwood samples and electrical resistivity of charcoal approached that of fossil fuels despite its twice lower bulk density.

Compacting pressure in SAF will increase by weight of the burden towards the carbon bed. Contact pressure and number of electrical contact points will increase with increasing pressure, resulting in a decreasing electrical resistivity [53,54]. Li et al. investigated electrical resistivity for high compacting pressure and concluded that electrical resistivity decreased up to a pressure of 19 GPa [55]. At compressive pressure of ~8 MPa an electrical resistivity of 2 Ω·m can be achieved [1]. However, with increasing pressure the particles can break at the contact point and create a new or better contact between the particles, resulting in a more even current density between particles at the cost of gas permeability [32].

A major issue for replacing metallurgical coke by renewable bed material is the mechanical property. In general, charcoal provides a lower mechanical stability and higher CO_2 reactivity than coke and coal. The current investigation has shown that the durability of charcoal particles improved after heat treatment at 1600 °C, indicating improved mechanical properties at the hearth of the furnace. The compression strength of heat-treated charcoal varied between 20 to 80 N, corresponding to a maximum pressure of 400 to 1600 kPa before breaking and the generation of large amounts of fines. A maximum pressure of 150 kPa is assumed by the burden in SAF based on the assumptions stated in Appendix E in the supplemental material. The thermal shrinkage indicates that between 15 to 20% of fines can be generated by bed movement and thermal stress in the burden, increasing an increased risk of clogging and slag boiling [19]. Thus, durability of carbonaceous material is considered more important for improvement than its mechanical stability towards compression.

Metallurgical coke and charcoal are assumed to last for a long time period as bed material. Xiao et al. reported that electrical resistivity of charcoal decreased with increasing residence time [56].

The current investigation showed that decreasing electrical resistivity was primarily correlated to heat treatment temperature, bulk density, and temperature profile inside the bulk, whereas prolonging the residence time did not affect electrical resistivity. At 1600 °C, the charcoal structure comprises amorphous and nanocrystalline graphite, in which prolonging residence time to 12 h led to the formation of ring graphitic structures [21]. However, the increase in crystallite growth is compensated by a decrease in excess electrons [1], which resulted in a quasi-constant electrical resistivity. Thus, increasing the residence time at 1600 °C has less of an effect on electrical resistivity than increasing the residence time below 950 °C [1].

Measurement at room temperature resulted in an overestimation or underestimation of electrical resistivity up to 50%. Bed shrinkage and bed movement led to the collapse and formation of current paths, affecting the measurement for heat treated materials. Measurements under load demonstrated a large impact of the initial compaction on electrical resistivity at low pressure. Thus, in situ measurement at elevated temperature in the same set-up is superior to multiple measurements of pre-heat-treated material. Deviation between hot temperature and room temperature measurement was especially large for charcoal with grain sizes less than 2 mm. Results at room temperature for particles larger than 2 mm scattered around electrical resistivity at high temperature with a deviation of ~25%. The results indicate that electrical resistivity can be determined from pre-heat-treated material with acceptable accuracy for grain sizes larger than 2 mm. However, cold measurements of charcoal fines underestimated electrical resistivity by more than 50%.

6. Conclusions

Electrical properties of charcoal, coal, semi-coke, and metallurgical coke have been investigated at elevated temperature and compared to measurements at room temperature under load. The results showed that the electrical resistivity of charcoal approached that of coal and metallurgical coke at temperatures above 1400 °C and was similar to semi-coke at temperatures higher than 1050 °C. Heat treatment prior to electrical resistivity measurement can be carried out to determine electrical properties at room temperature at acceptable accuracy for particles larger than 2 mm, whereas electrical resistivity of powder and smaller particles is overestimated. The mechanical load of the burden mostly affected electrical conductivity at compaction pressure less 50 kPa, whereas electrical resistivity only slightly decreased at higher load. The results showed that electrical resistivity is mainly dependent on heat treatment temperature, compacting pressure, and bulk density, whereas particle size and residence time provide only minor effects. Mechanical durability of charcoal after heat treatment at 1600 °C increased by 4% points to 96% and mechanical strength improved up to 80 N (corresponding 400 kPa), which is sufficient to withstand compression from the burden. In summary, charcoal can be used as an alternative carbon source for carbon bed in submerged arc furnaces with improved electrical properties.

Author Contributions: Conceptualization, G.R.S. and M.T.; methodology, G.R.S.; software, T.A.P.; validation, G.R.S., T.A.P., and A.C.; formal analysis, G.R.S., T.A.P., and A.C.; investigation, G.R.S.; resources, M.T.; data curation, T.A.P.; writing—original draft preparation, G.R.S.; writing—review and editing, M.T. and J.P.B.; visualization, G.R.S.; supervision, M.T. and J.P.B.; project administration, M.T. All authors have read and agreed to the published version of the manuscript.

Funding: This research was funded by Research Council of Norway grant number 280968. The APC was funded by Norwegian University of Science and Technology

Acknowledgments: The authors gratefully acknowledge the financial support from Research Council of Norway (Project Code: 280968) and our industrial partners Elkem ASA, TiZir Titanium & Iron AS, Eramet Norway AS, Finnfjord AS and Wacker Chemicals Norway AS for financing the *KPN Reduced CO$_2$ emissions in metal production* project. The support from the Norwegian Research Council through the project "Metal production—education, competence and research" (Project Code: 261692/H30) is also acknowledged for encouragement and funding cooperation between South Africa and Norway.

Conflicts of Interest: The authors declare no conflicts of interest. The funders had no role in the design of the study; in the collection, analyses, or interpretation of data; in the writing of the manuscript; or in the decision to publish the results.

Abbreviations

The following abbreviations are used in this manuscript.

A	Cross-sectional area
ar	as received
db	dry basis
EAF	Electric arc furnace
I	Current
IF	Induction furnace
l	length
NTNU	Norwegian University of Science and Technology
SAF	Submerged arc furnace
TS	Thermal shrinkage
T_{bw}	Temperature at the bottom Mo-wire
T_C	Temperature in the center of the bulk
T_{iw}	Temperature at the inner wall of the Al-tube
T_{ow}	Temperature at the outer wall of the Al-tube
T_{tw}	Temperature at the top Mo-wire
V	Voltage

Appendix A. Composition of Heat Treated Material

Table A1 summarizes the product yield, proximate, and ultimate analyses of the heat-treated material. Ash content of the products remained only slightly changed with heat treatment at 1600 °C, whereas volatile matter content was reduced from 13.1, 13.8, and 15.9 wt.% to less than 0.3 wt.% for charcoal, from 38.8 wt.% to less than 0.3 wt.% for coal, and from 5.4 and 0.9 wt.% to approximately 0.1 wt.% for semi-coke and metallurgical coke, respectively. Ultimate analysis was carried out at ASG Analytik-Service Gesellschaft mbH. After heat treatment, hydrogen and nitrogen contents were determined for all samples below detection limit to less than 0.1 wt.% and less than 0.5 wt.%, confirming previous results [36]. Sulfur content decreased from 0.57 to 0.5 wt.% for metallurgical coke, and from 0.35 to 0.3 wt.% for semi-coke and coal. The low volatile matter, hydrogen and oxygen contents confirm that solid residue at 1600 °C is mainly composed of carbon and ash.

Table A1. Product yield, proximate, and ultimate analyses of samples after heat treatment.

Particle Size /mm	Solid Yield /wt.%, db	Ash	Fixed Carbon	Volatile Matter /wt.%, db	C	H	N	O	S
			Metallurgical coke						
<2.0 mm	94.0 ± 0.3	11.4	88.5	0.1 ± 0.02	88.5	<0.1	<0.5	<0.1	0.5
$2.0 \leq d \leq 4.75$	94.0 ± 0.3	11.4	88.5	0.1 ± 0.02	88.4	<0.1	<0.5	<0.1	0.5
$4.75 \leq d \leq 9.5$	93.8 ± 0.3	11.9	88.0	0.1 ± 0.02	88.1	<0.1	<0.5	<0.1	0.5
			Semi-coke						
<2.0 mm	91.4 ± 0.3	2.3	97.6	0.1 ± 0.02	97.9	<0.1	<0.5	<0.1	0.3
$2.0 \leq d \leq 4.75$	90.1 ± 0.3	2.4	97.5	0.1 ± 0.02	97.3	<0.1	<0.5	<0.1	0.3
$4.75 \leq d \leq 9.5$	89.9 ± 0.3	2.3	97.6	0.1 ± 0.03	97.1	<0.1	<0.5	<0.1	0.3
			Coal						
<2.0 mm	61.9 ± 0.3	1.7 ± 0.3	98.0	0.3 ± 0.04	99.2	<0.1	<0.5	<0.1	0.4
$2.0 \leq d \leq 4.75$	62.8 ± 0.3	1.8 ± 0.2	97.9	0.3 ± 0.03	98.3	<0.1	<0.5	<0.1	0.3
$4.75 \leq d \leq 9.5$	62.9 ± 0.3	1.7 ± 0.3	98.0	0.3 ± 0.06	98.0	<0.1	<0.5	<0.1	0.3
			Charcoal A						
<2.0 mm	82.5 ± 0.3	2.0 ± 0.3	97.7	0.3 ± 0.03	97.9	<0.1	<0.5	<0.1	-
$2.0 \leq d \leq 4.75$	80.4 ± 0.3	2.5 ± 0.2	97.2	0.3 ± 0.04	97.5	<0.1	<0.5	<0.1	-
$4.75 \leq d \leq 9.5$	81.0 ± 0.3	3.0 ± 0.3	96.7	0.3 ± 0.02	97.1	<0.1	<0.5	<0.1	-
			Charcoal B						
<2.0 mm	79.7 ± 0.3	1.4 ± 0.2	98.4	0.2 ± 0.05	98.5	<0.1	<0.5	<0.1	-
$2.0 \leq d \leq 4.75$	81.5 ± 0.3	1.5 ± 0.1	98.3	0.2 ± 0.04	98.4	<0.1	<0.5	<0.1	-
$4.75 \leq d \leq 9.5$	78.9 ± 0.3	1.8 ± 0.3	97.9	0.3 ± 0.05	98.1	<0.1	<0.5	<0.1	-
			Charcoal C						
<2.0 mm	83.3 ± 1.3	1.9 ± 0.1	97.9	0.2 ± 0.03	98.0	<0.1	<0.5	<0.1	-
$2.0 \leq d \leq 4.75$	84.0 ± 0.9	1.7 ± 0.2	98.1	0.2 ± 0.05	98.1	<0.1	<0.5	<0.1	-
$4.75 \leq d \leq 9.5$	81.4 ± 1.5	2.1 ± 0.3	97.7	0.2 ± 0.05	97.9	<0.1	<0.5	<0.1	-

Appendix B. Electrical Resistivity at Low Heat Treatment Temperature

Electrical conductivity of semi-coke and metallurgical coke in the temperature range from 20 to 1650 °C is shown in Figure A1. Electrical resistivity slightly increased from room temperature to 300 °C by thermal expansion and bed alignment. The electrical resistivity of semi-coke decreased after surpassing 600 to 900 °C, which is the pyrolysis temperature in the production. Similarly, electrical resistivity of metallurgical coke decreased at temperatures higher 1100 °C, which can be accounted as the heat-treatment temperature in production [52]. Thus, electrical properties can also be used to identify the prior heat treatment temperature.

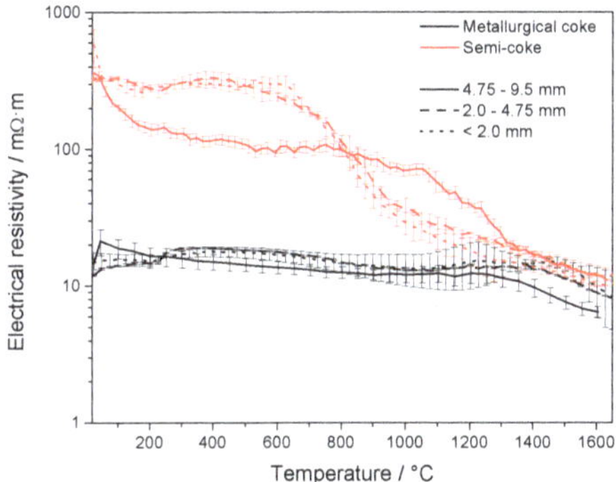

Figure A1. Electrical resistivity of semi-coke and metallurgical coke from 20 to 1650 °C.

Appendix C. Residence Time

Samples were kept at 1600 °C for 60 min to ensure an even temperature profile in the alumina tube. Electrical resistivity for charcoal and fossil-reducing agent samples is shown in Figure A2. It can be noted that electrical resistivity decreased in the first 30 min after reaching the set point temperature and continued to decrease slightly afterwards. The difference in electrical resistivity was pronounced in the fine fractions (d < 2 mm), whereas fraction of large particles was less affected by increasing residence time.

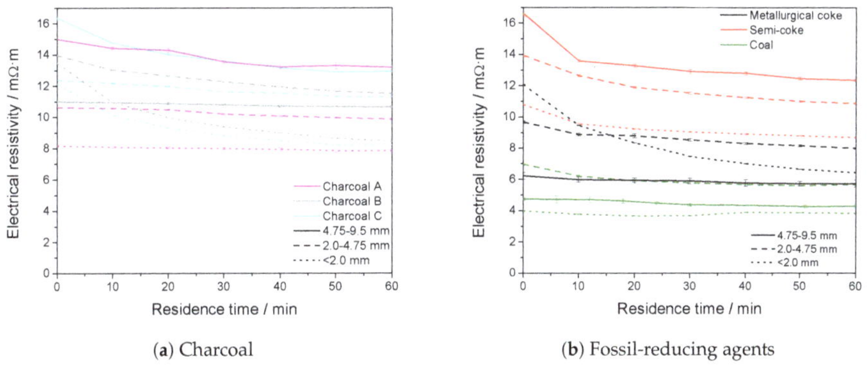

(**a**) Charcoal (**b**) Fossil-reducing agents

Figure A2. Electrical resistivity vs. temperature after reaching set point temperature for (**a**) charcoal and (**b**) fossil-reducing agents.

The decrease in electrical resistivity was related to the temperature gradient in the bulk. While temperature gradient between center temperature and lower wire was stable approximately 10 min after reaching the set point temperature, top wire temperature continued to increase slightly, schematically shown in Figure A3. It can be noted that temperature gradient between bottom and top wire decreased from about 300 °C at the beginning of residence time to 250 °C after 60 min and kept nearly constant afterwards.

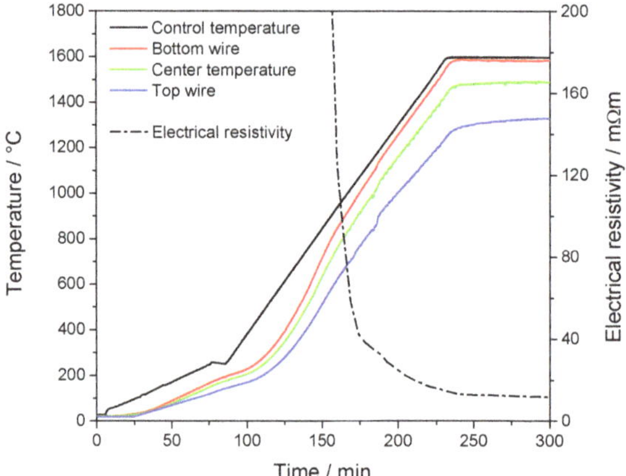

Figure A3. Temperature over time for control temperature, bottom wire, center, and top wire temperature for the heating program in IF75. Electrical resistivity over time is given for charcoal B as reference.

Appendix D. Electrical Resistivity under Load

Figure A4 shows the electrical resistivity and bulk density versus compression pressure for the cycle of compression and pressure release at room temperature in the 4-probe setup. Bulk density increased with increasing compaction pressure and decreases with a lower slope for pressure release, in which bulk density increased by 1.5 to 5.5% at 250 kPa compacting force, as shown in Figure A5. Compaction after releasing the pressure was measured to a range between 0.7 and 3.5%. The compaction after the cycle indicates that void space was filled by particle ordering in the bulk.

Electrical resistivity at 6 kPa compacting pressure is summarized for both set-ups in Table A2. Results are compared at final temperature in IF75 and pressure release in 4-probe setup in Figure A6. In the latter, electrical resistivity was higher for the compression route than the pressure release route due to ordering of the bulk. Charcoal and semi-coke particles larger than 2 mm showed a lower electrical resistivity in the 4-probe set-up compared to that at elevated temperature using IF75. On the other hand, electrical resistivity was higher for coal and metallurgical coke particles using the 4-probe setup at atmospheric temperature. The lower electrical resistivity was attributed to the higher bulk density and compaction for carbon material. The lower electrical resistivity of coal particles in IF75 was attributed to caking and agglomeration of particles, exemplary shown in Figure A7, whereas loose particles were used in 4-probe setup. It is known from the literature that contact resistance is higher than material resistance for dry coke beds [32]. The electrical resistivity of fines fraction remained higher at room temperature compared to experiments at elevated temperature for all investigated samples.

The results at room temperature indicate that electrical resistivity increases with increasing particle size as reported in the literature [19,32–34]. However, measurements at elevated temperature have shown that bulk density has just as great an influence on electrical resistivity as particle size for dry carbon beds. Thus, electrical resistivity measurements at room temperature can approve the tendency of materials, but quantification should be investigated at elevated temperature for bed material used in submerged arc furnaces.

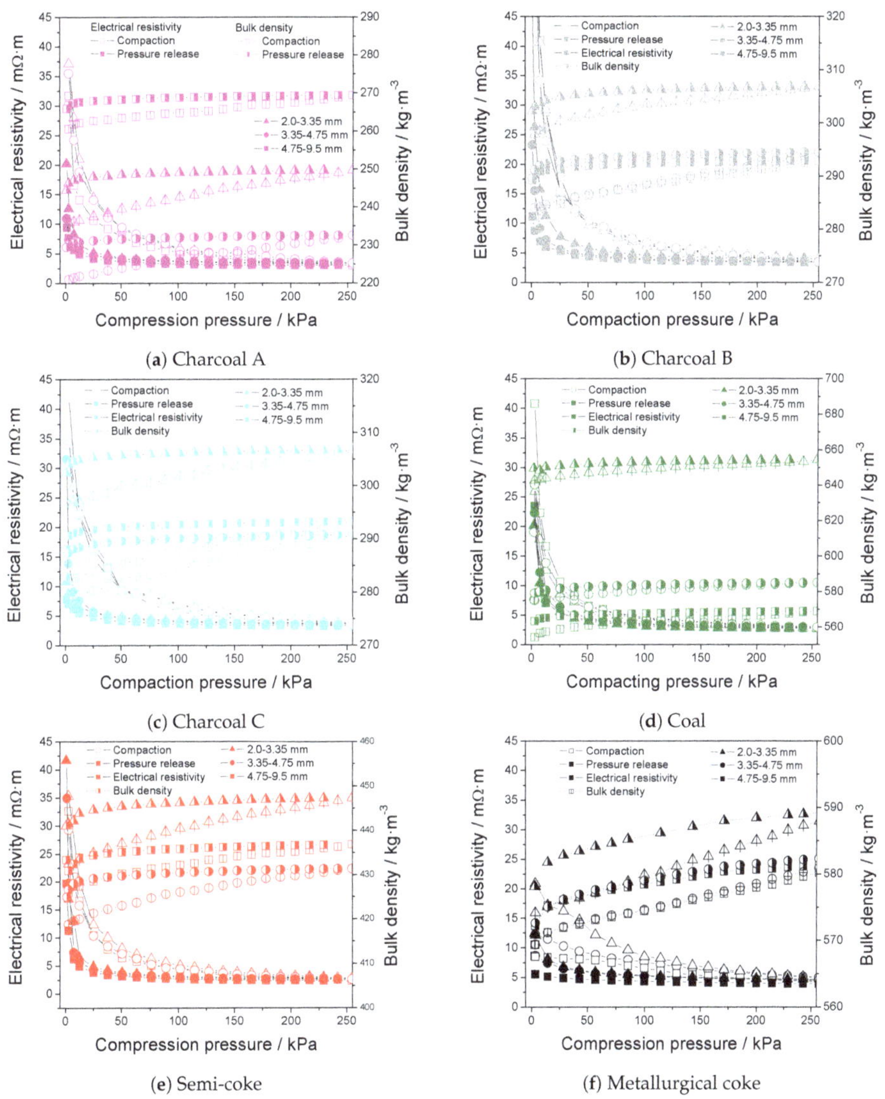

Figure A4. Electrical resistivity and bulk density versus compaction pressure for the cycle of compacting and pressure release of (**a**) charcoal A, (**b**) charcoal B, (**c**) charcoal C, (**d**) coal, (**e**) semi-coke, and (**f**) metallurgical coke.

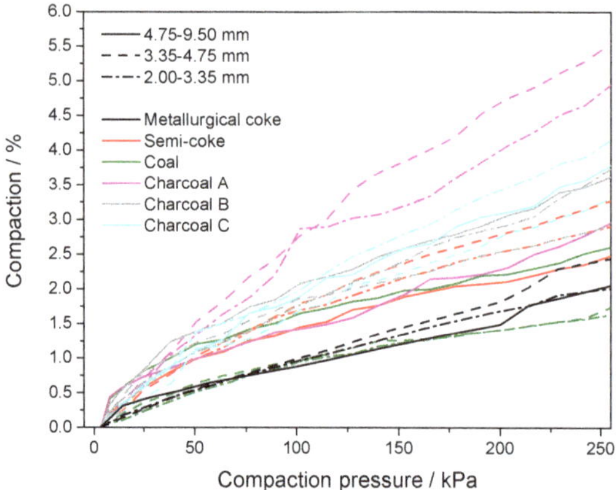

Figure A5. Compaction of carbon bulk material versus pressure for the fractions of small and large particles.

Table A2. Electrical resistivity of both setups at 6 kPa compaction pressure.

| | IF75 | | | 4-Probe Setup | | |
| | Electrical Resistivity /mΩ·m | | Bulk Density /kg·m^{-3} | Electrical Resistivity /mΩ·m | | Bulk Density /kg·m^{-3} |
	Hot *	After Cooling		Compaction	Pressure Release	
Metallurgical coke						
Fine	7.32	6.50	920	11.97	8.84	550
Small fraction	7.87	6.47	616	13.88	8.37	560
Large fraction	6.03	6.38	600	10.24	6.72	560
Semi-coke						
Fine	8.68	10.77	620	20.28	11.54	660
Small fraction	9.41	9.02	420	19.42	7.44	430
Large fraction	11.05	9.48	390	20.94	6.25	430
Coal						
Fine	3.48	5.98	590	36.03	12.43	750
Small fraction	6.01	9.01	520	19.02	12.24	580
Large fraction	5.00	8.37	480	22.33	8.84	560
Charcoal A						
Fine	7.84	12.94	240	41.72	18.70	350
Small fraction	9.73	11.66	230	24.27	6.12	270
Large fraction	11.71	12.24	230	16.61	5.66	270
Charcoal B						
Fine	7.64	19.00	385	43.08	17.10	400
Small fraction	11.23	12.13	265	34.14	9.15	290
Large fraction	11.39	13.08	255	40.99	6.97	290
Charcoal C						
Fine	8.22	16.75	360	45.76	11.62	410
Small fraction	10.97	11.80	280	30.99	9.13	290
Large fraction	12.48	10.49	260	24.83	6.13	290

* Final measurement before furnace was switched off for cooling.

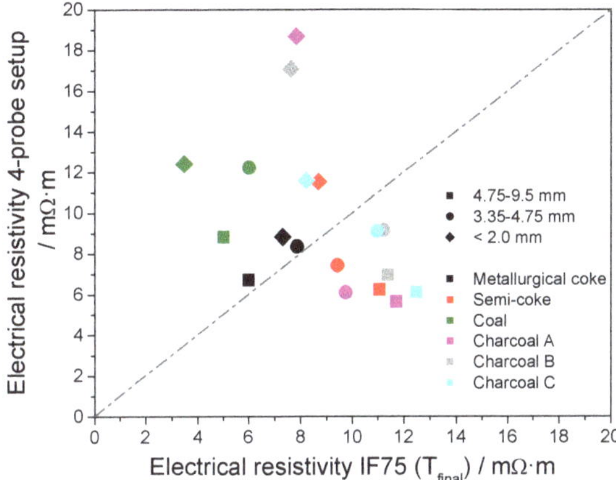

Figure A6. Electrical resistivity at 6 kPa in room temperature in 4-probe set,up versus measurements in IF75 at elevated temperature.

(a) Small particles   (b) Large particles

Figure A7. Images of coal char agglomerate from pyrolysis at 1600 °C using (**a**) small particles and (**b**) large particles.

Appendix E. Estimation of the Burden Load on the Carbon Bed

Measurements in the 4-probe set-up have shown that a load on the bed material has a large impact on electrical resistivity. In industrial furnaces, the load on carbon bed is dependent on the size and operation of the electric arc furnace and induced by the raw materials of the burden. Feedstock for manganese production is comprised of manganese ore, flux melting agents, and metallurgical coke, respectively, silica and reducing agents for silicon production. Maximum pressure is obtained for manganese ore and coke blend due to the higher density of the material. Bulk density of manganese ore is in the range of 2950 to 4600 kg·m^{-3} with a theoretical maximum of 5230 kg·m^{-3} [57] and the density of metallurgical coke is approximately 780 kg·m^{-3} [58] with an ash content of 10%. Approximately 15 wt.% of carbon is required to reduce MnO_2 to its metal form, corresponding to a volume fraction of 45 to 55%. Thus, average density of the feedstock without considering void fraction is in the range of 1950 to 2475 kg·m^{-3}. A maximum compression pressure of 150 kPa was calculated according to the following equation,

$$\text{Maximum compression} = \left(\rho_{\text{reducing agent}} \cdot w_{\text{reducing agent}} + \rho_{\text{manganese ore}} \cdot w_{\text{manganese ore}} \right) \cdot g \cdot h \quad \text{(A1)}$$

where ρ is the density, w is the mass fraction, g is the gravitational acceleration, and h is the burden height, which was assumed as 6 m. However, due to the void fraction between particles, additional flux agents, and larger amounts of reducing agents added, load is reduced to a range between 80 to

100 kPa. In addition, with increasing volume fraction required by charcoal, compaction pressure is further reduced to less than 55 kPa. In all cases, mechanical stability of charcoal and fossil reducing agents is sufficient to withstand the compaction pressure of the burden.

References

1. Mochidzuki, K.; Soutric, F.; Takokoro, K.; Antal, M.J.; Toth, M.; Zelei, B; Varhegyi, G. Electrical and Physical Properties of Carbonized Charcoals. *Ind. Eng. Chem. Res.* **2003**, *42*, 5140–5151. [CrossRef]
2. Hoffmann, V.; Rodriguez Correa, C.; Sautter, D.; Maringolo, E.; Kruse, A. Study of the electrical conductivity of biobased carbonaceous powder materials under moderate pressure. *GCB Bioenergy* **2019**, *11*, 230–248. [CrossRef]
3. Antal, M.; Grønli, M. The art, science, and technology of charcoal production. *Ind. Eng. Chem.* **2003**, *42*, 1619–1640. [CrossRef]
4. Lehmann, J.; Rillig, M.; Thies, J.; Masiello, C.A.; Hockaday, W.C.; Crowley, D. Biochar effects on soil biota—A review. *Soil Biol. Biochem.* **2011**, *43*, 1812–1836. [CrossRef]
5. Vorob'ev, V.; Golunov, A.; Ignat'ev, A. Carbon Reductants for the Production of Manganese Ferroalloys. *Russ. Metall.* **2009**, *8*, 752–755. [CrossRef]
6. Griessachenr, T.; Antrekowitsch, J.; Steinlechner, S. Charcoal from agricultural residues as alternative reducing agent in metal recycling. *Biomass Bioenergy* **2012**, *39*, 139–146. [CrossRef]
7. Adrados, A.; De Marco, I.; López-Urionabarrenchea, A.; Solar, J.; Caballero, B.; Gastelu, N. Biomass Pyrolysis Solids as Reducing Agents: Comparison with Commercial Reducing Agents. *Materials* **2016**, *9*, 1–18. [CrossRef]
8. Monsen, B.; Tangstad, M.; Midtgaard, H. Use of charcoal in silicomanganese production. In Proceedings of the International Ferro-Alloys Congress X, Cape Town, South Africa, 1–4 Febuary 2004; pp. 392–404.
9. Surup, G.; Nielsen, H.; Großarth, M.; Deike, R.; Van den Bulcke, J.; Kibleur, P.; Müller, M.; Ziegner, M.; Yazhenskikh, E.; Beloshapkin, S.; et al. Effect of operating conditions and feedstock composition on the properties of manganese oxide or quartz charcoal pellets for the use in ferroalloy industries. *Energy* **2020**, *193*, 116736. [CrossRef]
10. Surup, G.; Hunt, A.; Attard, T.; Budarin, V.L.; Forsberg, F.; Arshadi, M.; Abdelsayed, V.; Shekhawat, D.; Trubetskaya, A. The effect of wood composition and supercritical CO_2 extraction on charcoal production in ferroalloy industries. *Energy* **2020**, *193*, 116696. [CrossRef]
11. World Steel Association. *Steel's Contribution to a Low Carbon Future and Climate Resilient Societies—World Steel Position Paper*; World Steel Association: Brussels, Belgium, 2017.
12. Olsen, S.; Monsen, B.; Lindstad, T. Emissions from the Production of Manganese and Chromium Alloys in Norway. In Proceedings of the 56th Electric Furnace Conference, New Orleans, LA, USA, 15–18 November 1998; pp. 1–7.
13. Lindstad, T.; Olsen, S.; Tranell, G.; Færden, T.; Lubetsky, J. Greenhouse gas emissions from ferroalloy production. In Proceedings of the International Ferro-Alloys Congress XI, New Delhi, India, 18–21 February 2007; pp. 457–466.
14. Kero, I.; Eidem, P.; Ma, Y.; Indresand, H.; Aarhaug, T.; Grådahl, S. Airborne Emissions from Mn Ferroalloy Production. *JOM* **2019**, *71*, 349–365. [CrossRef]
15. Westfall, L.; Davourie, J.; Ali, M.; McGough, D. Cradle-to-gate life cycle assessment of global manganese alloy production. *Int. J. Life Cycle Assess.* **2016**, *21*, 1573–1579. [CrossRef]
16. Mathieson, J.; Rogers, H.; Somerville, M.; Jahanshahi, S.; Ridgeway, P. Potential for the use of biomass in the iron and steel industry. In Proceedings of the Chemeca 2011, Engineering a Better World, Sydney, Australia, 18–21 September 2011; pp. 1–12.
17. Surup, G.; Nielsen, H.; Heidelmann, M.; Trubetskaya, A. Characterization and reactivity of charcoal from high temperature pyrolysis (800–1600 °C). *Fuel* **2019**, *235*, 1544–1554. [CrossRef]
18. Lindstad, T.; Monsen, B.; Osen, K. How the ferroalloys industry can meet greenhouse gas regulations. In Proceedings of the International Ferro-Alloys Congress XII, Helsinki, Finland, 6–9 June 2010; pp. 63–70.
19. Eidem, P.; Tangstad, M.; Bakken, J.; Ishak, R. Influence of coke particle size on the electrical resistivity of coke beds. In Proceedings of the International Ferro-Alloys Congress XII, Helsinki, Finland, 6–9 June 2010; pp. 349–358.

20. Hussein, A.; Larachi, F.; Ziegler, D.; Alamdari, H. Effects of heat treatment and acid washing on properties and reactivity of charcoal. *Biomass Bioenergy* **2016**, *90*, 101–113. [CrossRef]

21. Surup, G.; Foppe, M.; Schubert, D.; Deike, R.; Heidelmann, M.; Timko, M.T.; Trubetskaya, A. The effect of feedstock origin and temperature on the structure and reactivity of char from pyrolysis at 1300–2800 °C. *Fuel* **2019**, *235*, 306–316. [CrossRef]

22. Eidem, P. Electrical Resistivity of Coke Beds. Ph.D. Thesis, NTNU, Trondheim, Norway, 2008.

23. Surup, G. Renewable Reducing Agents for the Use in Ferroalloy Industries. Ph.D. Thesis, University of Agder, Kristiansand, Norway, 2019.

24. Suopajärvi, H.; Pongrácz, E.; Fabritius, T. The potential of using biomass-based reducing agents in the blast furnace: A review of thermochemical conversion technologies and assessments related to sustainability. *Renew. Sustain. Energy Rev.* **2013**, *25*, 511–528. [CrossRef]

25. Suopajärvi, H.; Kemppainen, A.; Haapakangas, J.; Fabritius, T. Extensive review of the opportunities to use biomass-based fuels in iron and steelmaking processes. *J. Clean. Prod.* **2017**, *148*, 709–734. [CrossRef]

26. Suopajärvi, H.; Umeki, K.; Mousa, E.; Hedayati, A.; Romar, H.; Kemppainen, A.; Wang, C.; Phounglamcheik, A.; Tuomikoski, S.; Norberg, N.; et al. Use of biomass in integrated steelmaking—Status quo, future needs and comparison to other low-CO_2 steel production technologies. *Appl. Energy* **2018**, *213*, 384–407. [CrossRef]

27. Dijs, H.; Smith, D.J. Factors affecting the resistivity and reactivity of carbonaceous reducing agents for the electric-smelting industry. *J. S. Afr. Inst. Min. Metall.* **1980**, *80*, 286–296.

28. Adinaveen, T.; Vijaya, J.J.; Kennedy, L. Comparative Study of Electrical Conductivity on Activated Carbons Prepared from Various Cellulose Materials. *Arab. J. Sci. Eng.* **2014**, *41*, 55–65. [CrossRef]

29. Espinola, A.; Miguel, P.; Salles, M.; Pinto, A. Electrical properties of carbons - resistance of powder materials. *Carbon* **1986**, *24*, 337–341. [CrossRef]

30. Sánchez-González, J.; Macías-García, A.; Alexandre-Franco, M.; Gómez-Serrano, V. Electrical conductivity of carbon blacks under compression. *Carbon* **2005**, *43*, 741–747. [CrossRef]

31. Tsybul'kin, G. On the Calculation of the Electrical Resistivity of Conducting Loose Materials. *Elektrometallurgiya* **2010**, *2011*, 594–598. [CrossRef]

32. Eidem, P.; Tangstad, M.; Bakken, J. Measurement of material resistivity and contact resistance of metallurgical coke. In Proceedings of the International Ferro-Alloys Congress XI, New Delhi, India, 18–21 February 2007; pp. 561–571.

33. Buryak, V.; Vasil'chenko, G.; Chirka, T.; Konstantinov, S. Specific electrical resistance of carbon materials. *Refract. Ind. Ceram.* **2013**, *54*, 215–219. [CrossRef]

34. Tangstad, M.; Beukes, J.; Steenkamp, J.; Ringdalen, E. 14—Coal-based reducing agents in ferroalloys and silicon production. In *New Trends in Coal Conversion Combustion, Gasification, Emissions, and Coking*; Woodhead Publishing: Cambridge, UK, 2019; pp. 405–438.

35. Monsen, B.; Tangstad, M.; Solheim, I.; Syvertsen, I.; Ishak, R.; Midtgaard, H. Charcoal for manganese alloy production. In Proceedings of the International Ferro-Alloys Congress XI, New Delhi, India, 18–21 February 2007; pp. 297–310.

36. Surup, G.; Vehus, T.; Eidem, P.; Trubetskaya, A.; Nielsen, H. Characterization of renewable reductants and charcoal-based pellets for the use in ferroalloy industries. *Energy* **2019**, *167*, 337–345. [CrossRef]

37. Antal, M.; Mok, W.; Varhegyi, G.; Szekely, T. Review of methods for improving the yield of charcoal from biomass. *Energy Fuels* **1990**, *4*, 221–225. [CrossRef]

38. Olsson, J.; Jaglid, U.; Pettersson, J.; Hald, P. Alkali metal emission during pyrolysis of biomass. *Energy Fuels* **1997**, *11*, 779–784. [CrossRef]

39. Riaza, J.; Mason, P.; Jones, J.; Gibbins, J.; Chalmers, H. High temperature volatile yield and nitrogen partitioning during pyrolysis of coal and biomass fuels. *Fuel* **2019**, *248*, 215–220. [CrossRef]

40. Efika, E.; Onwudili, J.; Williams, P. Products from the high temperature pyrolysis of RDF at slow and rapid heating rates. *J. Anal. Appl. Pyrolysis* **2015**, *112*, 14–22. [CrossRef]

41. Wakchaure, G.; Mani, I. Thermal and Storage Characteristics of Biomass Briquettes with Organic Binders. *J. Agric. Eng. Res.* **2011**, *48*, 43–53.

42. Wang, L.; Barta-Rajnai, E.; Hu, K.; Higashi, C.; Skreiberg, Ø.; Grønli, M.; Czégény, Z.; Jakab, E.; Myrvågnes, V.; Várhegyi, G.; et al. Biomass Charcoal Properties Changes during Storage. *Energy Procedia* **2017**, *105*, 830–835. [CrossRef]

43. Koursaris, A.; See, J. The resistivity of mixtures of Mamatwan manganese ore and reducing agents. *J. S. Afr. Inst. Min. Metall.* **1980**, *80*, 229–238.

44. Mrozowski, S. Semiconductivity and Diamagnetism of Polycrystalline Graphite and Condensed Ring Systems. *Phys. Rev.* **1952**, *85*, 609–620. [CrossRef]

45. Golden, T.; Jenkins, R.; Otake, Y.; Scaroni, A. Oxygen Complexes on Carbon Surfaces. In Proceedings of the Workshop on The Electrochemistry of Carbon, Cleveland, OH, USA, 17–19 August 1983; Volume 84, pp. 61–78.

46. Pantea, D.; Darmstadt, H.; Kaliaguine, S.; Sümmchen, L.; Roy, C. Electrical conductivity of thermal carbon blacks Influence of surface chemistry. *Carbon* **2001**, *39*, 1147–1158. [CrossRef]

47. JoséSánchez-González, J.; Stoeckli, F.; Centeno, T. The role of the electric conductivity of carbons in the electrochemical capacitor performance. *J. Electroanal. Chem.* **2011**, *657*, 176–180. [CrossRef]

48. Razd'yakonova, G. The electrical conductivity of carbon-black filled vulcanisates. *Int. Polym. Sci. Tech.* **2013**, *40*, 11–15. [CrossRef]

49. Panshin, A.; De Zeeuw, C. *Textbook of Wood Technology: Structure, Identification, Properties, and uses of the Commercial Woods of the United States and Canada*; Number Bd. 1 in McGraw-Hill Series in Forest Resources; McGraw-Hill: New York, NY, USA, 1980.

50. Trubetskaya, A.; Jensen, P.; Jensen, A.; Steibel, M.; Spliethoff, H.; Glarborg, P. Influence of fast pyrolysis conditions on yield and structural transformation of biomass chars. *Fuel Process. Technol.* **2015**, *140*, 205–214. [CrossRef]

51. Weber, K.; Quicker, P. Properties of biochar. *Fuel* **2018**, *217*, 240–261. [CrossRef]

52. Díez, M.; Alvarez, R.; Barriocanal, C. Coal for metallurgical coke production: Predictions of coke quality and future requirements for cokemaking. *Int. J. Coal. Geol.* **2002**, *50*, 389–412. [CrossRef]

53. Babu, S.; Santella, M.L.; Feng, Z.; Riemer, B.; Cohron, J. Empirical model of effects of pressure and temperature on electrical contact resistance of metals. *Sci. Technol. Weld. Join.* **2001**, *6*, 126–132. [CrossRef]

54. Rani, A.; Nam, S.; Oh, K.; Park, M. Electrical Conductivity of Chemically Reduced Graphene Powders under Compression. *Carbon Lett.* **2010**, *11*, 90–95. [CrossRef]

55. Li, X.; Mao, H. Solid Carbon at High Pressure: Electrical Resistivity and Phase Transition. *Phys. Chem. Miner.* **1994**, *21*, 1–5.

56. Xiao, G.; Xiao, R.; Jin, B.; Zuo, W.; Liu, J.; Grace, J. Study on Electrical Resistivity of Rice Straw Charcoal. *J. Biobased Mater. Bioenergy* **2010**, *4*, 426–429. [CrossRef]

57. Emerson, D.; Schmidt, P. Pyrolusitic supergene manganese oxides: inductive properties, EM conductivity and magnetic susceptibility. *Preview* **2018**, *197*, 42–50.

58. Montiano, M.; Díaz-Faes, E.; Barriocanal, E.; Alvarez, R. Influence of biomass on metallurgical coke quality. *Fuel* **2014**, *116*, 175–182. [CrossRef]

Review

Life Cycle Assessment of Renewable Reductants in the Ferromanganese Alloy Production: A Review

Gerrit Ralf Surup [1,*], Anna Trubetskaya [2] and Merete Tangstad [1]

[1] Department of Materials Science and Engineering, Norwegian University of Science and Technology, 7491 Trondheim, Norway; merete.tangstad@ntnu.no

[2] Department of Chemical Sciences, University of Limerick, V94 T9PX Limerick, Ireland; anna.trubetskaya@ul.ie

* Correspondence: gerrit.r.surup@ntnu.no; Tel.: +47-934-73-184

Abstract: This study examined the literature on life cycle assessment on the ferromanganese alloy production route. The environmental impacts of raw material acquisition through the production of carbon reductants to the production of ferromanganese alloys were examined and compared. The transition from the current fossil fuel-based production to a more sustainable production route was reviewed. Besides the environmental impact, policy and socioeconomic impacts were considered due to evaluation course of differences in the production routes. Charcoal has the potential to substantially replace fossil fuel reductants in the upcoming decades. The environmental impact from current ferromanganese alloy production can be reduced by $\geq$20% by the charcoal produced in slow pyrolysis kilns, which can be further reduced by $\geq$50% for a sustainable production in high-efficient retorts. Certificated biomass can ensure a sustainable growth to avoid deforestation and acidification of the environment. Although greenhouse gas emissions from transport are low for the ferromanganese alloy production, they may increase due to the low bulk density of charcoal and the decentralized production of biomass. However, centralized charcoal retorts can provide additional by-products or biofuel and ensure better product quality for the industrial application. Further upgrading of charcoal can finally result in a CO_2 neutral ferromanganese alloy production for the renewable power supply.

Keywords: charcoal; life cycle assessment; sustainable biomass growth; mining; metallurgical coke

Citation: Surup, G.R.; Trubetskaya, A.; Tangstad, M. Life Cycle Assessment of Renewable Reductants in the Ferromanganese Alloy Production: A Review. *Processes* **2021**, *9*, 185. https://doi.org/10.3390/pr9010185

Received: 6 December 2020
Accepted: 13 January 2021
Published: 19 January 2021

Publisher's Note: MDPI stays neutral with regard to jurisdictional claims in published maps and institutional affiliations.

1. Introduction

Climate change caused by anthropogenic CO_2 emissions is considered one of the most prominent issues of the present time. About 5–10% of anthropogenic CO_2 emissions are emitted by metallurgical industries [1,2]. Most of these emissions are direct emissions generated in the smelting furnace, such as blast, electric arc, and submerged arc furnaces. The indirect emissions originate from coal dust, ore mining, metallurgical coke, and power production. Ferroalloys, such as ferromanganese (FeMn), silicomanganese (SiMn), or ferrochromium, are mainly produced in submerged arc furnaces [3], where between 40 and 70% of the required thermal energy is provided by electrical dissipation, emphasizing the importance of renewable power production [4,5]. The annual CO_2 emissions for the production of ferroalloys accumulate to $\approx$87 Mt (million tonnes) [6].

The differences in greenhouse gas (GHG) emissions generated by ferroalloy manufacturers vary in dependency on the source of energy and reductant material (biomass, coal, or metallurgical coke) [7,8]. The renewable hydropower makes Norway one of the most environmental friendly ferroalloy producer worldwide. However, the ferromanganese alloy production still relies on the use of fossil-based fuel reductants, e.g., metallurgical coke. The transition from fossil fuel-based to renewable reductants is hampered by the differences in properties, such as the low mechanical strength of renewable reductants and its high gas reactivity. In addition, the physicochemical properties of classical charcoal

175

can vary from batch to batch because of the undefined process conditions in charcoal kilns. However, the renewable charcoal-based reductant together with the carbon capture and storage (CCS) or carbon capture and utilization (CCU) are intended to eliminate the anthropogenic emissions in Norway by 2050 [9,10].

Life cycle assessment (LCA) is an analytical tool that supports users with the reduction of CO_2 emissions and sustainability challenges using ISO standards 14,001 and 14,040. Traditional cost and process optimization can be combined by linking environmental calculations for each process step within life cycle analysis. The important aspect of LCA is the system boundary, including processes and production routes which are considered and where the system begins and ends [11]. While a complete LCA evaluates the overall life time of a product (cradle-to-grave), cradle-to-gate approaches are commonly used to describe production of raw materials [8,11–13]. The cradle-to-gate approach includes the transition phases such as feedstock growth, pretreatment, transport, storage, and postprocessing by metallurgical smelters. However, not all aspects of sustainability can be considered within LCA [11]. Technical issues caused by the different properties and the economics of renewable reductants can play a significant role in replacing fossil fuel-based reductants [14,15]. In addition, renewable reductants may require more energy for handling and transport. For example, the milling of biomass requires more energy than that of coal samples [16], which will affect the overall energy demand in the pretreatment process. A systematic effort is needed to overcome technical and economical hurdles to realize a shift from fossil fuel-based to renewable reductants [17]. The low mechanical strength and high gas reactivity of charcoal hamper the direct replacement of fossil-based reductants in metallurgical industry [18,19]. However, charcoal is used as a major carbon source in mini blast furnaces in Brazil [17]. The low burden height of small size blast furnaces can result in a compaction pressure that makes the mechanical properties of reductants of less importance [14].

To the knowledge of authors, this review is the first attempt to examine the literature on life cycle assessment of ferromanganese alloy production route. This work aims to evaluate possible GHG emission sources and savings for the production routes of fossil fuel-based and renewable reductants used in ferromanganese alloy production. Post-treatment processes to adjust properties of charcoal to the fossil fuel-based counterparts were neglected, as the industrial process route for the replacement of metallurgical coke with renewable charcoal reductants is unknown. Three cases were reviewed in the present study: (1) the base case with the metallurgical coke as the reductant, (2) charcoal produced using classical charcoal kilns, and (3) charcoal formed by a sustainable production.

2. Ferromanganese Alloy Production

The total ferroalloy production has increased from 18 Mt in the 1990s to 36 Mt in 2008 and to 40 Mt in 2020 [7,20,21]. About one-third of the total ferroalloy production are ferromanganese alloys in a form of high-carbon ferromanganese (HC FeMn), silicomanganese, and refined ferromanganese alloys [7,8]. Ferroalloys are often used in high-quality steel production for the improvement of product strength and hardness [8,22]. The high-quality steel contains on average 3% of silicon, manganese, aluminum, or chrome as alloying elements [6]. Ferromanganese is mainly produced from manganese-ore by carbothermal reduction in submerged arc furnaces (SAF), in which metallurgical coke is used as reductant. About 59 Mt of manganese ore is estimated to be produced in 2020, with more than 50% reduced and refined in China [23].

The fossil fuel-based production route of ferromanganese alloy is shown in Figure 1. When this study was initiated, metallurgical coke has been mainly used as a reductant in closed hearth SAF to produce HC FeMn and SiMn. Metallurgical coke is produced from blends of coking coals in coke oven batteries at the temperature range 1100–1400 °C [24]. Charcoal is mainly produced from wood in kilns and medium sized retorts for the ferroalloy purposes [25,26]. The low bulk density of charcoal increases the transport and storage demand compared to fossil fuels, resulting in more complex logistics, higher emissions, and overall process cost. The efficiency for reducing manganese-ore to pure manganese is assumed to depend on the used furnace and is similar for fossil fuel-based and renewable reductants. Thus, similar emissions from mining during manganese-ore production occur using the different process routes.

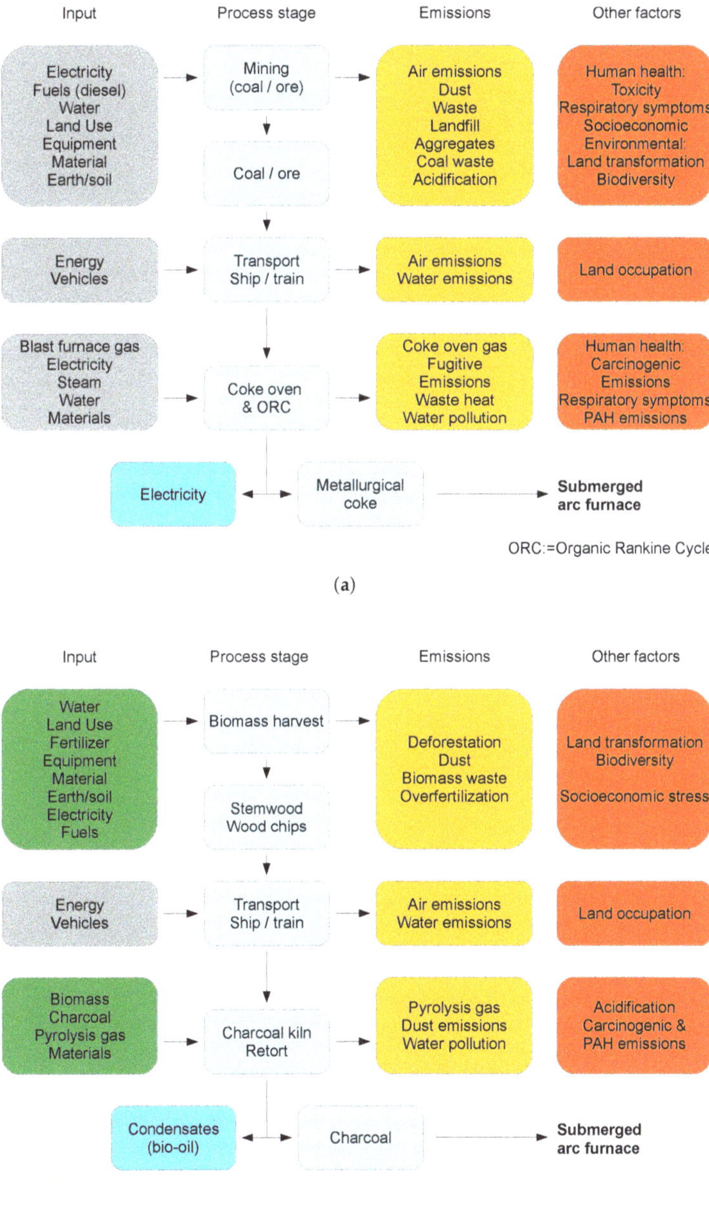

Figure 1. Production route of ferromanganese alloys using (**a**) metallurgical coke or (**b**) charcoal as reductant. Each process step can be accounted as a possible source for greenhouse gas (GHG) emissions. (**a**) Fossil fuel-based reductant. (**b**) Renewable reductant.

The system boundary is mostly set as a cradle-to-gate approach on the basis of 1 t final ferromanganese alloy. The reviewed articles concerning LCA for the ferromanganese alloy production route are summarized in Table 1. A bottom-up, layered approach should be chosen to evaluate sustainability for a charcoal use in a ferroalloy reduction process [17]. The fixed carbon yield of charcoal can be used to calculate the useable carbon content for metallurgical applications [27]. The physicochemical properties of renewable reductants are inferior to those of metallurgical coke and must be adjusted to the application in SAF. Previous studies have shown that charcoal can be post-treated to achieve properties which approach those of metallurgical coke [28,29]. However, classical charcoal and sustainably produced charcoal are considered as possible reductants for ferromanganese production, in which post-treatment processes, such as acid leaching or a secondary heat treatment, can close the gap between renewable to fossil fuel-based reductants [30,31]. In addition, renewable reductants can improve metal quality and productivity by their low ash content and ash composition [32,33].

Power generation and manufacturing of reductants are the main parameters which affect the overall GHG emissions in ferromanganese alloy production [8]. As power supply is considered CO_2 neutral in Norway, fossil fuel-based reductants are the main sources of GHG emissions. The overall emissions can be reduced by the recovery of CO off-gas and the on-site utilization [8], which is already performed by smelters in Norway. Between 50 and 85% of the total GHG emissions in ferroalloy production are generated by coal and coke [7]; 25–35% of the GHG emissions can be ascribed to the on-site air emissions [8]. GHG emissions for the production of metallurgical coke, mining, and transportation and possible reduction by renewable reductants are estimated from the steel industry [7,34]. About two-thirds of particulate matter emissions are associated with mining, handling, and preparation of the raw feedstocks, whereas only one-third are direct emissions from the SAF [35]. Direct emissions from tapping can be reduced by simple measures, such as the addition of curtains and ventilation of the tapping area [36].

Previous studies have shown that the emission factor of ferromanganese alloy production was between 1.04 and 6.0 kg CO_2 per kg FeMn, respectively, 2.8 kg CO_2 per kg SiMn and 3.4 kg CO_2 per kg FeSi [7,8,37]. The acidification potential for ferromanganese alloy production was stated to be 45 g SO_2-eq. per kg FeMn, while the photochemical ozone creation potential was determined to be 3 g C_2H_4-eq. per kg FeMn [8]. However, no complete LCA study has been found for ferromanganese alloy production.

Table 1. LCA studies reviewed to evaluate GHGe for ferromanganese production.

Process Stage	Region	Economics	Human Health	Environment	Other Factors	Source
BM, coal mining	US				land occupation and regeneration time	[38]
BM	North America, Europe		toxicity	GHG	land occupation, acidification	[39]
BM		considered	considered	considered	policy and societal impacts	[17]
BM,CC	South America			gas, water		[40]
BM,CC	IN, East Africa	considered		gas		[41]
BM (harvest,pyro)	US		TRACI	TRACI	activation of charcoal	[42]
Mining	CN			air, water	land occupation	[43]
Coal mining	PL			IPCC	human health, ecosystems and resources	[44]
Coke (Ferro)			considered	considered		[45]
BM,coke	FR	feedstock, CO_2 tax		GHG	transport (regional < 100 km)	
Coke production	CN		considered	considered	air and water emissions	[46]
Coke production	UK(steel),AU(coal)	considered	considered	GHG,gas,water		[47]
Coke production						[48]
Coke production	TR		considered	considered	by-product utilization	[49]
Mn-alloy production	AU, CN, FR, IN, ZA, US			GHG, SO_2e, C_2H_4e	LCIA, energy demand	[8]
Mn-alloy production	AU			GHG		[7]

LCIA:= Life cycle impact assessment

2.1. Mining

Mining of both manganese ore and coal contributes to the GHG emissions by ferromanganese alloy production. The emissions depend mostly on the country of origin and if surface mining or underground mining is applied. Coal mining can lead to air pollution (especially dust formation), surface and ground water pollution, solid waste land occupation, and destruction of local ecological environment [38,43,50]. This includes biodiversity, which has attracted more focus in the recent years [51]. Underground mining requires significantly more electrical energy, and thus GHG emissions from power production can become predominating. Methane emissions from the shaft system and auxiliary vents are additional air pollutants and GHG emissions for underground mining of coal [44,47]. Acidification is the main problem for water pollution at long-established mining operations. An advanced water management can improve the discharged water quality by prevention and innovative water treatment technologies. Other direct impacts of land use include landscape transformation, vegetation removal, and soil destruction [38].

Most of the energy required for surface mines is provided by diesel-powered mobile equipment [8,47], resulting in GHG emissions from combustion of the required fuel. Underground mines can also require more electricity, but produce less waste rocks than surface mines [8]. Overall, the environmental impact of power supply for surface mining is small compared to the particulate matter (PM) or dust formation, that has the largest impact on the environment ($\approx$37%), followed by global warming ($\approx$29%) or acidification ($\approx$23%) [8,43]. Other studies have shown that the largest impact on the GHG emissions from underground mining is related to the energy demand, methane emissions, and processing of wastes [44]. The effect of methane emissions for a time frame of 20 and 100 a (85 and 121 kg CO_2-eq. per kg) represents about two-third of all factors, while electricity production is the main GHG factor (representing 50%) for the timeframe of 500 years [44].

For a cleaner production, the main aim is to improve the environmental performance of mining, whereas sustainable development requires minimization of environmental costs [51,52]. In addition, the negative health effects due to mining, handling, and processing should be taken into consideration [44]. The treatment of mining water, especially acidic water or water with high contamination (heavy metals, total suspended solids, total dissolved solids and oils), will effectively decrease the environmental impact from coal and manganese ore mining [53]. The CO_2 emission factors for underground coal mining per tonne of coal were $\approx$15 kg CO_2 for electricity production, $\approx$11 kg CO_2 for coal processing, $\approx$9 kg CO_2 for haulage and hoisting coal, $\approx$7 kg CO_2 for ventilation work, and $\approx$5 kg CO_2 for exploitation [44]. However, secondary effects on the biodiversity and land recultivation after post-mining have not been considered in previous LCA studies [38,54].

2.2. Reductants

Direct GHG emissions in SAF are mainly produced from fossil fuel-based reductants, e.g., metallurgical coke. This currently used fossil fuel-based reductants can be replaced by renewable reductants, such as bio-coke and charcoal. However, the inferior mechanical and chemical properties of bio-coke and charcoal hamper the direct replacement of metallurgical coke [55]. Most likely, metallurgical coke will be partly replaced by bio-coke in the short-term [17] and by tailor-made charcoal in the long-term [15]. Other technologies, such as electrolytic manganese metal production, COREX® or FINEX® in combination with renewable reductants may also provide a CO_2 neutral production route as an alternative to the SAF [56,57].

Charcoal as an alternative reductant was mainly investigated for steel making, including pyrolysis by-product utilization [17,58]. However, renewable reductants have been also studied in ferroalloy, silicomanganese, and manganese alloy context [59–61]. The main obstacles for using charcoal in metallurgical industry are its high costs, inferior properties of charcoal, and less investigated efficiency of the tailor-made conversion process. Cost of the feedstock followed by the consecutive economics of renewable reductants and by-products are the key parameters to evaluate the performance of renewable reductants in ferromanganese alloy production [17]. On LCA basis, production and conversion routes of fossil fuel-based reductants must be compared to renewable ones. Socioeconomic factors are especially important for the biomass and charcoal production in developing countries, in which biomass and charcoal are the main fuels for heating and cooking [62–65]. While a larger charcoal demand positively affects the economy of the producers [65], low- and middle-income countries rely on affordable and sustainable cooking fuel supply [64].

Less than 30% of the GHG emissions are related to coal mining, handling, and preparation, whereas $\approx$60% of GHG emissions are emitted by the metallurgical coke production [66]. The remaining emissions are generated by the combustion of diesel fuel (transport) and electricity production. A similar ratio is given for charcoal production, where $\approx$61% of GHG emissions are emitted during charcoal production, and the remaining 39% occurred during its usage [40]. Unsustainable biomass growth can decrease savings related to CO_2 emissions.

2.3. Fossil Fuel Reductants

The main GHG emissions from fossil fuel-based reductants beside its usage occur from mining, transport, and processing of the coal. The energy demand for surface mining, washing, and transport of coal is stated to be small [66], whereas emissions such as dust formation or acidification can have a significant impact from LCA perspective. Coal conditioning, such as coking or metallurgical coke formation, can increase the GHG emissions by the required energy demand and volatile matter release. The coke production by dry quenching can result in up to 15% less air emissions compared to the traditional wet quenching [46].

2.3.1. Coal

Similar to ore mining, coal can be obtained by surface and underground mining. For underground mining, the average power consumption per washed tonne of coal is stated to $\approx$25 kW h [66], resulting in $\approx$64 kg CO_2 emissions [67]. Power consumption for coal transportation is 1–2 kW h/(t km) [68] and is often neglected due to the low impact [66]. However, the on-site preparation of coal at the coke production can highly influence the environmental performance of the coke production process [46]. As a result, coal mining and coke production have the highest impact on the depletion of fossil fuels, whereas the energy demand for surface mining and coke formation is low [49]. Domestic long-range transport is carried out by railway, where GHG emissions are based on the power supply of the train grid system. For example, the average CO_2 emission factor for EU railway transport in 2009 was 370 g CO_2 per kW h [69], whereas average CO_2 emissions factor in China was 627 g CO_2 per kW h [67].

2.3.2. Metallurgical Coke

Metallurgical coke is produced from blends of several coal types in coke oven batteries. About two-thirds of the total production are formed in China by classical technologies from the 1990s [46,70]. Previous studies have shown that the production of metallurgical coke is one of the main GHG emission factors for pig iron production [45]. The upstream coal mining in combination with the coke production are the main sources of GHG emissions, in which CO_2 and CH_4 significantly contribute with fractions up to $\approx$61% and $\approx$32%, respectively [46].

About 2% of the energy demand of coke production is covered by electricity (16–43 kW h/t) [66,71]. Most of the energy demand for coke production (>90%) is covered by natural gas, blast furnace gas, or coke oven gas [47]. The overall emission factor for coke production was estimated to $\approx$0.8 kg CO_2 per kg coke [47]. Energy demand for gasification coke is about 25% larger than for metallurgical coke production due to the lower yield of reductant [66].

The emission factor is reduced to 0.5 kg CO_2 per kg coke when the metallurgical coke yield is increased by 10% or by the usage of natural gas if no coke oven gas is used [47]. One possibility to increase the solid yield per tonne of coal is the production of bio-coke, where charcoal is added to the coal blend. Previous studies have shown that 2–15% of charcoal can be added to the coal mixture without any negatively impact on the coke properties [72–74]. A 20% replacement of fossil fuel-based coke with the renewable charcoal can result in a reduction of 15% GHG emissions [75]. As an alternative feedstock, low rank coals may be utilized in the coking process to improve economics and energy efficiency of the overall reduction process [66].

Using the coke oven gas for heating and power generation can improve the environmental impact of coke production. For example, the organic Rankine Cycle (ORC) technology for coke oven and blast furnace gas can reduce GHG emissions by $\approx$6% [47]. Coal charging and pushing can release additional emissions to the air and freshwater [46]. The increased coke production results in an increase of PAH and benzo[a]pyrene emissions [76], which have an adverse impact on local ecosystems and human health [77,78]. The post-treatment of airborne emissions and wastewater in combination with strict control policies can significantly reduce these emissions [46].

Current coke oven batteries have an emission factor of $\approx$0.4–1.27 kg CO_2-eq. per kg of coke, resulting in $\approx$2.3 kg CO_2-eq. per kg liquid steel [7,45,66]. Other fossil fuel-based reductants, such as gasification coke showed about 17% greater emission factor [66]. Overall, the release of organics and fine dust affecting human health, as well as the consumption of resources and the environmental impact by emissions have the largest impact from metallurgical coke production [49]. To minimize fossil fuel depletion and anthropogenic CO_2 emissions, alternative reducing agents can replace coal and metallurgical coke partially nowadays and fully in the future [1]. The replacement of metallurgical coke by anthracite or charcoal can decrease GHG emissions by $\approx$3.1 and 3.2 kg CO_2 per t of metal, as reported for the steel manufacturing [75].

2.4. Biomass Growth

Sustainable biomass growth is essential to reduce GHG emissions from metallurgy. Deforestation, soil degradation, as well as air and water pollution are the main challenges to avoid unsustainable charcoal production [79]. To avoid deforestation in developing countries, only wood from secondary forestry or biomass waste should be used as feedstock material in charcoal production. The charcoal production in Brazil showed the transition from primary to secondary forestry in the last decades [80,81], whereas African countries rely on the production of charcoal from primary forestry due to the high demand and low price [82,83]. Additional measures are carried out in Brazil to improve biomass growth in a frame of circular economy [84]. Biomass certification, such as the Programme for the Endorsement of Forest Certification (PEFC), Forest Stewardship Council (FSC), or European Biochar Certificate (EBC), can support the production of charcoal as a renewable reductant from sustainable source.

Between 2 and 7% of the anthropogenic CO_2 emissions are attributed to the production and usage of fuelwood and charcoal [85]. Most of the GHG emissions are caused by deforestation and combustion-related pollutants [85]. Poor plantation management, fuels for harvesting and transport can result in an emission factor of 105–120 kg CO_2 per t of charcoal [14]. Thus, the combination of GHG emissions from charcoal production by kilns in combination with deforestation can result in a net increase of global warming potential [86]. When a sustainable biomass production is ensured, the main sources of GHG emissions occur from thermochemical conversion and raw feedstock processing [11].

Non-sustainable land use and soil degradation are expected to occur mainly in regions with high poverty and uncontrolled state management, resulting in deforestation and degradation of the environment [87]. Such unsustainable biomass growth can result in an emission factor of $\approx$40 kg CO_2 per t of dry biomass, which can increase to $\approx$80 kg CO_2 per t if additional biomass treatment (e.g., chipping) is required [14]. A land usage with maximizing the biomass growth may negatively affect biodiversity and soil properties [39]. Thus, a sustainable biomass production has a larger land requirement to maintain the biodiversity in the region [38]. The land requirements can be divided into the size of transformed land to produce the biomass and the land occupation for the time the land is used to produce the biomass. Growth rates of biomass mainly depend on biomass species, climate, irradiation, and soil and are between 10 and 20 t/(ha a) [14,38,88].

Charcoal produced from both sustainable biomass and waste by-products is beneficial to reduce GHG emissions from metallurgy and fossil fuel depletion [1]. Biodiversity is the basis of ecosystem health [89] and has attracted more attention in recent years. The LCA for wood from secondary forestry should therefore comprise the environmental, social, and economic impacts of the whole value chain [90]. Forest residues and waste streams from wood industry can support a sustainable biomass production without transforming natural forests to secondary forestry [79]. The GHG emissions from indirect land use can increase GHG emissions by factor of $\approx$13 [17].

2.5. Biomass Pretreatment

Wood is mainly used as feedstock in the classical charcoal production and will be the most reliable feedstock for metallurgy. Stemwood is the best feedstock material based on the low ash content. Previous studies have investigated the forest management, harvest, transport, and processing of biomass, as well as the biomass pyrolysis [42]. The emissions from harvest and transport of biomass are low for short distance transportation and can be neglected if these emissions are biogenic [75].

Most of the carbon losses occur by the thermochemical conversion process, where about 50% of the carbon is lost as CO_2 and volatile gases [75]. Biomass sizing is necessary to provide a particle size which is required for the pyrolysis process. The processing of biomass and charcoal can require less energy than coal, resulting in lower GHG emissions for power production. For example, milling of dry biomass requires up to 50% less energy than wet biomass [91]. On the other hand, bio-oil production can result in GHG emission factor of up to 84 kg CO_2-eq. [75].

2.6. Classical Charcoal Production

Charcoal in Africa and Asia is often produced in earthmound kilns and pits at a low efficiency (batch processes), whereas charcoal in industrialized countries is produced in continuous retorts with by-product utilization [13–15,79]. Classical charcoal production results in the emission of incomplete combusted volatile matter, such as particulate matter, volatile organic carbon, organic acids, or polycyclic aromatic hydrocarbons (PAHs) [86,92], which are considered hazardous to health and environment [93,94]. One kg of smoldering wood can pollute $\approx$700 m^3 of air [41]. Thus, the LCA of charcoal production should include different environmental impact categories, such as GHG emissions, acidification, eutrophication, as well as by-products and waste management [13,43].

At least 80% of the unburnt volatiles can be fully combusted by using an after-burner [92], resulting in a 33–40% reduction in the environmental impact of charcoal production [13]. Although CO_2 emissions from biomass and its derivatives are accounted as CO_2 neutral, toxicity, and land acidification of volatile matter can result in net GHG emissions. Based on the emissions, the selection and used technique of the carbonization process is as important as the sustainable biomass production [11]. Previous studies have shown that GHG savings from charcoal fines in metal production can be similar to lump charcoal ($\approx$1780 kg CO_2-eq. per t hot metal) [75].

Classical charcoal production can result in an emission factor of up to 9 kg CO_2 per kg of charcoal [83]. The high emission factor is based on the low charcoal yield ($w < 20\%$) and the incomplete combustion of volatiles. The release of volatile matter (e.g., tars and organic acids) can increase overall GHG emissions, making the replacement of fossil fuel-based reductants with the renewable materials redundant. For example, the emission factor for charcoal as an energy carrier can be 10 times larger than that of the direct combustion of wood due to the excess of methane and CO_2 release in the pyrolysis [41]. In worst case, production of charcoal by earthmound kilns can result in net GHG emissions [11].

The gross GHG emission factor of charcoal production was 1.6–4.7 kg CO_2-eq. per kg charcoal [13,14,86] with a 100-year global warming potential of up to 5.685 kg CO_2-eq. [95]. The lowest reported emission factor for biochar production was 0.22 kg CO_2 per kg of biochar [7]. These emissions can be accounted as CO_2 neutral if the volatiles are completely combusted in the process. However, the biomass feedstock, transport distance and plant size must be considered for the LCA and pyrolysis location [17,96]. The main factors that have to be considered in LCA for classical charcoal production are climate change, photochemical oxidant formation, and human toxicity due to the release of unburnt hydrocarbons [13]. The impact on climate change can be between 2700–4700 kg CO_2 per t of charcoal for classical charcoal production and can be reduced by 33–40% for the advanced charcoal production [13].

2.7. Sustainable Charcoal Production

Sustainable charcoal production is assumed as the currently best available case scenario. The charcoal production in industrial retorts results in a twice to three times larger solid yield compared to classical charcoal production [25,97,98], while the volatile hydrocarbons are recirculated and utilized in the process [13]. Both benefits improve the economics of the process, decrease the GHG emissions, and reduce the risk of local pollution [13]. However, these processes have high capital expenditures (capex), and thus are not used in Africa or Asia. In France, $\approx$50 kt of charcoal are produced in industrial retorts for households, barbeque or catering, and thus cannot cover the demand from metallurgical industry for high quality charcoal [75]. However, modern technologies charged with licensed biomass from secondary forestry shall avoid deforestation and ensure a defined income for the biomass producers in developing countries [85].

Industrialized charcoal production requires higher capex than classical charcoal production, and thus is economically feasible only for large scale applications [13]. Decentralized biomass growth can be combined with a centralized charcoal production to exploit by-products and use other synergistic effects [99]. The amount of by-products depends on the feedstock quality and process conditions, such as heating rate, gas residence time, and final temperature [14]. The liquid by-products from pyrolysis can be condensed and post-treated to biofuels or chemical feedstocks [100]. New technologies, such as multi-stage pyrolysis units, can decrease the energy demand for the process and enable the utilization of the by-products [101]. These by-products can replace other fossil fuel-based feedstocks and further reduce anthropogenic CO_2 emissions.

In the long-term, sustainable biomass growth and high efficient charcoal production will have a much lower area demand, stable economics for producer and consumer, and greater environmental savings than classical charcoal production. The conditioning of liquid by-products can open new markets for biofuels, chemicals, or pharmaceutics and should reduce the environmental impact to minimum [100,102]. Although GHG emissions from biomass growth and charcoal production are considered CO_2 neutral, handling and transportation of reductants will rely on fossil fuels for the next decades.

2.8. Transport

The transport of ore, coal, coke, and charcoal is another source of GHG emissions and land occupation. Previous studies have shown that the GHG emissions are negligible (<5%) for coal and coke transportation in steel industries [11,46,103]. However, charcoal has a twice lower density than coal and coke that can affect the cost and environmental impact of charcoal transportation, especially since volume limitations occur for these low bulk density [14]. The transport of whole stem wood is most favorable for the charcoal production chain, as bulk density is highest and handling and sizing can be carried out at the charcoal retort [104].

The metallurgical smelters in Norway are located on the coast and international transport is mostly operated using seaways. Thus, only biomass transport from harvest to the pyrolysis plant and charcoal transport to the harbor are expected to use road ground transportation. Previous studies have shown that an increased transport distance from 50 to 100 km has mainly an effect on the economics, whereas the effect on total energy consumption and GHG emissions was small [66]. Loose biomass, such as forest residues or straw can be mechanically densified to improve transport efficiency [104,105]. Wood or forestry residues can also be used as a transportation biofuel to reduce GHG emissions by up to ≈45–75% [106,107].

3. Environmental Impact

Most LCA studies are carried out for the environmental impact, e.g., GHG emissions, acidification, and land occupation, whereas some also include human health issues and economics. Solid waste, eutrophitcation, and dust formation are also environmental impacts which should be considered due to the feedstock growth [43]. While the standard approach for LCA is "cradle-to-grave", the "cradle-to-gate" approach is most suitable for the metallurgical industry. This approach can be used according to ISO 14040 for silicomanganese or ferromanganese alloys [8]. The emission factors of ferroalloy production vary between 1.04 and 1.15 kg CO_2 per kg ferromanganese [37], 1.4 and 6.9 kg CO_2 per kg silicomanganese, 2.5 and 4.8 kg CO_2 per kg ferrosilicon [8,108,109], and 6.0 kg CO_2 per kg manganese [8]. The large differences in the emission factors were mostly observed due to the electric power supply by hydropower or coal.

Mining and biomass growth have the largest impact on the land consumption and local biodiversity. The direct and indirect GHG emissions from mining expands over time and are often driven by global factors which are uncontrollable by the local management or policies [110]. The renaturation of mining areas to the full recovery can take several decades to hundreds of years [38] and is concomitant with changes in the landscape, in which the change in landscape may result in environmental and social impacts [110]. Secondary forestry and other biomass growth scenarios have a large area consumption to provide sufficient biomass for renewable reductant production. The increased global biomass demand will open new markets and rise the risks of rapid changes in local landscapes and ecosystems [87] and the local impacts in the biomass production countries [52].

Industrialized countries will ensure by policies a sustainable biomass growth and charcoal production, whereas developing countries may have challenges to execute the policies at district or regional level [79]. However, markets and policies are driven by global factors and create the opportunities and constraints for the new land users [87]. To ensure the sustainable ferromanganese alloy production, global factors, as well as natural and social sciences in foreign countries have to play an increasing role in the process chain from cradle-to-gate. Fuel and power supply in developing countries will be a source of GHG emissions for mining and railway transportation. However, the energy demand for both process stages is small compared to the metallurgical coke and ferromanganese production. Replacement ratios of fossil fuels by bioenergy can significantly reduce the emission factors in each process stage [17].

4. Socioeconomic Effects

Charcoal is mainly produced in Africa and Asia by households with low income. In some countries the charcoal market accounts for more than 2% of the GDP [111], and the biomass demand is greater than the sustainable biomass growth [82]. To supply biomass for the industrial scale production, ecological and ethical criteria must be followed to avoid social tension. Local policy, structured governance, and well-funded instructions can improve the long-term success of sustainable charcoal production in these countries [79]. The charcoal production costs in Brazil are ≈200 EUR/t of charcoal, whereas production costs in Finland or Austrilia are between 270–480 EUR/t without considering the by-products as value-added compounds [17]. Bio-oil from flash pyrolysis ranges at 200–300 EUR/t [17], indicating an economical value of the liquid by-products.

Simple retorts can be constructed for ≈300 EUR [41], whereas industrial retorts such as Lambiotte retorts have an capex of 0.5–2 million EUR [112]. The high capex of industrial retorts is not economically reasonable for local farmers in developing countries. From socioeconomic perspective, centralized charcoal production units can be provided as a development assistance to minimize emissions from charcoal production, create local jobs, ensure high conversion rates and stable charcoal properties for a long period. The production in industrial retorts would increase the available charcoal by a factor of ≈ 2 without increasing the demand in the raw feedstock.

Biomass certification such as the Programme for the Endorsement of Forest Certification (PEFC), Forest Stewardship Council (FSC), or European Biochar Certificate (EBC) can support local farmers with the licensing of sustainable wood sourcing. The sustainable biomass growth in combination with the industrial charcoal production will maximize CO_2 emission saving potential and concomitant reduce local toxic emissions formed by classical charcoal production. An adequate income will be provided to farmers for a sustainable biomass growth, and the additional charcoal yield will at least partially cover the carbon demand of ferromanganese alloy production without increasing social tension in the charcoal producing countries.

5. Discussion

Previous studies have shown that charcoal has the potential to reduce the GHG emissions for the integrated steel making route by 31–74%, respectively, up to a CO_2 neutral production by EAF [14]. The power supply in Norway makes it most likely that ferromanganese alloys can be produced CO_2 neutral by compensating direct emissions from SAFs by renewable reductants. Other ferromanganese alloy-producing countries can increase the renewable energy production in their energy mix to reduce indirect emissions from power supply [9,10]. When the properties of carbon reductants inhibit a complete replacement of fossil fuel-based reductants, CCS and CCU can be used to compensate remaining emissions. Previous studies have shown that CO_2 neutral reductants can be used for silicon and silicomanganese production [113], whereas closed hearth SAF still rely on metallurgical coke due to the low volatile matter content. Post-treatment of charcoal may improve physicochemical properties to replace fossil fuel reductants in the future [28,29,55]. The possible impacts of a classical and sustainable charcoal production on the LCA of ferromanganese alloy production are summarized in Table 2.

Mining and metallurgical coke production are the main GHG emitters in the process chain upstream of the metallurgical industry. Environmental effects such as dust formation and acidification are often not considered in the life cycle analysis, and will further improve the application of renewable reductants in ferromanganese alloy production. Land consumption for mining, biomass growth, or road construction for transportation have received more attention in recent studies, but the environmental impact is considered negligible compared to the coke production for the metallurgical processes. While renaturation of coal mining can take centuries, sustainable biomass growth can provide biodiversity in the regions. Short rotation coppice or classical forestry can provide biomass with low ash content. Transport of biomass and coal has only a minor impact on the GHG emissions which are often neglected in LCA studies.

Biomass growth and charcoal production are the key process variables to decrease GHG emissions during ferromanganese alloy reduction. Sustainable biomass growth requires large areas, and large areas of monocultures should be avoided for plant diversity. A stem wood production in Southern Norway was estimated to $18\,m^3/(ha\ a)$ of solid wood [114], approximately $8\,t/(ha\ a)$. An average growth rates of 10–$15\,t/(ha\ a)$ (dry basis) can be realized in temperate climate for short rotation coppice [88,115]. However, the lower material density of short rotation coppice such as willow may limit the transport distance and storage time of this biomass [116]. Thus, GHG emissions by transport may be covered by the additional biomass growth in southern European region for selected biomass species.

Table 2. Impact of classical and sustainable charcoal production on the LCA in the ferromanganese alloy production.

Process Stage	Current Situation and Main Impacts	Changes by Classical Charcoal Production	Changes by Sustainable Charcoal Production
Ore mining	Dust emissions, water pollution (acidification), land transformation, and destruction of local ecological environment.	No changes expected by renewable reductants.	
Coal mining	Dust, CH_4 and CO_2 emissions, water pollution (acidification), fossil fuel depletion, land transformation, and destruction of local ecological environment	Reduced coal demand to a fully replacement of coal by charcoal as a renewable reductant results in a decreased fossil fuel depletion, reduced air and water pollution and avoids land transformation.	
Biomass production	Partly deforestation and soil degradation	Additional land occupation, deforestation and reduced biodiversity.	Reduced demand of additional biomass by an efficient and sustainable charcoal production, as well as consideration of socioeconomic factors.
Transport (local)	Mainly conveyor belt and railroad transport for coal, respectively truck transport for biomass and charcoal.	Increased diesel consumption by truck transport for biomass and charcoal transport expected.	
Coke production	Local emissions, resulting in air and water pollution.	Charcoal can replace up to 20% in bio-coke production, resulting in a reduced volatile matter.	Bio-cokes and tailor made charcoal may fully replace metallurgical coke in long-term.
Charcoal production	Incomplete combustion and release of volatile matter, resulting in air and water pollution, photochemical oxidant formation and human toxicity.	Increased biomass demand can result in an increased non-sustainable production and additional local emissions.	Improved conversion technologies result in an increased conversion efficiency, by-product utilization and improved charcoal quality. The greater conversion efficiency can compensate the increased land demand for sustainable biomass production.
Transport (international)	Emissions by ship and railroad transport ($<5\%$ of total emissions)	Emissions may increase due to the lower bulk density of charcoal and the volume limited transport (emissions may increase by a factor up to 2)	By-products may be utilized as fuel for transport, making long-distance transport more sustainable
Smelting (SAF)	CO_2, CO and dust emissions	CO_2 emissions from charcoal are considered CO_2 neutral, additional gas cleaning required for high volatile matter content	

About 50% of the carbon content in the raw biomass can be converted to charcoal in industrial retorts [117], resulting in a carbon yield of $\approx$2.5–5 t/(ha a). The low conversion efficiency of earthmound kilns and pits in combination with the incomplete combustion of the volatile matter makes these technologies counterproductive for the large replacements of fossil fuel-based reductants in ferromanganese alloy production. Replacing 20% of the metallurgical coke by bio-coke can reduce the GHG emissions by 10–15% for classical charcoal production, whereas the sustainable charcoal production would further reduce local emissions and environmental impact by acidification and water contamination. In addition, the sustainable charcoal production would further reduce the resource consumption by the twice to three times greater charcoal yield compared to classical charcoal production.

An ensured price range by certified biomass can convince local farmers to produce biomass sustainable. Centralized industrial retorts can create high-quality local jobs and improve the quality of charcoal concomitant to the increased charcoal yield. The current charcoal production of $\approx$55 Mt [85] would be increased to >75 Mt without the consumption of additional biomass. The high capex of industrial retorts can be covered by government subsidies or financial aid.

6. Conclusions

Renewable reductants can decrease the direct and indirect GHG emissions from ferromanganese alloy production. The sustainable charcoal production can reduce the indirect GHG emissions due to the improved feedstock growth and additional area occupation, whereas the direct emissions can be decreased by the integration of energy efficient pyrolysis retorts. GHG emissions generated due to the transportation of feedstocks and charcoal are expected to be greater than emissions from transportation of metallurgical coke and fossil fuels due to their low bulk density. The utilization of pyrolysis by-products with the concurrent production of biofuels can further decrease the emissions related to transportation. In the next few decades, the integration of renewable charcoal reductants is expected due to the increased governmental requirements and policies towards development of environmental and sustainable processes in metallurgical sector within a circular economy. Overall, the authors believe that ferromanganese alloys can be manufactured in a CO_2 neutral way using carbothermal process with the addition of charcoal that has properties striving metallurgical coke and electricity provided by renewable sources.

Author Contributions: Conceptualization, G.R.S., A.T., and M.T.; methodology, G.R.S.; validation, G.R.S., A.T., and M.T.; writing—original draft preparation, G.R.S.; writing—review and editing, G.R.S., A.T., and M.T.; visualization, G.R.S.; supervision, M.T.; project administration, M.T.; funding acquisition, M.T. All authors have read and agreed to the published version of the manuscript.

Funding: This research was funded by Research Council of Norway grant number 280968. The APC was funded by Norwegian University of Science and Technology.

Institutional Review Board Statement: Not applicable.

Informed Consent Statement: Not applicable.

Acknowledgments: The authors gratefully acknowledge the financial support from Research Council of Norway (Project Code: 280968) and our industrial partners Elkem ASA, TiZir Titanium & Iron AS, Eramet Norway AS, Finnfjord AS and Wacker Chemicals Norway AS for financing *KPN Reduced CO_2 emissions in metal production* project.

Conflicts of Interest: The authors declare no conflict of interest.

Abbreviations

The following abbreviations are used in this manuscript:

BM	Biomass
capex	capital expenditures
CC	Charcoal

CCS	carbon capture and storage
CCU	carbon capture and utilization
EBC	European Biochar Certificate
FeMn	ferromanganese
FSC	Forest Stewardship Council
GHG	greenhouse gas
LCA	life cycle assessment
ORC	organic Rankine Cycle
PAHs	polycyclic aromatic hydrocarbons
PEFC	Programme for the Endorsement of Forest Certification
SAF	submerged arc furnace
SiMn	silicomanganese
tonne	metric ton

References

1. Vadenbo, C.O.; Boesch, M.E.; Hellweg, S. Life Cycle Assessment Model for the Use of Alternative Resources in Ironmaking. *J. Ind. Ecol.* **2013**, *17*, 363–374. [CrossRef]
2. World Steel Association. *Steel's Contribution to a Low Carbon Future and Climate Resilient Societies-Worldsteel Position Paper*; World Steel Association: Brussels, Belgium, 2017.
3. Office of Air Quality Planning and Standards. *Background Report-AP-42 Section 12.4 Ferroalloy Production*; Report Number: OAQPS/TSD/EIB; Research Triangle Park, NC 27711; U.S. Environmental Protection Agency: Washington, DC, USA, 1992; p. 40.
4. Schei, A.; Tuset, J.K.; Tveit, H. *Production of High Silicon Ferroalloys*; Tapir Forlag: Trondheim, Norway, 1998.
5. Tangstad, M. Ferrosilicon and Silicon Technology. In *Handbook of Ferroalloys*; Elsevier: Amsterdam, The Netherlands, 2018.
6. Holappa, L. Basics of Ferroalloys. In *Handbook of Ferroalloys-Theory and Technology*; Butterworth-Heinemann: Oxford, UK, 2013; pp. 9–28.
7. Haque, N.; Norgate, T. Estimation of greenhouse gas emissions from ferroalloy production using life cycle assessment with particular reference to Australia. *J. Clean. Prod.* **2013**, *39*, 220–230. [CrossRef]
8. Westfall, L.A.; Davourie, J.; Ali, M.; McGough, D. Cradle-to-gate life cycle assessment of global manganese alloy production. *Int. J. Life Cycle Assess.* **2016**, *21*, 1573–1579. [CrossRef]
9. Industri, N. The Norwegian Process Industries' Roadmap-Combining growth and zero emissions by 2050. 2016. Available online: https://www.norskindustri.no/siteassets/dokumenter/rapporter-og-brosjyrer/the-norwegian-process-industries-roadmap-summary.pdf (accessed on 15 August 2020).
10. Dalaker, H.; Ringdalen, E.; Kolbeinsen, L.; Mårdalen, J. Road-Map for Gas in the Norwegian Metallurgical Industry: Greater Value Creation and Reduced Emissions. 2017. Available online: http://hdl.handle.net/11250/2488736 (accessed on 15 August 2020)
11. Matuštík, J.; Hnátková, T.; Kočí, V. Life cycle assessment of biochar-to-soil systems: A review. *J. Clean. Prod.* **2020**, *259*, 120998. [CrossRef]
12. Nuss, P.; Eckelmann, M.J. Life Cycle Assessment of Metals: A Scientific Synthesis. *PLoS ONE* **2014**, *9*, e101298. [CrossRef]
13. San Miguel, G.; Méndez, A.M.; Gasecó, G.; Quero, A. LCA of alternative biochar production technologies. In Proceedings of the 15th International Conference on Environmental Science and Technology, Rhodes, Greece, 31 August–2 September 2017.
14. Norgate, T.; Haque, N.; Somerville, M.; Jahanshahi, S. Biomass as a Source of Renewable Carbon for Iron and Steelmaking. *ISIJ Int.* **2012**, *52*, 1472–1481. [CrossRef]
15. Surup, G.R.; Trubetskaya, A.; Tangstad, M. Charcoal as an Alternative Reductant in Ferroalloy Production: A Review. *Processes* **2020**, *8*, 1432. [CrossRef]
16. Trubetskaya, A.; Beckmann, G.; Poyraz, Y.; Weber, R. Secondary comminution of wood pellets in power plant and laboratory-scale mills. *Fuel* **2017**, *160*, 675–683. [CrossRef]
17. Suopajärvi, H.; Pongrácz, E.; Fabritius, T. The potential of using biomass-based reducing agents in the blast furnace: A review of thermochemical conversion technologies and assessments related to sustainability. *Renew. Sustain. Energy Rev.* **2013**, *25*, 511–528. [CrossRef]
18. Eidem, P.A. Electrical Resistivity of Coke Beds. Ph.D. Thesis, Norges Teknisk-Naturvitenskaplige Universitet (NTNU), Trondheim, Norway, 2008; p. 279.
19. Surup, G.R.; Heidelmann, M.; Kofoed Nielsen, H.; Trubetskaya, A. Characterization and reactivity of charcoal from high temperature pyrolysis (800–1600 °C). *Fuel* **2019**, *235*, 1544–1554. [CrossRef]
20. Schulte, R.F.; Tuck, C.A. Ferroalloys. In *U.S. Geological Survey Minerals Yearbook*; US Geological Survey: Reston, VA, USA, 2016; pp. 25.1–25.14.
21. Ferroalloys Market Size, Share & Trends Analysis Report by Product (FeCr, FeSiMn), by Application (Carbon & Low Alloy Steel, Cast Iron) by Region, (Asia Pacific, Europe, North America), and Segment Forecasts, 2020–2027. Available online: https://www.grandviewresearch.com/industry-analysis/ferroalloys-market (accessed on 16 November 2020).
22. Olsen, S.E.; Tangstad, M.; Lindstad, T. *Production of Manganese Ferroalloys*; Tapir Akademisk Forlag: Trondheim, Norway, 2007; p. 247.

23. Rozhikhina, I.D.; Nokhrina, O.I.; Yolkin, K.S.; Golodova, M.A. Ferroalloy production: State and development trends in the world and Russia. *IOP Conf. Ser. Mater. Sci. Eng.* **2019**, *866*, 012004. [CrossRef]
24. Díez, M.A.; Alvarez, R.; Barriocanal, C. Coal for metallurgical coke production: Predictions of coke quality and future requirements for cokemaking. *Int. J. Coal Geol.* **2002**, *50*, 389–412. [CrossRef]
25. Emrich, W. *Handbook of Charcoal Making: The Traditional and Industrial Methods*; Solar Energy R&D in the Ec Series E; Springer: Berlin, Gremany, 1985.
26. Kajina, W.; Junpen, A.; Garivait, S.; Kamnoet, O.; Keeratiisariyakul, P.; Rousset, P. Charcoal production processes: An overview. *JSEE* **2019**, *10*, 19–25.
27. Antal, M.J.; Allen, S.G.; Dai, X.; Shimizu, B.; Tam, M.S.; Grønli, M. Attainment of the Theoretical Yield of Carbon from Biomass. *Ind. Eng. Chem. Res.* **2000**, *39*, 4024–4031. [CrossRef]
28. Hussein, A.; Larachi, F.; Ziegler, D.; Alamdari, H. Effects of heat treatment and acid washing on properties and reactivity of charcoal. *Biomass Bioenergy* **2016**, *90*, 101–113. [CrossRef]
29. Surup, G.R.; Vehus, T.; Eidem, P.A.; Trubetskaya, A.; Nielsen, H.K. Characterization of renewable reductants and charcoal-based pellets for the use in ferroalloy industries. *Energy* **2019**, *167*, 337–345. [CrossRef]
30. Surup, G.R.; Foppe, M.; Schubert, D.; Deike, R.; Heidelmann, M.; Timko, M.T.; Trubetskaya, A. The effect of feedstock origin and temperature on the structure and reactivity of char from pyrolysis at 1300–2800 °C. *Fuel* **2019**, *235*, 306–316. [CrossRef]
31. Surup, G.R.; Leahy, J.J.; Timko, M.T.; Trubetskaya, A. Hydrothermal carbonization of olive wastes to produce renewable, binder-free pellets for use as metallurgical reducing agents. *Renew. Energy* **2020**, *155*, 347–357. [CrossRef]
32. Babich, A.; Senk, D.; Fernandez, M. Charcoal behaviour by its injection into the modern blast furnace. *ISIJ Int.* **2010**, *50*, 81–88. [CrossRef]
33. Surup, G.R.; Nielsen, H.K.; Großarth, M.; Deike, R.; Van den Bulcke, J.; Kibleur, P.; Müller, M.; Ziegner, M.; Yazhenskikh, E.; Beloshapkin, S.; et al. Effect of operating conditions and feedstock composition on the properties of manganese oxide or quartz charcoal pellets for the use in ferroalloy industries. *Energy* **2020**, *193*, 116736. [CrossRef]
34. Norgate, T.; Haque, N.; Somerville, M.; Jahanshahi, S. The Greenhouse gas footprint of charcoal production and of some applications in steelmaking. In Proceedings of the 7th Australian Life Cycle Assessment Conference, Melbourne, Australia, 9–10 March 2011.
35. Davourie, J.; Westfall, L.; Ali, M.; McGough, D. Evaluation of particulate matter emissions from manganese alloy production using life-cycle assessment. *Neurotoxicology* **2017**, *58*, 180–186. [CrossRef] [PubMed]
36. Ravary, B.; Grådahl, S. Improving environment in the tapping area of a ferromanganese furnace. In Proceedings of the INFACON XII, Helsinki, Finland, 6–9 June 2010; pp. 99–107.
37. Olsen, S.E.; Monsen, B.E.; Lindstad, T. CO$_2$ Emissions from the Production of Manganese and Chromium Alloys in Norway. In Proceedings of the 56th Electric Furnace Conference, New Orleans, LA, USA, 15–18 November 1998; pp. 363–369.
38. Fthenakis, V.; Kim, H.C. Land use and electricity generation: A life-cycle analysis. *Renew. Sustain. Energy Rev.* **2009**, *13*, 1465–1474. [CrossRef]
39. D'Amato, D.; Gaio, M.; Semenzin, E. A review of LCA assessments of forest-based bioeconomy products and processes under an ecosystem services perspective. *Sci. Total Environ.* **2020**, *706*, 135859. [CrossRef] [PubMed]
40. De Oliveira Miranda Santos, S.D.F.; Piekarski, C.M.; Ugaya, C.M.L.; Donato, D.B.; Júnior, A.B.; De Francisco, A.C.; Carvalho, A.M.M.L. Life Cycle Analysis of Charcoal Production in Masonry Kilns with and without Carbonization Process Generated Gas Combustion. *Sustainablity* **2017**, *9*, 1558. [CrossRef]
41. Adam, J.C. Improved and more environmentally friendly charcoal production system using a low-cost retort-kiln (Eco-charcoal). *Renew Energy* **2009**, *34*, 1923–1925. [CrossRef]
42. Gu, H.; Bergman, R.; Anderson, N.; Alanya-Rosenbaum, S. Life cycle assessment of activated carbon from woody biomass. *Wood Fiber Sci.* **2017**, *50*, 229–243. [CrossRef]
43. Zhang, L.; Wang, J.; Feng, Y. Life cycle assessment of opencast coal mine production: a case study in Yimin mining area in China. *Environ. Sci. Pollut. Res.* **2018**, *25*, 8475–8486. [CrossRef]
44. Burchart-Korol, D.; Fugiel, A.; Czaplicka-Kolarz, K.; Turek, M. Model of environmental life cycle assessment for coal mining operations. *Sci. Total Environ.* **2016**, *562*, 61–72. [CrossRef]
45. Burchart-Korol, D. Evaluation of environmental impacts in iron-making based on life cycle assessment. In Proceedings of the 20th Anniversary International Conference on Metallurgy and Materials, Brno, Czech Republic, 18–20 May 2011; pp. 1–6.
46. Liu, X.; Yuan, Z. Life cycle environmental performance of by-product coke production in China. *J. Clean. Prod.* **2016**, *112*, 1292–1301. [CrossRef]
47. Walsh, C.; Thornley, P. The environmental impact and economic feasibility of introducing an Organic Rankine Cycle to recover low grade heat during the production of metallurgical coke. *J. Clean. Prod.* **2012**, *34*, 29–37. [CrossRef]
48. Iosif, A.M.; Hanrot, F.; Birat, J.P.; Ablitzer, D. Physicochemical modelling of the classical steelmaking route for life cycle inventory analysis. *Int. J. Life Cycle Assess.* **2010**, *15*, 304–310. [CrossRef]
49. Olmez, G.M.; Dilek, F.B.; Karanfil, T.; Yetis, U. The environmental impacts of iron and steel industry: A life cycle assessment study. *J. Clean. Prod.* **2016**, *130*, 195–201. [CrossRef]
50. Milder, A.I.; Fernández-Santos, B.; Martínez-Ruiz, C. Colonization patterns of woody species on lands mined for coal in Spain: preliminary insights for forest expansion. *Land Degrad. Dev.* **2013**, *24*, 39–46. [CrossRef]

51. Bridge, G. Contested Terrain: Mining and the Environment. *Annu. Rev. Environ. Resour.* **2004**, *29*, 205–259. [CrossRef]
52. Fearnside, P.M.; Figueiredo, A.M.R.; Bonjour, S.C.M. Amazonian forest loss and the long reach of China's influence. *Environ. Dev. Sustain.* **2013**, *15*, 325–338. [CrossRef]
53. Tiwary, R.K. Environmental Impact of Coal Mining on Water Regime and Its Management. *Water Air Soil Pollut.* **2001**, *132*, 185–199. [CrossRef]
54. Awuah-Offei, K.; Adekpedjou, A. Application of life cycle assessment in the mining industry. *Int. J. Life Cycle Assess.* **2011**, *16*, 82–89. [CrossRef]
55. Surup, G.R. Renewable Reducing Agents for the Use in Ferroalloy Industries. Ph.D. Thesis, University of Agder, Grimstad, Norway, 2019; p. 350.
56. Sithole, N.A.; Bam, W.G.; Steenkamp, J.D. Comparing Electrical and Carbon Combustion Based Energy Technologies for the Production of High Carbon Ferromanganese: A Literature Review. In Proceedings of the INFACON XV, Cape Town, South Africa, 25–28 February 2018.
57. Zhang, R.; Ma, X.; Shen, X.; Zhai, Y.; Zhang, T.; Ji, C.; Hong, J. Life cycle assessment of electrolytic manganese metal production. *J. Clean. Prod.* **2020**, *253*, 119951. [CrossRef]
58. Suopajärvi, H.; Fabritius, T. Effects of biomass use in integrated steel plant—Gate-to-gate life cycle inventory method. *ISIJ Int.* **2012**, *52*, 779–787.
59. Pistorius, P.C. Reductant selection in ferro-alloy production: The case for the importance of dissolution in the metal. *J. S. Afr. Inst. Min. Met.* **2002**, *1*, 33–36.
60. Monsen, B.; Tangstad, M.; Solheim, I.; Syvertsen, M.; Ishak, R.; Midtgaard, H. Charcoal for manganese alloy production. In Proceedings of the INFACON XI, New Delhi, India, 18–21 February 2007; pp. 297–310.
61. Monsen, B.; Tangstad, M.; Midtgaard, H. Use of charcoal in silicomanganese production. In Proceedings of the INFACON X, Cape Town, South Africa, 1–4 February 2004; Volume 68, pp. 155–164.
62. Oladeji, S.O.; Ologunwa, O.P.; Tonkollie, B.T. Socio-economic Impact of Traditional Technology of Charcoal Production in Kpaai District-Bong County Liberia. *Environ. Manag. Sustain. Dev.* **2018**, *7*, 86–107. [CrossRef]
63. Wood, T.S.; Baldwin, S. Fuelwood and charcoal use in developing countries. *Ann. Rev. Energy* **1985**, *10*, 407–429. [CrossRef]
64. Lohri, C.R.; Rajabub, H.M.; Sweeney, D.J.; Zurbrügg, C. Char fuel production in developing countries—A review of urban biowaste carbonization. *Renew. Sust. Energ. Rev.* **2016**, *59*, 1514–1530. [CrossRef]
65. Harindintwari, J.; Gaspard, N. Socio-economic and environmental impact of charcoal production in Rangiro, Cyato and Bushekeri sectors, Nyamasheke District. *GSJ* **2019**, *7*, 1206–1223.
66. Li, Y.; Wang, G.; Li, Z.; Yuan, J.; Gao, D.; Zhang, H. A Life Cycle Analysis of Deploying Coking Technology to Utilize Low-Rank Coal in China. *Sustainability* **2020**, *12*, 4884. [CrossRef]
67. Gao, D.; Qiu, X.; Zhang, Y.; Liu, P. Life cycle analysis of coal based methanol-to-olefins processes in China. *Comput. Chem. Eng.* **2018**, *109*, 112–118. [CrossRef]
68. García-Álvarez, A.; Pérez-Martínez, P.; González-Franco, I. Energy Consumption and Carbon Dioxide Emissions in Rail and Road Freight Transport in Spain: A Case Study of Car Carriers and Bulk Petrochemicals. *J. Intell. Transp. Syst.* **2013**, *17*, 233–244. [CrossRef]
69. International Energy Agencys. *Railway Handbook 2012-Energy Consumption and CO_2 Emissions*; International Energy Agency: Paris, France, 2012.
70. Yang, J.X.; Xu, C.; Wang, R.S. *Methodology and Application of Life Cycle Assessment*; China Meteorological Press: Beijing, China, 2002.
71. McKenna, R.C.; Norman, J.B. Spatial modelling of industrial heat loads and recovery potentials in the UK. *Energy Policy* **2010**, *38*, 5878–5891. [CrossRef]
72. MacPhee, J.A.; Gransden, J.F.; Giroux, L.; Price, J.T. Possible CO_2 mitigation via addition of charcoal to coking coal blends. *Fuel Process. Technol.* **2009**, *90*, 16–20. [CrossRef]
73. Ng, K.W.; MacPhee, J.A.; Giroux, L.; Todoschuk, T. Reactivity of bio-coke with CO_2. *Fuel Process. Technol.* **2011**, *92*, 801–804. [CrossRef]
74. Khanna, R.; Li, K.; Wang, Z.; Sun, M.; Zhang, J.; Mukherjee, P.S. Biochars in iron and steel industries. In *Char and Carbon Materials Derived from Biomass: Production, Characterization and Applications*; Elsevier: Amsterdam, The Netherlands, 2019; pp. 429–446.
75. Fick, G.; Mirgaux, O.; Neau, P.; Patisson, F. Using biomass for pig iron production: A technical, environmental and economical assessment. *Waste Biomass Valoriz.* **2014**, *5*, 43–55. [CrossRef]
76. Xu, S.; Liu, W.; Tao, S. Emission of Polycyclic Aromatic Hydrocarbons in China. *Environ. Sci. Technol.* **2006**, *40*, 702–708. [CrossRef] [PubMed]
77. Parodi, S.; Stagnaro, E.; Casella, C.; Puppo, A.; Daminelli, E.; Fontana, V.; Valerio, F.; Vercelli, M. Lung cancer in an urban area in Northern Italy near a coke oven plant. *Lung Cancer* **2005**, *47*, 155–164. [CrossRef]
78. Wang, L.C.; Wang, Y.F.; Hsi, H.C.; Chang-Chien, G.P. Characterizing the Emissions of Polybrominated Diphenyl Ethers (PBDEs) and Polybrominated Dibenzo-p-dioxins and Dibenzofurans (PBDD/Fs) from Metallurgical Processes. *Environ. Sci. Technol.* **2010**, *44*, 1240–1246. [CrossRef] [PubMed]
79. Neuberger, I. *Policy Solutions for Sustainable Charcoal in Sub-Saharan Africa*; The World Future Council: Hamburg, Germany, 2015.
80. Muylaert, M.S.; Sala, J.; de Freitas, M.A.V. The charcoal's production in Brazil—Process efficiency and environmental effects. *Renew. Energy* **1999**, *16*, 1037–1040. [CrossRef]

81. Coomes, O.T.; Burt, G.J. Peasant charcoal production in the Peruvian Amazon: Rainforest use and economic reliance. *Forest Ecol. Manag.* **2001**, *140*, 39–50. [CrossRef]
82. Msuya, N.; Masanja, E.; Temu, A.K. Environmental Burden of Charcoal Production and Use in Dar es Salaam, Tanzania. *J. Environ. Prot.* **2011**, *2*, 1364–1369. [CrossRef]
83. Njenga, M.; Karanja, N.; Munster, C.; Iiyama, M.; Neufeldt, H.; Kithinji, J.; Jamnadass, R. Charcoal production and strategies to enhance its sustainability in Kenya. *Dev. Pract.* **2013**, *23*, 359–371. [CrossRef]
84. Da Silva, F.A.; Simioni, F.J.; Hoff, D.N. Diagnosis of circular economy in the forest sector in southern Brazil. *Sci. Total Environ.* **2020**, *706*, 135973. [CrossRef]
85. African Forestry and Wildlife Commission. *Sustainable Charcoal Production for Food Security and Forest Landscape Restoration*; Report Number FO:AFWC/2020/4.2; Food and Agriculture Organization of the United Nations: Rome, Italy, 2020; pp. 1–6.
86. Pennise, D.M.; Smith, K.R.; Kithinji, J.P.; Rezende, M.E.; Raad, T.J.; Zhang, J.; Fan, C. Emissions of greenhouse gases and other airborne pollutants from charcoal making in Kenya and Brazil. *J. Geophys. Res.* **2001**, *106*, 24143–24155. [CrossRef]
87. Lambin, E.F.; Turner, B.L.; Geist, H.J.; Agbola, S.B.; Angelsen, A.; Bruce, J.W.; Coomes, O.T.; Dirzo, R.; Fischer, G.; Folke, C.; et al. The causes of land-use and land-cover change: Moving beyond the myths. *Glob. Environ. Chang. Hum. Policy Dimens.* **2001**, *11*, 261–269. [CrossRef]
88. Capareda, S. *Introduction to Biomass Energy Conversions*; Taylor & Francis: New York, NY, USA, 2013.
89. Rockström, J.; Steffen, W.; Noone, K.; Person, A.; Chapin, S.F., III; Lambin, E.F.; Lenton, T.M.; Scheffer, M.; Folke, C.; Schellnhuber, H.J.; et al. A safe operating space for humanity. *Nature* **2009**, *461*, 472–475.
90. Lindner, M.; Suominen, T.; Palosuo, T.; Garcia-Gonzalo, J.; Verweij, P.; Zudin, S.; Päivinen, R. ToSIA—A tool for sustainability impact assessment of forest-wood-chains. *Ecol. Model.* **2010**, *221*, 2197–2205. [CrossRef]
91. Miao, Z.; Grift, T.E.; Hansen, A.C.; Ting, K.C. Energy requirement for comminution of biomass in relation to particle physical properties. *Ind. Crop. Prod.* **2011**, *33*, 504–513. [CrossRef]
92. United States Environmental Protection Agency. Charcoal. In *AP 42, Fifth Edition, Volume I Chapter 10: Wood Products Industry*; United States Environmental Protection Agency: Washington, DC, USA, 1995.
93. Oanh, N.T.K.; Nghiem, L.H.; Phyu, Y.L. Emission of Polycyclic Aromatic Hydrocarbons, Toxicity, and Mutagenicity from Domestic Cooking Using Sawdust Briquettes, Wood, and Kerosene. *Environ. Sci. Technol.* **2002**, *36*, 833–839. [CrossRef] [PubMed]
94. Andersson, J.T.; Achten, C. Time to Say Goodbye to the 16 EPA PAHs? Toward an Up-to-Date Use of PACs for Environmental Purposes. *Polycyl. Aromat. Comp.* **2015**, *35*, 330–354. [CrossRef]
95. Van Dam, J. *The Charcoal Transition: Greening the Charcoal Value Chain to Mitigate Climate Change and Improve Local Livelihoods*; FAO: Rome, Italy, 2017.
96. De Wit, M.; Junginger, M.; Faaij, A. Learning in dedicated wood production systems: Past trends, future outlook and implications for bioenergy. *Renew. Sustain. Energy Rev.* **2013**, *19*, 417–432. [CrossRef]
97. Reumerman, P.J.; Frederiks, B. Charcoal production with reduced emissions. In Proceedings of the 12th European Conference on Biomass for Energy, Industry and Climate Protection, Amsterdam, The Netherlands, 17–21 June 2002; pp. 1–4.
98. Balat, M. Mechanisms of Thermochemical Biomass Conversion Processes. Part 1: Reactions of Pyrolysis. *Energ Source Part A* **2008**, *30*, 620–635. [CrossRef]
99. Funke, A.; Niebel, A.; Richter, D.; Abbas, M.M.; Müller, A.K.; Radloff, S.; Paneru, M.; Maier, J.; Dahmen, N.; Sauer, J. Fast pyrolysis char—Assessment of alternative uses within the bioliq® concept. *Bioresource Technol.* **2016**, *200*, 905–913. [CrossRef]
100. Mostafazadeh, A.K.; Solomatnikova, O.; Drogui, P.; Tyagi, R.D. A review of recent research and developments in fast pyrolysis and bio-oil upgrading. *Biomass Convers. Biorefin.* **2018**, *8*, 739–773. [CrossRef]
101. Oyedun, A.O.; Lam, K.L.; Hui, C.W. Charcoal Production via Multistage Pyrolysis. *Chin. J. Chem. Eng.* **2012**, *20*, 455–460. [CrossRef]
102. Surup, G.R.; Attard, T.M.; Trubetskaya, A.; Hunt, A.J.; Budarin, V.L.; Arshadi, M.; Abdelsayed, V.; Shekhawat, D.; Trubetskaya, A. The effect of wood composition and supercritical CO2 extraction on charcoal production in ferroalloy industries. *Energy* **2019**, *193*, 116696. [CrossRef]
103. Gupta, R.C. Woodchar as a sustainable reductant for ironmaking in the 21st century. *Miner. Process. Extr. Metall. Rev.* **2003**, *24*, 203–231. [CrossRef]
104. Suurs, R. *Long Distance Bioenergy Logistics—An Assessment of Costs and Energy Consumption for Various Biomass Energy Transport Chains*; Copernicus Institute Report NWS-E-2002-01; Universiteit Utrecht: Utrecht, The Netherlands, 2002; p. 79.
105. Trubetskaya, A.; Beckmann, G.; Poyraz, Y.; Weber, R.; Velaga, S.; Wadenbäck, J.; Holm, J.K.; Velaga, S.P.; Weber, R. One way of representing the size and shape of biomass particles in combustion modeling. *Fuel* **2017**, *206*, 675–683. [CrossRef]
106. Zhang, F.; Johnson, D.M.; Wang, J. Life-Cycle Energy and GHG Emissions of Forest Biomass Harvest and Transport for Biofuel Production in Michigan. *Energies* **2015**, *8*, 3258–3271. [CrossRef]
107. Tsalidis, G.A.; Discha, F.E.; Korevaar, G.; Haije, W.; de Jong, W.; Kiel, J. An LCA-based evaluation of biomass to transportation fuel production and utilization pathways in a large port's context. *Int. J. Energy Environ. Eng.* **2017**, *8*, 175–187. [CrossRef]
108. Lindstad, T.; Olsen, S.E.; Tranell, G.; Færden, T.; Lubetsky, J. Greenhouse gas emissions from ferroalloy production. In Proceedings of the INFACON XI, New Delhi, India, 18–21 February 2007; pp. 457–466.
109. Kero, I.T.; Eidem, P.A.; Ma, Y.; Indresand, H.; Aarhaug, T.A.; Grådahl, S. Airborne Emissions from Mn Ferroalloy Production. *JOM* **2019**, *71*, 349–365. [CrossRef]

110. Sonter, L.J.; Moran, C.J.; Barrett, D.J.; Soares-Filho, B.S. Processes of land use change in mining regions. *J. Clean. Prod.* **2014**, *84*, 494–501. [CrossRef]
111. Chidumayo, E.N. Is charcoal production in *Brachystegia-Julbernardia* woodlands of Zambia sustainable? *Biomass Bioenergy* **2019**, *125*, 1–7. [CrossRef]
112. Garcia-Nunez, J.A.; Pelaez-Samaniego, M.R.; Garcia-Perez, M.E.; Fonts, I.; Abrego, J.; Westerhof, R.J.M.; Garcia-Perez, M. Historical Developments of Pyrolysis Reactors: A Review. *Energy Fuels* **2017**, *31*, 5751–5775. [CrossRef]
113. Myrhaug, E.H. Non-Fossil Reduction Materials in the Silicon Process-Properties and Behaviour. Ph.D. Thesis, Norges Teknisk-Naturvitenskapelige Universitet (NTNU), Trondheim, Norway, 1 July 2003.
114. Andersson, F.; Brække, F.H.; Hallbäcken, L.; Nordic Council of Ministers. *Nutrition and Growth of Norway Spruce Forests in a Nordic Climatic and Deposition Gradient*; Tema Nord Series; Nordic Council of Ministers: Copenhagen, Denmark, 1998.
115. Bassam, N.E. *Handbook of Bioenergy Crops*; Earthscan Ltd.: London, UK, 2010.
116. Volk, T.A.; Verwijst, T.; Tharakan, P.J.; Abrahamson, L.P.; White, E.H. Growing fuel: A sustainability assessment of willow biomass crops. *Front. Ecol. Environ.* **2004**, *2*, 411–418. [CrossRef]
117. Quicker, P.; Weber, K. *Biokohle-Herstellung, Eigenschaften und Verwendung von Biomassekarbonisaten*; Springer Vieweg: Wiesbaden, Germany, 2016.

Review

Charcoal as an Alternative Reductant in Ferroalloy Production: A Review

Gerrit Ralf Surup [1,*], Anna Trubetskaya [2] and Merete Tangstad [1]

[1]   Department of Materials Science and Engineering, Norwegian University of Science and Technology, 7491 Trondheim, Norway; merete.tangstad@ntnu.no
[2]   Department of Chemical Sciences, University of Limerick, V94 T9PX Castletroy, Ireland; anna.trubetskaya@ul.ie
*   Correspondence: gerrit.r.surup@ntnu.no; Tel.: +47-934-73-184

Received: 12 October 2020; Accepted: 3 November 2020; Published: 9 November 2020

Abstract: This paper provides a fundamental and critical review of biomass application as renewable reductant in integrated ferroalloy reduction process. The basis for the review is based on the current process and product quality requirement that bio-based reductants must fulfill. The characteristics of different feedstocks and suitable pre-treatment and post-treatment technologies for their upgrading are evaluated. The existing literature concerning biomass application in ferroalloy industries is reviewed to fill out the research gaps related to charcoal properties provided by current production technologies and the integration of renewable reductants in the existing industrial infrastructure. This review also provides insights and recommendations to the unresolved challenges related to the charcoal process economics. Several possibilities to integrate the production of bio-based reductants with bio-refineries to lower the cost and increase the total efficiency are given. A comparison of challenges related to energy efficient charcoal production and formation of emissions in classical kiln technologies are discussed to underline the potential of bio-based reductant usage in ferroalloy reduction process.

Keywords: charcoal; pyrolysis; bio-based reductant; ferroalloy industry; kiln

---

1. Introduction

Charcoal is one of the first man-made products which has been used for millennia. First application of charcoal were found from about 30,000 to 35,000 BC [1], and its first usage in metallurgy can be dated to about 6000 BC for copper production [2] respectively to about 1800 BC initiating the Iron Age [3]. Charcoal enabled high energy intense processes, such as glass production and metal smelting [4] and was used as the major energy source in industrial processes until the industrial revolution, when is was replaced by coal and its derivatives. Nowadays, charcoal is used as a fuel for heating and barbecue, and is considered as a reductant for ferrosilicon processes in Norway and Brazil [5]. However, metallurgical production continues to rely on fossil-based reductants due to limited knowledge of charcoal properties and high costs.

The global production of charcoal increased from 40.5 million tonnes in 2002 [6] to 53.2 million tons in 2018 [7], in which about two third were produced in Africa. Fuelwood and charcoal generate between 2 to 7% of global greenhouse gas (GHG) emissions, mainly caused by deforestation, and combustion-related pollutants [7]. Brazil is the largest producer of charcoal with an annual average production of 6.5 million tons between 1993–2017 [8,9]. It was reported that about 50% of charcoal were produced from planted Eucalyptus wood [10]. Over the last decades, charcoal production shifted from primary to secondary forestry, reducing anthropogenic CO_2 emissions by deforestation [8,11]. Large amounts of Brazil's charcoal are used in the metallurgical industry that consumes about 10 million tons of charcoal per year and replaces fossil fuels for pig-iron (30%), steel

(15%) and ferroalloy (98%) production [8]. In Europe, the production of charcoal from wood has almost vanished in the last centuries [12], leading to the charcoal import from countries like Brazil, Nigeria or Tunisia.

The metallurgical industry in one of the most energy intensive industrial sectors. About 10% of the annual anthropogenic CO_2 emissions are ascribable the direct and indirect GHG emissions from metallurgical industry [13–15]. These emissions are mainly produced by carbothermal reduction processes, such as blast furnaces (BF), electric arc furnaces (EAF) and submerged arc furnaces (SAF), and by power production. About 1 billion tons of metallurgical coal are consumed in global steel industry [14]. About half of it was used to produce 350 million tons of coke used in metallurgical processes [16]. The consumption of raw materials is highly dependent on the final product and used furnace type. For example, the consumption of reducing agents increases from 500–550 kg per metric ton high carbon ferrochrome (HC FeCr) in closed furnaces to 550–700 kg per metric ton in open furnaces, whereas only 410–500 kg per metric ton of high carbon ferromanganese (HC FeMn) are required in closed furnaces [17].

On average, special materials and high quality steels are composed of about 3% ferroalloys [17]. The ferroalloy industry refers to iron alloys with a high portion of additional elements, such as aluminium, chromium, manganese or silicon. Ferroalloys were initially produced in small-scale blast furnaces in the second half of the 19th century [17]. The scale of production increased with the introduction of electric arc furnaces, in which the major production nowadays takes place. The electrical consumption is in the range of 3000 to 3500 kWh per metric ton of ferroalloy, respectively up to 7000 to 8000 kWh per metric ton of high-silicon ferroalloy [17]. The electrical resistance of the charge material is used to heat the hearth of the electric furnace, in which a higher resistivity enables a deeper electrode tip position in the furnace [18]. Simulation have shown that a lower electrode tip positions can decrease current through the burden by about one third [19], improving the temperature profile inside the smelter.

Bio-based reductants have potential to replace the reductants made from fossil-based materials in metallurgy [20,21]. Between 800 and 1200 kg of wet woodchips and about 10% charcoal have been used for the silicon production in the EU countries [22,23]. The current demand of sustainable carbon for ferroalloy production is smaller than the production of charcoal, whereas the markets and quality of charcoal are very diverse. The production and usage of charcoal can have a socioeconomic impact on different regions and should be considered beside the technical requirements for metallurgical application. The Food and Agriculture Organization of the United Nations summarized the main sustainable development goals for charcoal to poverty reduction, food security and nutrition, health and availability of clean water, sustainable energy and sustainable forest management to ensure the conservation of biodiversity [7]. Some kilns used for the charcoal production do not include off-gas cleaning systems and Thus, can release unburned by-products to the environment. These emissions are composed of particulate matter (PM), volatile organic compounds (VOC) and combustion products. Volatile pyrolysis products can be reduced by at least 80% by using an afterburner [24], in which the CO_2 emissions from the process are increased. Modern technologies produce charcoal continuously or semi-continuously, in which the off-gases are combusted or recirculated to provide the energy for the pyrolysis process [1,25]. However, only limited improvements have been made for these technologies over the last decades. Thus, charcoal production should be based on a sustainable forestry with concomitant reduction in CO_2 emissions.

Norway has a significant potential of forest-based resources due to the high ratio of hydro-power in their energy supply and the availability of unexploited forests [26]. Biocarbon in combination with carbon capture and storage (CCS) is intended to reduce the CO_2 emissions from Norwegian metallurgical industry by 6.8 million tons CO_2-equivalence per year in 2050 [27,28]. To substitute the fossil fuel carbon reductants in the ferroalloy industry, biocarbon reductants with specific chemical, mechanical, and electrical properties are required [18,23]. In this review, current technologies and the resulting properties of charcoal are discussed to understand (i) how bio-based reductants can replace fossil fuel-based carbon and (ii) how charcoal production routes can be improved to enable an economical production in Europe.

2. The Process Chain of Renewable Reductants

The properties of carbon reductants are crucial for a stable operation of SAF [16,18]. Differences in chemical and physical properties compared to fossil fuel based reductants could lead to a range of technical challenges for the bioreductant use in ferroalloy production. Currently used fossil fuel reductants generally provide a high mechanical stability and low gas reactivity. The fixed carbon content of fossil fuel reductants is expected to be larger than 85%, whereas ash content should be less than 12%. Ash constituents such as alkali and alkali earth metals, phosphorous and sulfur should be low to minimize catalytic reactions and slag formation. The high volatile matter content of biomass and it derivatives can be problematic for especially closed hearth SAF. Selected properties of biomass, charcoal and fossil fuel coke are summarized in Table 1, while minimum requirements for metallurgical processes are summarized in chapter 4 (Table 5). The typical particle size of carbon reductants in SAF can vary from 5 to 30 mm [29,30]. It is obvious that raw biomass must be processed to fulfill the required properties in SAF. While wood has been used as the only feedstock over the millennia, herbaceous biomass waste [31–34] and algae [35] can be also considered as feedstocks nowadays. Possible steps for the charcoal production and use as a reductant in ferroalloy industries are shown in Figure 1.

Table 1. Comparison of biomass and charcoal properties to fossil fuel reductants used in ferroalloy production [36–39].

Property	Unit	Wood	Herbaceous Biomass	Industrial Charcoal	Coke
Fixed carbon	%	15–20	15–20	65–85	86–88
Ash content	%	0.1–1.0	1–12	0.4–4	10–12
Compressive strength	kg cm^{-2}	250–400	-	10–80	130–160
Bulk density	kg m^{-3}	100–850 *	80–310	180–350	500–550
Electrical resistance	Ω m	Very high	Very high	High	Low
CO$_2$ reactivity	-	High	High	Medium-high	Low-medium

* 100–310 for wood chips and 300–850 for stemwood.

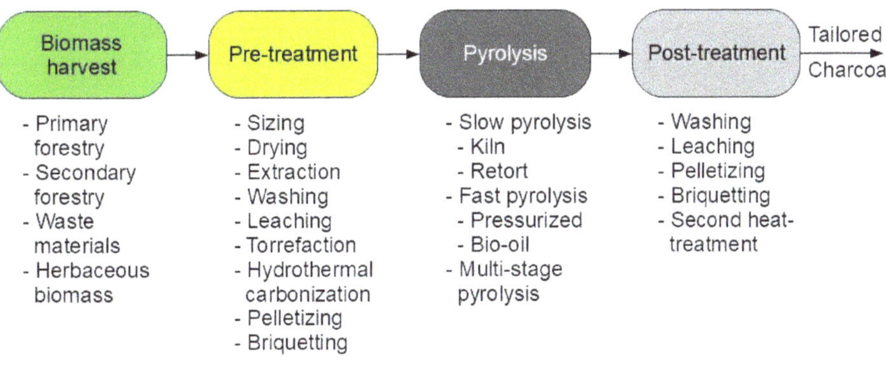

Figure 1. Schematic of the possible charcoal production steps.

Charcoal is defined as a solid residue from wood pyrolysis. Pyrolysis is the thermochemical decomposition of organic material at elevated temperatures, most often between 350 to 700 °C, in absence of oxygen. The required thermal energy is provided by the partial combustion of additional feedstock or pyrolysis by-products, i.e., gas and oil (volatile matter), which is released during pyrolysis. While charcoal kilns have been used over millenia and are still used in several countries, modern kilns and large-scale retorts are the main production processes in industrialized countries.

2.1. Biomass Composition

The most common biomass used as a renewable reductant is wood. However, due to the high price of woody biomass, alternative feedstock such as herbaceous biomass or waste have become more attractive in recent years. Biomass is a mixture of organic compounds and mineral matter. Woody biomass is mainly composed of cellulose, hemicellulose, lignin and minor amounts of extractives, in which softwood has a greater lignin content than hardwood, whereas the cellulose, hemicellulose and extractives are greater in hardwood [40]. The compositional analysis of herbaceous biomass as well as selected softwood and hardwood species is shown in Table 2. The analysis of specific feedstock can be found elsewhere [38,41], such as differences between bark and stemwood [42].

The biomass constituents have a major impact on the product properties, composition and yields in thermochemical conversion. The constituents react in broad temperature ranges using different mechanisms, pathways and rates [43] and Thus, affect the primary and secondary reactions in the kiln or retort. Pyrolysis reactions of cellulose are mainly endothermic, whereas decomposition of lignin and secondary decomposition of volatiles are exothermic [43,44]. Thus, biomass composition has a strong effect on the char yield and by-product composition.

The volatile composition is based on the biomass feedstock and heating rate of the process, in which cellulose, hemicellulose and lignin (CHL) decompose to characteristic monomer and monomer-related fragments [45]. Cellulose decomposes in the temperature range between 280 to 380 °C with a maximum reaction rate at 350 °C [46,47]. Hemicellulose decomposes in the wider temperature range between 190 to 380 °C with a maximum reaction rate at 260 and 310 °C [48–51]. Lignin decomposes in the widest temperature range between 200 to 500 °C with a maximum reaction rate at 360 and 400 °C [48,50,52]. Thus, thermal breakdown (degradation) of hemicellulose is larger than that cellulose and larger than that of lignin [43]. The pyrolysis of cellulose results in a formation of levoglucosan and its derivatives [46] with a solid residue of 7–20% [53]. The char yield from xylan can be twice greater than that of cellulose and three times greater than that of glucomannan [54,55]. Lignin has the greatest solid yield with about 50% at 500 °C, further decreasing to 40% at 900 °C [55,56].

Minerals in the ash can catalyze primary and secondary pyrolysis reactions. Potassium and silica showed a greater catalytic effect on the remaining radical concentration compared to the CHL composition in the biomass [57]. In addition, charcoal forming reactions under fast heating rates were much stronger affected by the potassium content than by CHL composition [58]. Sugar and sugar derivatives were increased at the cost of carbonyls and phenolic compounds after acid leaching [59]. To evaluate the conversion performance of biomass, five grades on basis of the compositional and proximate analyses were established [60].

Table 2. Composition of hardwood, softwood, wheat straw and rice husks.

	Cellulose	Hemicellulose	Lignin /wt.%	Extractives	Ash	Sources
Herbaceous biomass						
Wheat straw	37.4	25.5	16.00	6.2	8.6	[38]
Rice husks	33.7	22.0	22.8	9.8	11.0	[38]
Bagasse	38.0	27.5	18.5	10.6	10.9	[38]
Softwood						
Softwood	43.1	27.7	29.2		0.5	[43]
Stem wood	43 *	28 *	29 *			[61]
Pinewood	38.3–40.7	17.8–26.9	27.0–31.4	5.0–8.8	0.3	[42,62]
Spruce	37.8–42.0	25–27.3	27.4–28.6	2.0–7.8	0.8	[42,63]
Hardwood						
Hardwood	43.0	35.3	21.7		0.4	[43]
Stem wood	43 *	35 *	22 *			[61]
Eucalyptus	46.1–48.8	21.9–22.5	28.8–31.4	3.1–5.0	0.1–0.2	[64]
Beechwood	35.0	19.2	33.5	7.5	1.4	[62]
Birch	43.9	28.9	20.2	3.8		[42]
Oak	36.7	18.7	21.9	11	1.6	[63]
Birch						
Stem wood	43.9	28.9	20.2	3.8		[42]
Bark	10.7	11.2	14.7	25.6		[42]
Branches	33.3	23.4	20.8	13.5		[42]

* On dry ash free basis.

2.2. Pre-Treatment of Raw Feedstock and Charcoal

Ferroalloy industries require pre-treatment processes to have a low capital and operating cost and also to be effective on a wide range and loading of lignocellulosic material. To improve biomass properties for transport, storage and thermochemical conversion, biomass can be pretreated by resizing and drying. Fresh woody biomass provides a moisture content up to 60% [65] and ≈80% for fresh algae [66,67]. To reduce the energy demand in the first pyrolysis stage, biomass should be naturally dried to less than 25% [68]. The equilibrium moisture content of wood can vary from 4 to 18.1 wt.% [69], similar to that of dried herbaceous biomass such as wheat straw or rice husks. A high moisture content increases the energy demand for pyrolysis, resulting in a decreased energy conversion efficiency [70]. In addition, the amount of bio-oil would be increased by a larger water fraction, but with a decreased amount of organic compounds and heating value [71,72].

A stem wood length of 5 to 6 m is most often used for road transport in Europe. The feedstock size must be adjusted to the thermochemical conversion process, whereas a length of ≈30 cm and a thickness of 10–18 cm are required for pyrolysis retorts, respectively a length of 60–120 cm and a diameter of 12–20 mm for charcoal kilns [25,73], or milled to fines for fast pyrolysis processes [74]. Sawdust from sawmills is therefore useful for fast and flash pyrolysis units, or must be densified by pelletizing or briquetting. Herbaceous biomass must be milled and densified to be usable in classical charcoal production. However, fine biomass milling is energy-consuming, and fibrous particles with low bulk densities may cause feeding problems.

The use of a bioreductant that is more reactive in CO_2 than fossil-based coke may increase maintenance costs of the metallurgical process [75,76]. Ash constituents such as potassium act as a catalyst for CO_2 reactivity and should be minimized. Stem wood can be debarked to reduce the ash content of the feed material and increase the value of the woody biomass [77], in which debarking can be carried out with modified harvesting heads in the forests [78]. The removal of bark from wood logs reduces the concentration of alkali metals which can have a negative impact on the metal parts of the kiln leading to corrosion. One of the major challenges of agricultural residues is therefore related to its quality, such as a higher ash content of problematic alkali metal compounds. Washing or leaching are common methods to reduce the ash content of the parental feedstock.

Thermochemical processes such as torrefaction and hydrothermal carbonization can be used to improve physicochemical properties. However, and efficient and cost-effective biomass pre-treatment is very important for overcoming biomass limitations and hurdles in use of biocarbon-based reductants in the ferroalloy industry. Value-added compounds may be removed from the biomass prior to the pyrolysis process to improve the economics of the process chain. In contrast to classical charcoal production require extraction processes wood particles of small size for the high process efficiency. Thus, particle size of the material must be adjusted for each step in a process chain, and should target a final particle size that can vary from 5 to 30 mm for the further use as a reductant in ferroalloy industries [29,30]. Briquetting or pelletizing can be carried out after the extraction process.

2.3. Milling/Grinding

The particle size of the feedstock material must be adjusted for each step in the process chain, and should target a final particle size between 5 to 30 mm for the further use as renewable reductant [29,30]. While wood is sawed and split into logs prior to charcoal production, herbaceous biomass must be milled and grinded prior to its compaction. The pre-treatment of herbaceous biomass can be adopted from biomass combustion, where the lignocellulosic materials are first decentralized milled and pelletized, and second biomass pellets are then milled using coal roller mills prior to combustion [79]. A number of studies [80–88] have investigated the influence of mill type on both the particle size and shape. Momeni [80] showed that comminuting woody pellets in hammer and roller mills produced significantly different sized particles. In other investigations [81,83], higher fractions of fine particles were obtained after comminution in a hammer mill compared to milling using a knife mill. In agreement with this observation, the energy consumption of the knife mill was found in all cases to be smaller than that of the hammer mill [82,83]. The feedstock type (hardwood, straw, corn cobs and corn stover) affected the energy consumption of the hammer and disk mills [84].

The energy consumption for the comminution of dry pellets was lower for the hammer mill than for the disk mill, and the particle size distribution was broader with larger particle aspect ratios after comminution in the hammer mill [86]. In addition, it was reported that a high moisture content (>20%) increased the specific energy consumption by 50% [86]. It appeared that different feedstocks (switchgrass, corn and soybean) showed differences in the particle size and shape during comminution and generated particles with various morphological properties [87,88]. Milling of the thermally treated material showed that the energy efficiency can increase twice at temperatures up to 180 °C and decrease at carbonization temperatures above 270 °C reduced by factor of 4 [89].

Particles should be milled to an optimal size for the consecutive processes, in which a later densification must be ensured. Biomass can be compacted by pelletization and briquetting, in which pellets showed best mechanical stability for a broad particle size distribution and a low amount (≤20%) of particles less than 0.5 mm [90]. However, pellets produced from smaller particle sizes (≤300 µm) resulted in a higher yield stress and density than for larger particles (300–600 µm) [91]. Particles larger 5 mm can be compacted by briquetting.

2.4. Washing/Leaching

The alkali metals, particularly potassium, calcium and sodium catalyze the thermal degradation of biomass, increasing the yield of reaction water and decreasing the yield of tar and char [92,93]. Other soluble inorganic species can also be problematic because of their effects on the charcoal properties, such as increased reactivity, electrical conductivity and purity of base metals [94,95]. Water leaching removes alkali sulfates, carbonates, and chlorides, whereas HCl leaches carbonates and sulfates of alkaline earth and other metals [96]. For example, ash content of wheat straw and rice husks were reduced by nearly 50% by acid leaching [97–99], in which especially alkali and alkali-earth metals were removed. In addition, ammonia removes organic compounds of Mg, Ca, K and Na. Overall, leaching with dilute acids is known to remove only soluble metals which are not physiologically bound to the matrix of lignocellulosic feedstocks, while dilute alkali or other catalysts are required to disrupt cell walls and release ash components which are physiologically bound to the feedstock matrix [33]. However, previous studies have shown that leaching of biomass can also remove the organic fraction, especially reducing the lignin amount [92,100].

Leaching can also result in a loss of cellulose and hemicelluloses through hydrolysis pre-treatment [101]. Another challenge related to the biomass leaching with the acid catalysts is the remaining acid in the charcoal matrix that requires post-processing including calcination or disposal of catalysts and additional washing of charcoal with the deionized water. Concurrently, water washing may also be desirable to reduce volatilization of sulfur and chlorine products during pyrolysis [102,103]. In addition, leaching may increase the energy density up to 25% through increased heating values and pellet density [104]. Differences in carbon structure suggest that leaching of original biomass affect the charcoal properties after heat treatment at elevated temperatures, whereas temperature is the dominant process variable at temperatures above 2000 °C [62,105]. The alkali and alkali-earth metals are removed in a rapid phase, whereas other elements are removed in slow phases which can exceed several days [106]. Boiling-water leaching is inferior to acid leaching to remove ash elements, but superior on the basis of investment costs and chemical demand [98]. However, the removal of these elements is beneficial for the metallurgical industry [105], since these are catalyzing the Boudouard reaction.

2.5. Extraction

The separation of extractives (e.g., lipids/resin acids) from the biomass feedstock provides a possibility to considerably reduce off-gassing and provides valuable chemicals or biofuel for the energy sector and metallurgical industries [107]. The extracted fatty/resin acids can be utilized as primary feedstocks for chemicals and biorefinery applications [108–111], whereas the wood fraction after extraction is of high importance as a source of green carbon that could be utilized in metallurgical industries. Several methods exist for the extraction of high-value molecules from biomass including conventional organic solvent extraction, hydrodistillation, low-pressure solvent extraction and hydrothermal feedstock processing [112–114].

Supercritical fluids demonstrate properties between those of a liquid and a gas, with the viscosity of a supercritical fluid being an order of magnitude lower than a liquid, whereas the diffusivity is an order of magnitude higher and Thus, leading to the enhanced heat and mass transfer [115]. The properties of a solvent can be fine-tuned by varying the temperature and pressure. Conventional solvents traditionally utilized in wax extraction (such as hexane) are frequently viewed as being problematic due to the toxicological and environmental impacts [116]. Supercritical fluid extraction using CO_2 as a solvent has an easily accessible critical point, is non-flammable, has minimal toxicity and is widely available [117]. Supercritical CO_2 extraction ($scCO_2$) has been conducted on a commercial scale for over two decades for the extraction of high-value products from biomass [118]. Supercritical extraction process has been shown to improve the off-gassing of wood pellets, thus reducing the potential for uncontrolled auto-oxidation, while maintaining pellet properties [107,119]. Moreover, supercritical CO_2 extraction can also improve the physicochemical properties of solid char from pyrolysis at high temperatures, leading to greater electric conductivity and low reactivity of charcoal [120]. Supercritical CO_2 extraction increases the bending strength and stiffness of residual wood and, Thus, decreases the cost of process scaling up, wood storage and transportation [121].

The pre-treatment using $scCO_2$ extraction of wood removes more than half of value-added compounds without any significant influence on the physical properties of original wood and on the yield of solid charcoal [122]. Under properly selected treatment conditions (e.g., >1100 °C), charcoal samples can be produced from a mixture of different low quality wood fractions with reactivity and dielectric properties approaching that of fossil-based metallurgical coke and with the low content of liquid products, including naphthalene, PAHs, aromatic and phenolic fractions.

2.6. Torrefaction

Torrefaction is a mild pyrolysis process that converts biomass into a carbon material with increased energy density and decreased oxygen content, removing smoke-forming volatiles and result in a product yield of ≈45–70% of the initial weight [123,124]. Torrefaction contributes to dehydration, deoxygenation, partial degassing, and structural changes through breaking hemicellulose, lignin and cellulose chains at elevated temperatures [125], in which the lignin content is increased by ≈10–15% [124]. This leads to increased calorific value that improves biomass physiochemical properties during co-firing with coal [126,127]. For example, the feedstock becomes more uniform that improves pelletization and the flow properties of torrefied products [123] and the energy density is increased to more than 20 GJ m^{-3} [124].

The higher heating value of torrefied olive stones (28.8 MJ kg^{-1}) at 300 °C can be achieved in a rotary slow pyrolysis reactor to meet the requirements of metallurgical industries [128]. In addition, the torrefaction length did not have a strong influence on the CO_2 reactivity of olive stones, whereas the temperature and particle size had a significant influence on the composition and product yields [129]. Longer torrefaction times and greater heat treatment temperatures led to the improvement of higher heating value of olive stones leading to higher carbon and lower oxygen content. One way to improve torrefied biomass handling and combustion properties is by densification into briquettes or pellets which have many advantages over torrefied feedstock including reduction of dust, improved handling properties and higher bulk density (up to 66% greater) [130].

2.7. Hydrothermal Carbonization

In hydrothermal carbonization (HTC) feedstocks remains in contact with hot, compressed water increasing carbon and hydrogen contents and also producing hydrochar with a high stability [131]. HTC treatment generates three products: gases, liquid compounds, and hydrochar that contains 80–95% of the energy content of the raw feedstock and 55–90% of the original mass [132]. Gaseous products cover ≈10% by mass of the raw feedstock and liquid compounds, primarily hydroxymethylfurfural, furfural, phenol, pentoses, hexoses, comprise the remainder of the products [133]. HTC is carried out in a hot liquid water in the temperature range 180 to 250 °C at solids loading ranging from 7 to 25%, and reaction time ranging from a few minutes to several hours [134]. Specifically, HTC removes a significant fraction of undesired inorganic elements such as Na and K that would otherwise contribute to slag [135]. Previous studies showed that the alkali content can be further reduced by increasing the heat treatment temperature or by washing the hydrochar with the deionized water after the pre-treatment [136]. In addition, hydrochar has superior mechanical properties and pelletability compared with torrefaction or pyrolysis biochar [129,137]. Besides the complexity of structure-property relationship, the use of bioreductants is hindered by the price of feedstock and hydrochar yield [138,139]. Thus, a limited number of studies using lignocellulosic biomass and waste has been conducted to investigate the hydrochar properties for the use in ferroalloy industries.

2.8. Pelletizing and Briquetting

Densification processes such as pelletizing and briquetting can significantly reduce carbon losses due to dust formation. In addition, pelletizing and briquetting provide the product with a defined uniform shape and size [133]. Combined carbonization and pelletizing can be used to increase the fuel value of feedstock by increasing its energy density and improving its handling and transport processes in the reduction furnace [140]. One of the challenges in using renewable reductants in metallurgical processes is related to its fragility with the generation of large amounts of fine particles during transportation and storage [141]. The mechanical strength of biomass pellets can be improved through pelletization or briquetting and is slightly larger than that of charcoal pellets [142]. However, recent studies have shown that pelletizing charcoal fines increase the usable carbon yield in ferroalloy production [143].

In ferroalloy industries, manganese ore pellets must be sintered at higher temperatures to provide a mechanical stability that is similar to that of iron ore pellets. The addition of wood dust and dolomite are known to increase the required sintering temperature of the ore pellets and Thus, generally used in iron and manganese alloy production [16,144]. The use of pellets from charcoal-ore blends is known to reduce the electricity demand and increase the yield of elemental manganese [145]. Current metallurgical production is based on the use of fossil-based fuels because the use of biomass-ore and charcoal-ore pellets in the reduction process can increase the overall power consumption by 72–152 kWh per tonne of FeMn and will increase the cost of the reduction process [146].

Previous studies showed that the durability of high quality pellets is required to be >97.5% to fulfill the criteria of the European Standard Committee CET/TC 335 and be used in ferroalloy production [147]. Pelletizing of HTC char prepared at 200–240 °C provided mechanically stable bioreductants compared to the dried torrefied-based pellets without an additional binder [135,148,149]. An additional heat treatment improved the agglomeration of the hydrochar particles, increasing the durability of hydrochar pellets to >95% at temperatures above 300 °C [129]. A maximum durability of 98.5% was measured during the heat treatment of hydrochar at 450 °C showing similar properties to charcoal pellets with the pre-mixed bio-oil binder [63,150]. Torrefied biomass particles are loose and nonuniform due to decreased hemicellulose content [126]. In comparison, mechanical strength and pelletability of torrefied charcoal are less compared to that of HTC chars [135,137,151]. Therefore, the pre-treatment of biomass under the HTC conditions might be more suitable for the production of reductants for the ferroalloy industries than torrefaction.

2.9. Slow Pyrolysis

Slow pyrolysis is the thermal decomposition of organic feedstocks i.e., biomass, coal in an inert atmosphere at low heating rates (up to $20\,°C \cdot min^{-1}$). Slow pyrolysis processes are used to produce solid products like biochar or charcoal from biomass [152,153]. The feedstock is firstly dried by driving off the moisture and pore water of the organic feedstock. The dried feedstock undergoes a thermal decomposition of the organic matrix with increasing temperature, in which volatile compounds are released. The product composition is based on the feedstock and process conditions, such as heating rate, heat treatment temperature and residence time. Biomass is heated up to about 300 to 700 °C, and up to 50 wt.% of the feedstock are recovered as solid residue [56,154]. Pyrolysis at temperatures less than 1200 °C is defined as carbonization process [40], in which the solid yield decreases with increasing heat treatment temperature. The volatile fraction is composed of condensates (e.g., bio-oil) and pyrolysis gases, such as CO, CO_2, H_2 and CH_4. The solid pyrolysis product is rich in carbon [155], whereas the liquid by-products from biomass pyrolysis have a high oxygen content. The final yield and quality of the pyrolysis products are affected by the feedstock composition, pre-treatment, heat treatment temperature, heating rate, reaction gas atmosphere, particle residence time, pressure and catalysts (e.g., alkali and alkali earth metals) [152,154,156].

In general, the pyrolysis of biomass follows a three-step mechanism as described above: 1. dehydration, 2. primary pyrolysis reactions and 3. secondary pyrolysis reactions. Primary pyrolysis reactions include dehydrogenation, depolymerization, and fragmentation reactions of the organic matrix, and can be considered as homogenous reactions. Secondary (and ternary) pyrolysis reactions summarize the reactions of intermediate decomposition products and gas reactions, as well as the heterogeneous reactions with the formed char [56]. Moreover, secondary reactions of bio-oil compounds can increase the solid yield by more than 4%-points [63]. Previous studies have shown that secondary reactions can be induced by long residence times, large particles and increased pressure [157,158], and are affected by the ash yield and composition.

Charcoal yields are maximized at low heating rates, long residence times and high pressures [157,158], where increased pressure leads to an increase in a char yield (up to 50%) with the decreased residence time [159]. Classical charcoal is produced at a temperature below 700 °C at low heating rates. An operation cycle can vary from 7 days to more than 30 days [160,161] and is accounted as slow pyrolysis. At temperatures of 400–450 °C, about 37–50% charcoal, 34–47% bio-oil (4–11% tar, 30–36% aqueous phase), and 14–29% permanent gases are formed from wood [8,10,162]. However, due to the partial combustion of feedstock and undefined process conditions in kilns, charcoal yield is highly influenced by the skills of the operator, and is generally less than 20–25% on dry basis [160,163,164]. Industrial retorts used in Europe provide a charcoal yield of ≈35% on dry basis [153]. Char yields of 15–43 wt.% can be obtained for the slow pyrolysis of microalgae [35], similar as reported for agricultural residues [25,32].

Recent studies have shown that multi-stage pyrolysis processes can improve the solid yield and reduce its energy demand. The fixed carbon yield can be increased by 11% to 25.7% by a three-stage pyrolysis process with different heating rates, resulting in an energy saving of 20–30% [165,166]. This approach incorporates the temperature regions of endothermic and exothermic reactions and Thus, enables a better control of product yield and quality. However, the multi-stage pyrolysis requires carefully design to obtain the optimized product yield and energy savings [167] and no multi-stage process was installed on industrial scale when this review was initiated.

2.10. Fast Pyrolysis

Fast and flash pyrolysis of biomass are mainly used to produce liquid biofuels (e.g., bio-oil) at high heating rates (>100 °C s^{-1}) and reduced to atmospheric pressure. At a heat treatment temperature of 500 °C, about 24 wt.% of solid residue (including ash) can be recovered [168], whereas high yields of charcoal are formed at increased pressure [152,157]. The greater charcoal yield and the larger throughput in pressurized pyrolysis can improve the economics of the overall process, in which the fixed carbon yield can be increased to 70–85% of the theoretical value [157]. However, the fixed carbon content of the produced charcoal was less than 85% [152,157] and Thus, increases the demand for enhanced off-gas cleaning systems due to the reduced quality of the off-gases [169,170] or additional post-treatment.

Properties of flash pyrolysis char and soot differ from that of industrial charcoal and may be considered as an alternative resource in metallurgy. The solid residue is strongly affected by the feedstock composition and final temperature in the process [171]. Charcoal from fast pyrolysis exhibits a higher volatile matter content compared to slow pyrolysis char produced at same temperature [172]. The charcoal yield of wood and herbaceous biomass decreases to less than 4% on dry ash free basis at temperatures above 1000 °C [173,174], whereas the formation of soot increases at temperatures larger than 900 °C with a maximum yield between 1100 to 1250 °C [97,175]. Biomass with a high lignin content will form larger soot yields compared to biomass which is enriched in cellulose or xylan content [171,176], in which the soot and tar formation is reduced by high concentrations of alkali metals. At temperatures larger than 1000 °C, beechwood and straw samples retained their original structure, while low ash pinewood underwent a morphological transformation with a highly molten surface [100]. Cellulose pyrolysis under high heating rates produces mostly permanent gases and a limited amount of bio-oil [171].

Fast pyrolysis processes are technically complex [177] and will require a profound understanding of the primary and secondary pyrolysis reactions [164] to provide a biocarbon with the required properties for metallurgical application. A major drawback of bio-oil from fast pyrolysis is the lack of integrated biorefinery concepts into metallurgy [178]. The bio-oil is generally upgraded into a valuable biofuel or chemical feedstock [179]. In addition, bio-oil can be used in the ferroalloy industry as a binder for carbon briquettes, pellets and agglomerates [63,150,180] or as feedstock for electrode material [181]. The large fractions of small aromatics (e.g., benzene and toluene) and phenols which are found mostly in lignin bio-oil are beneficial for the use as a binder [176]. However, both process chains would require an additional post-treatment of the collected bio-oil samples.

2.11. Emissions

The charcoal fuel cycle from classical charcoal production is one of the most GHG intensive energy sources used by mankind [5]. Liquid and gaseous products from biomass pyrolysis can be accounted as value-added by-products or emissions. Charcoal is commonly produced in simple kiln technologies and small scale retorts without off-gas cleaning system or flares, releasing the unburnt by-products as emissions. These kilns generate multiple hazardous emissions by the incomplete combustion of the feedstock, such as PM, VOC, NO_x and CO [24,182] as summarized in Table 3. Some advanced charcoal kilns partly condense the VOC by a chimney and use these as a value-added by-product [25]. VOC can be subdivided into methane and non-methane organic compounds [182]. PM are increased from incomplete combustion of especially dried feedstock, whereas feedstock with high moisture content increases smoke formation [183]. Highest PM emissions occur in the first hours after the under-stoichiometric combustion is initiated [10]. Large PM emissions in combination with deforestation can result in a net increase in global warming potential [182] and must therefore be inhibited.

About 3–3.6% of the biomass carbon is emitted as condensable liquid emissions from simple kilns [182]. Polycyclic aromatic hydrocarbons (PAH) are one class of these products which are partly considered as hazardous to health and environment [184,185], and are mainly emitted in the first hours of the wood carbonization process [10]. 5.0 to 7.6% of total PAHs or 82 to 100% of the aromatics with 5 to 6 rings are found within the PM [10,184]. The number of PAHs in pyrolysis off-gases is higher than that of complete biomass combustion [10]. The concentration of dust including PM can range from 1.77 to 38.9 mg m^{-3} at highest concentration of total dust [186]. About 2.8 mg m^{-3} of particle bound PAH is smaller than the total emission of gaseous PAHs, which accounted to 23.6 mg m^{-3} [10]. Maximum concentration of 64–100 mg m^{-3} have been detected in the initial phase of the pyrolysis process [10,186].

NO_x emissions are dependent on the nitrogen content of the original feedstock. 60–80% of the nitrogen involved in the pyrolysis exists as protein nitrogen in the biomass, resulting in the formation of N-containing compounds at lower heating rates [187]. A high concentration of lignin promote to formation of NO_x [187]. The amount of NO_x emissions can range between 0.016 to 11 g per kg of charcoal [24,182]. Sulfur of the parental biomass will be partly released as SO_x emissions. About one-third of the sulfur is released as emissions with the volatiles in carbonization of coal [16]. Sulfur content of biomass transited from mostly sulfate to organosulfur from 500 °C to 850 °C [188]. However, sulfur content in wood biomass is generally low.

CO and CO_2 emissions are formed by incomplete combustion and thermal decomposition of the biomass. Carbon monoxide emissions are reported in the range of 130 to 373 g kg^{-1} of charcoal [24,182]. Approximately 25% of the biomass carbon is converted to CO_2, resulting in an emission factor between 543–3027 g CO_2 per kg of charcoal for the different kiln technologies [182]. The maximum of 3.03 kg of CO_2-equivalent per kg charcoal is produced in earthmound kilns. The 100 year global warming potential is stated to 1.144–5.685 kg of CO_2-equivalent per kg charcoal [189]. However, CO_2 emissions can be accounted as CO_2 neutral for the sustainable biomass production. On average, 28 to 61% of the CO_2 emissions are referred to the charcoal production process, whereas 29 to 61% arise from the biomass production [189]. Thus, a sustainable biomass production and utilization of the by-products are essential to reduce anthropogenic CO_2 emissions in metallurgy, whereas an increased carbon conversion efficiency and recirculation of volatiles are the most important improvements to reduce the pyrolysis gas emissions into the atmosphere [68].

More than 80% of the volatile emissions from simple charcoal kilns can be reduced by using a chimney and afterburner [24], which are usually installed in advanced kilns such as Brazilian Beehive kiln and Missouri kiln [25]. Industrial retort systems such as Degussa retort, Lambiotte retort etc. are closed systems that use off-gas scrubber and other gas cleaning systems in addition. Thus, modern charcoal production retorts do not pollute the environment compared to their simple predecessor [190] and generally fulfill the regulations for industrial emissions in Europe and North-America.

Table 3. Possible emissions from charcoal production in kiln technologies.

Emission Source	PM	VOC	NO_x	CO	CO_2	Source
	$kg\ ton^{-1}$ of charcoal					
Uncontrolled	140	125	11	130	500	[24]
combustion	110 (PM+VOC)		-	-	-	[10]
	13–41	24–124	0.0014–0.13	143–373	500–3000	[182]
	-	81	0.3	210	-	[6]
	420–690	34–100	2.7–5.9	340–620	1400–3350	[191]
	20–200	10–99	0.1–0.7	19–89	2600–6000	[192]
Retort kiln	100–300	10–50	0.8–2.8	100–220	1750–2150	[191]
	<1	<1	0.2	0.1	-	[6]

2.12. Post-Treatment Processes

Charcoal produced from feedstock of differing particle size showed various chemical properties. Small feedstock particles increase the rate of volatile escape, and the charcoal has an increased ash content [92,193]. The charcoal (or biocarbon) can be post-treated to adjust its properties and to utilize additional by-product streams, such as fine material or bio-oil. The ash content of charcoal is about 3 to 4 times larger than that of biomass due to the devolatilization of the organic volatile matter. Post treatment processes like water- and acid-leaching can be used to reduce the final ash content of the charcoal as described above [59,99,194]. A large extent of fine particles is produced during transport and handling of charcoal. Briquetting, pelletizing and agglomeration can be used to compact charcoal fines and to produce a renewable reductant with increased properties [124], such as mechanical strength and density [150,180,195,196]. Densification processes vary by energy consumption and can impact the chemical and physical properties of the charcoal [124].

A balance between mechanical strength and porosity of charcoal has to be found for the use of bioreductants in ferroalloy industries. Charcoal is often porous (both micro- and macroporous), which is disadvantageous when carbon needs to be mechanically stable and not excessively reactive, as it is the case in manganese production. The micropores in charcoal can be closed by the deposition of carbon from methane that could ease the transition to the use of biocarbon as a means of metal ore reduction [28,197]. In addition, properties of the original feedstock such as density, moisture content, heating value, ash content, and compressive strength affect the quality of charcoal and pellets/briquettes made from charcoal. Therefore, process route of the raw feedstock and by-products is a dominating factor governing properties of charcoal and bioreductant-based pellets/briquettes in the reduction furnace.

The possible method for preparing charcoal-based reductants is to compact fine charcoal particles with the addition of a binder. The fines can be agglomerated to improve the mechanical properties of charcoal-based reductants by adding water and organic binder i.e molasse [172], starch [195], lignin [124,198] and bio-oil [63,150]. The binder is required due to the lack of plasticity of charcoal particles compared to untreated biomass. The binder forms solid bridges between particles, resulting in an improved mechanical stability of the densified product [172]. Starch in combination with water are superior to bio-oil based on its availability and low economical value [195]. Lignin softens at temperature above 140 °C and forms agglomerates between charcoal particles [124,198,199], in which hardwood lignin is superior to softwood lignin due to the greater amount of methoxy content [199]. Lignin is commercially available as a by-product from second generation bio-fuels [150] and from pulping industry [199].

Mechanical properties are affected by the added binder, its content and compacting pressure. The mechanical strength of carbon briquettes can be increased by increasing the binder content to 6.5–8.6 wt.% of softwood and hardwood lignin [199]. An increased bio-oil ratio from 10 to

40% resulted in an increased compression strength by approximately 50% after a second heat treatment [150]. The second heat treatment reduced the volatile matter from the pellets and resulted in cross-linking of the bio-oil and charcoal particles and improved the mechanical strength of charcoal pellets and briquettes [105,200,201]. Lignin and bio-oil were successfully tested as binders for heat treatment temperatures up to 1100 °C [150,199]. To ensure a high mechanical stability at higher temperatures, metallic silicon can be added as an additive to the binder-charcoal mixture [199]. Overall, charcoal producents strive for the use of small charcoal particles in pelletizing process without further milling to keep the cost of the overall bioreductant process low. Further heat treatments can be used to reduce the volatile matter content and to improve mechanical strength of the lump charcoal, pellets and briquettes [105,200,201].

However, charcoal pellets can reduce a self-heating risk when the operating temperature in the furnace exceeds 600 °C [180]. In addition, porosity and surface area decreased after charcoal pelletizing using the pyrolysis bio-oil, whereas the use of coal tar as a binder did not affect the combustion and physicochemical properties of pellets made from fossil-based reductants [202]. Coal tar is a type of toxic, hazardous and carcenogenic solid waste generated in the process of coal gasification or coking and composed of heavy tar oil, pulverised coal and particles in gases produced in coal pyrolysis [203,204]. Therefore, renewable and carbon-neutral binders i.e., bio-oil with the addition of paraffin oil, castor oil, mineral oil and linseed oil can be an alternative to fossil-based binders due to the increased energy density of charcoal pellets with the low self-heating risk [205].

3. Charcoal

Charcoal is comprised of the unconverted organic solids, the non-volatile mineral matter and the carbonized products. Fixed carbon content of charcoal is generally greater than 70% [5,73], and it is considered greater than 85% for metallurgical grade charcoal [5,152]. The feedstock origin and the pyrolysis conditions are the main influencing parameter on the physicochemical properties [156,206]. The obtained structure of charcoal is defined by the original feedstock, particle size and heating rate, in which pyrolysis at slow heating rates and larger particles results in the similar charcoal structure compared to its feedstock [206]. Charcoal produced from dense hardwood provide a greater mechanical stability than charcoal produced from softwood, which is preferable in the ferroalloy industry [152]. The electrical properties of charcoal are also important for the use as a reductant in SAF.

The charcoal yield of classical kiln production processes is in the range of 5 to 20 wt.% on dry basis due to the limited control of process conditions (e.g., air supply) [8]. These production processes release often large quantities of unburnt hydrocarbons to the atmosphere, and provide a fixed carbon content in the range of 65 to 80% [25]. Industrial retorts on the other hand provide charcoal yields of ≈35% [207,208], whereas charcoal yields up to 62% were reported for pilot scale experiments [153,209]. Thus, the conversion efficiency of the carbonization process is one of the key issues for a sustainable process chain [7].

3.1. Yields

The charcoal yield from industrial charcoal production is summarized in Table 4. Carbon efficiency of modern retorts is about twice greater than that of earth mound kilns, and achieve a yield of about 35 wt.% on dry basis [6,43]. Well operated Brazilian kilns with tar recovery provided a similar charcoal yield with up to 36.4% and carbon yield of 69% [182]. In charcoal kilns, the required thermal energy is provided by the partial combustion of wood, its by-products and other fuels i.e., gas, oil, etc. [25]. The combustion of by-products and wood can decrease the charcoal yield to less than 20%. The modernized energy efficient kilns include the bio-oil condensation units and can obtain higher charcoal yields than it was reported for traditional kilns [6,182].

Charcoal yield from pyrolysis of "dead" biomass is similar to "living" biomass, whereas the bio-oil yield is approximately 1% greater for dead biomass at the cost of pyrolysis gases [210]. Thus, charcoal yield from sapwood at short rotation forestry (≤20 years) is similar to stem wood (sapwood and heartwood) from classical forestry (50–150 years cycle). The low solid yield in fast pyrolysis indicates a limited potential of charcoal use in ferroalloy industries [74,101,177,179]. However, bio-oil utilization as a binder or an electrode material using fast pyrolysis opens new opportunities for the process upscale in ferroalloy industries.

To estimate the usable carbon fraction for metallurgical application, yield and fixed carbon content of charcoal were combined in the fixed carbon yield. The fixed carbon yield is defined as: $\gamma_{FC} = \gamma \cdot (FC) \cdot (1 - a)^{-1}$, where γ_{FC} is the fixed carbon yield, γ is the charcoal yield, FC is the fixed carbon content of the charcoal and a is the ash content of the biomass [1,209,211]. The fixed carbon yield of charcoal varies from 20 to 30 wt.% on dry basis [155,208]. Bio-oil utilization can increase γ_{FC} by 4–5%-points [63,105]. Herbaceous biomass such as sugarcane has a lower conversion efficiency than wood pyrolysis, resulting in a lower bio-oil yield and lower exergy efficiency than that of wood pyrolysis on area basis despite its greater growth rate [212].

Table 4. Charcoal yield for the pyrolysis of biomass.

Biomass	Pyrolysis Process	Charcoal Yield	Temperature	Firing Time	Source
		/wt.%	/°C	/h	
Wood	Small scale pyrolyzer	18–40	400–700	0.1–6	[40,155,213]
Biomass waste	Small scale pyrolyzer	14–45	390–900	0.5–3	[214–216]
Herbaceous biomass	Small scale pyrolyzer	18–48	400–700	0.1–35	[155,217]
Biomass	Pressurized small scale reactor	35–40	500–650	<0.5	[1,218]
Biomass	Fast pyrolysis (small scale)	23–35	450–650	<0.01	[35,217]
	Fast pyrolysis (industrial)	combusted	500–550	<0.001	[219]
	Fluidized bed	18–25	n.s.	<0.001	[25]
Hardwood	Charcoal kilns	9–34.2	450–600	80	[6,25,73,182]
Blend	Rectangular kiln	27–31	450–650 (700)	96–120	[220]
Cord wood	Kiln	8–9	n.s.	<6	[221]
Hardwood	Missouri kiln	20–30	450–550	80–144	[25,161,182]
Hardwood	Brick kiln	15–25	450–600	80	[25,73]
	Keyan earthmound kiln	21–34	n.s.	170–235	[182].
	Hot-tail kiln	38	n.s	40–50	[182].
Beechwood	Retort	30–38	450–550	11–22	[25,73]
	Twin-retort	33	500	12	[6]
	CML process	22–24	500	6–8	[161]
Biomass	DPC technology	34	500	6–20	[161]

3.2. By-Products

Volatile pyrolysis products are accounted as by-products from charcoal production. Volatiles are composed of light pyrolysis gases, such as CO, CO_2, CH_4, H_2 and light hydrocarbons [222,223], and condensable liquids, such as organic acids, methanol, heavy hydrocarbons, oxygenates, etc. [43,61]. The liquid products are defined as bio-oil, which is composed of more than 300 compounds with an oxygen content between 35 and 40% and heating value of 11 to 26 MJ kg^{-1} [223,224]. Water content of slow pyrolysis bio-oil is more than twice larger than that of fast pyrolysis oil [225] and strongly affected by the initial moisture content of the biomass feedstock. The high water content of slow pyrolysis bio-oil 50% [63,225] can lead to a phase separation, which generally occurs for bio-oils with a water content larger than 30–35% under storage [222]. This emphasizes the importance of dried biomass feedstock, since the moisture content increases the bio-oil yield and decreases the concentration of the organic fraction and its heating value. A moisture content less than 45–50% is required for an autothermal operation [226], respectively less than 20–25% for an efficient and stable operation of industrial scale retorts, such as the Lambiotte process [6,68].

The major organics in bio-oil are organic acids, alcohols (e.g., methanol: 1.4–7.9%), oxygenates (e.g., acetone: 0.3–1.6%) and tars (11.8–26.6%) [61]. Most of the compounds are water soluble, while 15–35% of bio-oil is water insoluble [227,228]. Chemical composition analysis of the water insoluble fraction showed a carbon content of 50–71%, hydrogen content of 5.8–6.7% and oxygen content of 23–43% [227,229]. The high oxygen content facilitates ageing reactions in the bio-oil, leading to an increased viscosity and average molecular weight from 530 to 990 g mol^{-1} [230]. In addition, the low pH-value (2–3) of bio-oil increases the maintenance cost for the condenser and storage system [224]. PAHs were found in bio-oil within a range of 10 to 50 mg kg^{-1} [231], which are accounted as harmful to the environment [185]. However, only one-third to two-third of the biooil compounds can been identified by GC-MS and HPLC [227,232].

About 10–20% of pyrolysis gas is formed by pyrolysis of biomass at temperatures up to 800 °C [43,225,233]. The yield of pyrolysis gas increases with increasing heat treatment temperature and can be 75% at 1200 °C [154]. Its composition is strongly affected by heat treatment temperature and operation of the kiln or retort. The pyrolysis gas in charcoal production is mainly composed of H_2 ($\leq$ 20.1%); CH_4 (5–22.0%); CO (30.1–58%); CO_2 (22.5–43%) and light hydrocarbons C_xH_y (2.3%) [210,222,223]. CO and CO_2 concentrations are increased in classical charcoal production by the partly combustion of additional feedstock, whereas H_2 concentration would increase at high heat treatment temperature by dehydrogenation reaction and carbonization of the charcoal [234]. Unburnt pyrolysis gas can therefore be accounted as a strong pollutant based on the high CO and CH_4 concentration [5]. However, the gas can have a heating value of 8000 kJ m^{-3} [5], respectively 2.3 MJ kg^{-1} [161]. Therefore, the released volatiles during pyrolysis can be combusted to provide the thermal energy for the process [5].

3.3. By Product Utilization

Pyrolysis of biomass results in the formation of 50% of liquid and 20% of gaseous by-products. In current processes, by-products can be combusted to provide the thermal energy of the process, increasing the amount of (renewable) CO_2 emissions at the same time. Unburnt by-products reduce the process efficiency and lead to net GHG emissions. Industrial retorts on the other hand recirculate the volatile compounds and can condensate parts as value-added by-products for consecutive processes [25]. Recent studies have shown that the charcoal yield can be increased by up to 5%-points by the bio-oil recirculation [63]. Pressurized retorts decompose a large fraction of the bio-oil inside the charcoal particles [157,158], decreasing the yield of high molecular hydrocarbons in the bio-oil.

The high molecular bio-oil can be used as feedstock material for green electrodes, as a feedstock in gasification, as a binder for briquetting or pelletizing or as fuel for boilers according to EN 16900, i.e., as standardized, readily marketable product [180,181]. Briquetting and pelletizing enable a conditioning of charcoal fines at the pyrolysis plant and thus, increase the carbon yield suitable in ferroalloy industry. In addition, the homogeneity of charcoal briquettes and pellets is increased. Green electrodes have successfully been produced from fast pyrolysis bio-oil [181], but composition of bio-oil from fast pyrolysis differs from that of slow pyrolysis [225,235]. The high water content and the large fraction of organic acids, such as acetic and propionic acid, inhibit a direct usage as engine fuel [61]. Distillation of the tarry compounds will result in the formation of 35–50% of solid residues [224] and is only limited applicable for the bio-oil conditioning. Recent studies focus on the bio-oil upgrading with in-situ generation of hydrogen [236].

Gasification of bio-oil can provide a synthetic gas for fuel and chemical production [237,238]. Some bio-oil components have established markets which were supplied by coke oven batteries and other pyrolysis processes before products were synthesized by natural gas and other fossil fuels, such as acetic acid. Bio-oil conditioning may provide renewable chemicals which can directly substitute fossil fuels in the future. In short term, bio-oil as a binder can increase the charcoal yield and concurrently reduce waste streams.

4. Charcoal Properties

Properties of charcoal are crucial for the use as a reductant in metallurgical processes [16,18]. To minimize changes in a process design and optimization, charcoal properties are expected to be similar to those of fossil-based reductants, as summarized in Table 5. Fossil-based reductants show high mechanical stability and low to medium gas reactivity at a low price, whereas charcoal has a high gas reactivity and low mechanical stability. In addition, charcoal has a lower density than fossil-based reductants, that may change the mass flow and volume distribution of the raw materials in the SAF. Charcoal is produced in large proportions by kilns as a household fuel that does not require specific properties. However, the quality of charcoal produced in kilns can significantly vary from one batch to another.

Understanding mechanical, chemical, physical and electrical properties of charcoal is a key step that affects the use of charcoal in ferroalloy industries [16,18,29,239]. Mechanical strength is required to avoid collapse of the carbon bed by the load of the burden, while abrasive strength (or durability) is required to minimize generation of fines while handling and feeding of charcoal particles into the reactor [239]. Chemical properties such as carbon content and reactivity are essential to provide the carbon in the reaction zone, while ash content and composition of the feedstock can affect the product quality and energy demand [172]. Electrical properties are crucial for the quality of carbon bed material in the SAF to provide the thermal heat by electrical energy dissipation.

Table 5. Required properties of carbon reductants in the ferro and ferroalloy industry [29,239–245].

Property	Unit	Blast Furnace	Ferroalloy	
			Silicon	Manganese
Moisture	[%]	1–6	$\leq$6	$\leq$6
Fixed carbon	[%]	$\geq$85	$\geq$84	$\geq$85
Volatile matter	[%]	$\leq$1.5	$\leq$9.5	$\leq$3
Ash	[%]	$\leq$12	$\leq$12	$\leq$12
Phosphorous	[%]	$\leq$0.06	<0.02	<0.02
Sulfur	[%]	$\leq$0.9	$\leq$0.6	$\leq$0.6
Size	[mm]	47–70	5–40	5–40
Bulk density	[kg m^{-3}]		400–500	400–500
Ash fusion temperature	[°C]		1250–1450	1250–1450
CSR	[%]	$\geq$60	-	-
CRI	[%]	20–30	-	-

4.1. Chemical Analysis

While the charcoal yield is the most important variable for its production, specific charcoal properties are compulsory for its application. The chemical analysis comprises the proximate and ultimate analyses, as well as the structural properties. The carbonization degree increases with increasing heat treatment temperature, resulting in an greater fixed carbon content of the charcoal. The carbonization results in structural changes of the carbon matrix, affecting functional group and reactivity of the sample.

4.1.1. Proximate and Ultimate Analyses

Carbon content increases with increasing heat treatment temperature and residence time. Classical charcoal production results in a carbon content greater than 70% [161]. The proximate analysis of classical charcoal production is shown in Table 6. Based on the biomass feedstock, industrial retorts provide a carbon content of 80–94% [6,37]. Charcoal with a fixed carbon content greater than 83% is considered as barbecue charcoal according to EU requirements [6] and as metallurgical grade charcoal with a fixed carbon content greater than 85% [152]. A fixed carbon content greater than 85% is therefore recommended in ferroalloy production to minimize additional slag formation by ash constituents and off-gas cleaning by volatile matter [239]. Metallurgical coke used in a closed SAF has a volatile matter content less than 3 wt.% (generally 1–1.5% [246,247]), whereas more than 10% of volatiles are released by charcoal samples. Volatile matter content of charcoal pellets is increased to more than 24% by the addition of water and binder [105,172,247]. This high volatile matter content requires an improved off-gas cleaning system and increases the risk of slag foaming in secondary steelmaking routes [248]. One possibility to reduce volatile matter content similar to metallurgical coke is to use the secondary heat treatment at temperatures of 1300 °C [63].

The ash content of charcoal is correlated to the ash content of the feedstock. Therefore, the ash content is low for stem wood charcoal and relatively high for charcoal from herbaceous biomass i.e., wheat straw, rice husks, etc. [38,155]. Ash content with more than 8–12 wt.% should be avoided in SAF to reduce the amount of impurities and generated slag [239]. Biomass ash is rich in alkali and alkali earth metals compared to fossil fuels, but low in sulfur content. An increased input of alkali metals can cause corrosion issues in the off-gas system and change reduction stages in the burden [172].

Table 6. Proximate analysis of industrially produced charcoal.

Reactor	Biomass	Ash	FC wt.%	VM	Source
	Wood	0.4–4	65–85	15–35	[36,37]
	Wood	1.9	84.7 (*)		[249]
Kiln	Mangrove	1.66	67.53	29.15	[161]
Kiln	Rain Tree	1.79	78.52	18.22	[161]
Kiln	Rubber	4.67	78.98	15.24	[161]
Twin-retort	Hardwood and softwood	-	92% *	-	[6]

* based on ultimate analysis.

4.1.2. Structural Analysis

The carbon structure of charcoal is highly affected by the heat treatment temperature and the surrounding atmosphere, for example gasification agents or oxygen. The higher carbonization degree at higher heat treatment temperature results in a decreased CO_2 reactivity and increased electrical conductivity [1]. Torrefaction and low pyrolysis of biomass on the other hand only slightly increase the carbonization degree. The cellulose peak can be still observed by XRD analysis after a heat treatment at ≈340 °C [220].

Charcoal produced at 240 °C exhibited the same FT-IR bands as biomass, whereas specific bands of cellulose, hemicellulose and aromatic ring groups from lignin disappeared after a heat treatment temperature of 400 °C [250]. This change in chemical composition started at temperatures above 350 °C [251]. A uniform spectra with strong aryl carbon signals indicated poly-condensation of aromatic rings at temperatures above 500 °C [220], in which fast pyrolysis char showed a lower aromaticity compared to classical charcoal from slow pyrolysis. In addition, these results correspond to mass loss curves from the thermogravimetric analysis [155,172].

Charcoal structure is different from fossil fuel reductants and graphite. The charcoal structure is characterized by the greater surface area and absorptive capacity than carbon black or graphite [40]. Structural changes in coal graphitization occur mainly in four temperature ranges, 1000–1500 °C, 1500–2000 °C, 2000–2500 °C and 2500–3000 °C [252,253]. The basic structural units reorganize between 800 and 1500 °C and coalesce between 1600 to 2000 °C [253]. A nano-crystalline structure of charcoal was observed at heat treatment temperatures larger than 1300 °C [62,200]. This temperature range is above the production temperature of industrial charcoal, but structural changes will appear in the burden and carbon bed inside SAF. Charcoal produced at temperatures above 1000 °C exhibited a similar structure to petroleum coke [254]. The increased graphitization with increasing heat treatment temperatures improves the electrical properties of the charcoal.

4.2. Density and Porosity

The density and porosity are important properties for the handling, transport and storage of bulk and compacted charcoal. Density of charcoal is highly affected by the feedstock and process conditions, in which pyrolysis of softwood results in a charcoal with a lower density compared to hardwood. Density of parental wood increases with age [255], resulting in a superior charcoal from older wood. However, due to the easier and less labor-intensive handing of young growth fresh wood, secondary forestry is commonly used as a feedstock in charcoal production [11]. A greater particle density results in improved mechanical and electrical properties, which are important properties for the renewable reductant. The skeletal density of charcoal produced at temperatures up to 500 °C is similar to that of the parental wood ($\approx$1400–1550) kg m^{-3} [220,256]. Above 500 °C, skeletal density starts to increase with increasing heat treatment temperature [155,208,257] to $\approx$1750 kg m^{-3} at 700 °C [258] and to $\approx$2000 kg m^{-3} at 800 °C [220].

While true density increases with increasing heat treatment temperature, charcoal's bulk density is nearly constant in the temperature range from 450 to 650 °C [257,258], leading to a greater BET surface area and porosity compared to low temperature treated charcoal. Metallurgical grade charcoal has a bulk density of 180 to 350 kg m^{-3} [36,37,201], about half of that for metallurgical coke (450–600 kg m^{-3}) [36,37,239,246], leading to higher transport and storage costs. In addition, the lower density of charcoal increases the volume fraction of the carbon reductant in the SAF, possibly decreasing the maximum load of the furnace by its available construction height.

The porosity of charcoal increased from $\approx$50% at 300 °C to 50–70% at temperatures above 700 °C [155,259]. The porosity of spruce charcoal was about 10% points larger than that of oak charcoal [62] and similar to that of grass charcoal [258] and other woody biomass species [123]. On the other hand, total porosity of metallurgical coke ranges from 25 to 62%, increasing with larger particle size [18,62,239]. It is known from cokes that the porous structure can affect the mechanical strength of particles [260]. While a relationship between tensile strength and pore structure of coke was determined in previous studies [261], no clear correlation was found for the mechanical strength and porosity by other authors[262].

Bulk density of charcoal can be increased to 900 kg m^{-3} by agglomeration and compaction [172], in which an increased particle density is achieved by blending with an organic binder and increased compaction pressure [124,150]. The addition of organic binder such as lignosulphonate and bio-oil significantly improved the density and mechanical durability of charcoal pellets [150]. A further increase in a bulk density was reported by the increased residence time of pellets and briquettes during compaction [124]. An apparent density of about 1000 kg m^{-3} was reported for charcoal produced under a compressive pressure of 500 kPa, that was similar to that of metallurgical coke [259]. On the other hand, the addition of 5% sawdust to a coal blend resulted in a slightly decreased bulk density of coke, from 780 kg m^{-3} for metallurgical coke to 770 kg m^{-3} for bio-coke [262].

4.3. Surface Area

Operation of SAF is affected by the gas reactivity of the carbon material in the burden due to the reaction with CO_2 and SiO. The gas reactivity depends on the accessible surface area and presence of functional groups on the surface. The morphology of charcoal is highly affected by the heat treatment temperature, residence time and gas pressure [263]. The specific surface area increased from 5 m^2 g^{-1} at 300 °C to 80 m^2 g^{-1} at 500–600 °C [32,264–266] and further to 500 m^2 g^{-1} at 800–900 °C [99,213]. Above 1000 °C, the specific surface area of charcoal decreased to about 10 m^2 g^{-1} [62,99]. The increase in surface area was correlated to the increase in micro- and mesoporosity and the collapse of micropores thereafter [32,263].

Charcoal produced at intermediate pyrolysis pressure (10 bar) showed a smaller surface area than charcoal generated at low pressure (5 bar), respectively high pyrolysis pressure (20 bar) [263]. Yang et al. reported that the surface area decreased with increasing heating rate [35] and was generally lower for herbaceous charcoal than for woody charcoal [32]. The results indicate that the release and secondary pyrolysis reactions of volatile matter has a high impact on the surface area of charcoal. Fossil fuel reductants, such as metallurgical coke, bio-coke and petroleum coke, exhibit a surface area of less than 3 m^2 g^{-1} [21,63,267]. Metallurgical coke and bio-cokes are produced at temperatures above 1100 °C, while petroleum coke is the solidified residue from crude oil distillation. Differences in the accessibility were observed for the surface characterization in N_2 and CO_2. Surface area from oak and pine charcoal increases from less than 3 m^2 g^{-1} at 400 °C to 220–280 m^2 g^{-1} at 650 °C for measurements in N_2, respectively from 250 to 530 m^2 g^{-1} for oak and 360 to 650 m^2 g^{-1} for pine in

CO_2 [268]. Thus, the accessibility of gas species into the pore volume can also affect the reactivity in different atmospheres.

4.4. Mechanical Properties

Mechanical properties of charcoal are essential to avoid generation of fine particles by transport, handling and feeding into the SAF. The furnace operation and off-gas cleaning system are significantly affected by the particle size of original feedstock and charcoal [18]. In general, charcoal is more fragile than fossil fuels samples due to its high porosity and lower carbonization degree [253]. Global charcoal transport results in the generation of 5–20% fines [159,269–271] due to the loading and unloading of the charcoal [271]. This material is not directly suitable for a usage in SAF, since a particle size of 5–40 mm is required for ferroalloy industry to ensure a good gas permeability [29,239]. In particular, the silicomanganese industry uses fossil fuel reductants with a particle size ranging from 5 to 20 mm [29]. Thus, particles less than 5 mm are considered as fines and should be avoided.

Standard tests for measuring the mechanical properties are drum tests, such as ASTM D 294-64, ISO 556 and DIN 5171 for coke, and DIN EN ISO 17831-1 for pellets and briquettes. In addition, mechanical fragmentation is investigated by the drop shatter test for coal according to ASTM D440. Hot strength of metallurgical coke is investigated by the coke strength after reaction (CSR). The combination of hot strength and CO_2 reactivity are important parameters to evaluate the quality of reductants in the burden [37], especially for blast furnaces. However, most of the standards are not applicable for charcoal and have therefore been modified by the researchers [36,37].

While a very high compression strength is required in blast furnaces, mechanical strength is of less importance in SAF due to the low height of the burden [40,239]. Compressive strength of charcoal is 10–80 kg cm^{-2}, about half of that of metallurgical coke (130–160 kg cm^{-2}) [36,201]. Charcoal produced from hardwoods have generally a greater mechanical strength than charcoal from softwood or herbaceous biomass [152] and the mechanical strength is 3 to 4 times greater in the growing direction of wood fibers compared to its perpendicular direction [36]. The compression strength slightly increased after the sample was heat treated to elevated temperatures [201,272], indicating improved mechanical properties inside the SAF. Carbonized wood is about 28% stronger than its biomass precursor [273].

4.5. Charcoal Compaction

Charcoal produced at elevated temperatures has a lower tendency to agglomerate compared to raw biomass and requires the addition of a binder for compaction [123]. Charcoal fines can be compacted by pelletization and briquetting to increase bulk density and mechanical stability for the use in metallurgical industry [124,172,274]. Coal tar, coal tar pitch and bio-oil pitch have been used as binder for special purposes over decades [271]. Recent studies have confirmed that the mechanical properties can be improved for charcoal pellets and briquettes by the addition of bio-oil and bio-oil pitch and a second heat treatment [63,150,275]. The mechanical durability of the pellets increased to greater than 90% [150], similar to that of heat treated hydrochar pellets [129]. Pellets compacted at 116 MPa with an oil concentration of 34% provided optimal pelletizing conditions [180]. A slightly better mechanical durability was achieved by adding water and molasses at a compaction pressure of 100 MPa [172]. One of the major disadvantages of molasses is the addition of mineral matter (especially potassium) to charcoal pellets and briquettes. Other important parameter are particle size and moisture content, and the glass transition temperature for lignocellulosic binder [124]. Drop strength tests approved the results from mechanical durability test for charcoal pellets [172].

The compressive strength of charcoal pellets increased from less than 1 MPa to greater than 2 MPa after the secondary heat treatment [150]. A similar compressive strength was reported for charcoal briquettes using bio-oil binder [275]. The mechanical compression resistance of green pellets produced with molasses as a binder increased from 1 MPa to 2.6 MPa after storage for 12 days [172]. Similar values were found for fossil fuel briquettes compacted at 56 kPa [199], in which a higher compaction pressure resulted in improved mechanical properties and hardwood lignin developed twice the compressive strength than softwood lignin. Lignin as a binder was improved by adding tannic acid, in which the mechanical strength was also obtained as hot strength for temperatures up to 1400 °C [199]. A compressive strength of 4.2 MPa was reported for briquettes using starch binder [275], which was similar to that of charcoal pellets using bio-oil binder [150]. To ensure mechanical strength at higher temperatures, silicon metal can be added to the charcoal fines, forming a network of silicon carbide nanowires between the carbon matrix [276]. The carbide nanowires improved the resistance to compaction from 1 MPa to 3–4 MPa at 1200–1400 °C [199].

Pellets started to shrink after the primary heat treatment temperature was surpassed [142]. This additional size reduction must be considered if the particle size of charcoal pellets is close to the minimum particle size requirements. Pellets from fast pyrolysis and slow pyrolysis differ by its aromaticity and volatile matter content [172], leading to differences in a shrinkage at higher temperature. If lignin is selected as binder, a secondary heat treatment temperature above 450 °C should be chosen to form polyaromatic structures between the particles [277]. The additional heat treatment at elevated temperature also reduces the risk of self-heating under transport and storage [274] and decreases CO_2 reactivity [142]. In summary, the quality of pellets and briquettes is highly improved by the second heat treatment.

4.6. Electrical Properties

The electrical conductivity is one of the most important properties of carbon reductants in SAF and EAF [239,253,278]. The electrical conductivity of charcoal is affected by the heat treatment temperature (carbonization degree), density of the packed bed, particle size and volume fraction of carbon, as well as operating temperature and pressure inside the furnace [1,201,253,279,280]. Metallurgical coke used in SAF is expected to have an electrical resistivity of 7–10 mΩ m at room temperature and 1–10 mΩ m at 1400 °C for a particle size of 5–20 mm [37,239,280]. Charcoal on the other hand has an electrical resistivity above 10^6 Ω m at room temperature, decreasing to 14–23 mΩ m at 1000 °C and 9–18 mΩ m at 1400–1600 °C for 5–35 mm respectively 5–10 mm particles [37,201]. The electrical resistivity of charcoal particles from a heat treatment at 950 °C was similar to that of fossil fuel coal char ranging from 1.7 to 3.4 mΩ m [281].

The electrical resistivity of packed carbon beds is higher than that of particles due to larger contact resistance between particles [279,282]. Electrical resistivity of metallurgical coke decreased from 10–15 mΩ m at 1000 °C to 5–8 mΩ m at 1600 °C [37,201]. Packed beds of charcoal particles larger than 2 mm showed an electrical resistivity of 50 mΩ m at 1100 °C and 15–20 mΩ m at 1600 °C [37,201]. The electrical resistivity of metallurgical coke decreased by 50% when the particle size increased from 5–10 mm to 15–20 mm [280]. However, the previous results have also shown that only minimal differences were observed for small and large sizes of charcoal particles [201]. The compaction pressure of the burden has a large impact on the electrical resistivity. An increased compaction pressure decreases the air gap between particles and increases the number of contact points between the particles, resulting in a decreased electrical resistivity [18,201,279,283].

The loss in ohmic resistance can be attributed to the removal of oxygen groups by carbonization from the carbon matrix [252,284], especially at temperatures less than 950 °C. The high oxygen content and disordered carbon structure act as insulators in the carbon matrix of biomass, charcoal and coal [1,285–287], in which a decrease in electrical resistivity of 5 orders of magnitude was observed after the release of volatile matter [1,288]. Therefore, electrical resistivity of carbon reductants can be correlated to the ordering of the carbon matrix and its reactivity [281]. The higher electrical resistivity of charcoal inhibits the conduction of current through the burden inside the SAF [288,289], where ≈5–15% of current is conducted through [18]. In addition, the higher electrical resistivity of the charcoal enables a lower electrode tip position, which improves heat distribution in the lower part of SAF [280].

4.7. Gas Reactivity

Gas reactivity of carbon reductants is an important property for the burden in SAF. About 30% of the CO_2 emission from ferro-managnese and silicomanganese production are related to the Boudouard reaction ($C(s) + CO_2 \rightleftharpoons 2\ CO$) [290], indicating the importance of a low gas reactivity towards CO_2. A high reaction rate will increase the coke and power consumption and is therefore undesirable in silicomanganese and ferromanganese production [36]. The CO_2 reactivity of metallurgical coke is investigated according to coke reactivity index (CRI), where the mass loss is measured by difference after a 2 h heat treatment (1100 °C) in pure CO_2. The standard can not directly be adopted for charcoal samples due to its high gas reactivity. By AC-method, charcoals reactivity can be 30 to 40 times larger compared to metallurgical coke [36]. Many researchers investigate the CO_2 reactivity by thermogravimetric analysis at lower temperature, lower CO_2 concentration or by non-isothermal analysis [37,63,204,291,292]. In addition, knowledge of the CO_2 reactivity can be adopted from fixed bed biomass gasification for syngas production [293].

The CO_2 reactivities of metallurgical coke and charcoal samples produced at 1060 °C were compared. Carbon conversion of metallurgical coke was in the order $(0.2\text{–}0.5)\cdot10^{-2}\%C\ s^{-1}$, and thus, about 5 to 16 times lower than that for industrial and laboratory charcoal that ranged from $(2.1\text{–}3.2)\cdot10^{-2}\%C\ s^{-1}$) [37]. Similar results were obtained by other researchers [105,204], in which charcoals from forest residues and herbaceous biomass generally have a higher reactivity than those from stem wood [204]. CO_2 reactivity of charcoal samples post treated at temperatures larger than 1600 °C approached that of fossil fuel reductants [62,99]. Despite the high porosity and surface area of charcoal, CO_2 reactivity of charcoal samples decreased by about 20–50% when the particle size from 60 µm to 1–2 mm [294,295]. Previous studies have shown that up to 70% of the carbon conversion depends on the external surface area and the potassium content in original feedstock [291]. The higher reactivity of carbon fines may improve the stability of SAF by the consumption of small particles in the burden, resulting in an improved gas permeability of the bed.

Alkali and alkali earth metals are stated as catalysts for the Boudouard reaction [158,204,291]. The catalytic index is shown as the quotient of the sum of catalytic elements (K, Ca, Mg, Na, Fe) to inhibiting elements (Si, Al) [291]. Bio-based reductants with a low mineral matter content showed a similar reactivity compared to the fossil coke [20]. However, potassium impregnated coke was more reactive than low ash charcoal, but less reactive than potassium impregnated charcoal [36]. These results indicate a dominating role of potassium on the CO_2 reactivity, confirming the previous results [97]. In addition, alkali and alkali earth metals can recirculate in the burden, resulting in an accumulation of potassium in the SAF [36]. The CO_2 reactivity is also highly affected by CO_2 partial pressure [200,296–299]. In addition, the high concentration of CO is desirable for ferroalloy production to pre-reduce higher metal-oxides in the burden and as a value-added product gas for other applications [300,301]. Beside the Boudouard reaction, charcoal also reacts with SiO gas in silicon and ferrosilicon production. The SiO reactivity is determined by passing hot SiO gas (13.5% SiO, 4.5% CO) through a carbon bed according to the improved SINTEF procedure, which was developed by Tuset and Raaness in the 1970's and described elsewhere [302]. The SiO gas will react with the carbon reductant according to: $SiO(g) + 2\,C(s) \rightleftharpoons SiC(s) + CO(g)$, thereby forming a solid silicon carbide layer at the surface. Therefore, it can be assumed that the SiO reactivity is mainly depending on the porosity and accessible surface area of the carbon material, eventually blocking the reaction surface and decreasing SiO reactivity over time [94]. The general trend is SiO reactivity of charcoal $\geq$ coal $\geq$ metallurgical coke $\geq$ petcoke [94,303,304]. It is also known that about 80% of the silicon from the gas phase can be recovered in the ferro-silicon production, while 20% of the SiO gas is discharged by the off-gases [239]. Therefore, woodchips and charcoal can improve the gas permeability of the burden, prevent the crusting of the charge and increase the recovery of SiO gas [253,305].

4.8. Slag Reactivity

Reduction of metal-oxide to the pure metals takes place in the liquid state by carbon reductants in the lower part of SAF, for example by the reaction: $MnO + C \rightleftharpoons Mn + CO$. Carbon can be provided as solid carbon at the contact point between the slag and the "coke" bed, and by the dissolution of carbon into the liquid metal. The sessile drop wettability test [306] and thermogravimetric analysis (TGA) at $1600\,^{\circ}C$ [37] are techniques to evaluate the slag-carbon reactivity in ferroalloy industries. Thus, slag reactivity can be characterized by its dissolution and by the reaction kinetics [253]. In the former, the slag reactivity is depending on the dissolution rate of the carbon material into the liquid metal phase. Alkali metals in charcoal can decrease the carbon solubility into the liquid metal. It is known from carbon dissolution into liquid iron that S, P, Si, Al, Ni and Co decrease the solubility of carbon, whereas Mn and Cr slightly increase it [253]. Charcoal and fossil coke can be equally used as reductants in silicomanganese production [36], in which the dissolution rate constant is significantly improved by increasing heat treatment temperature [37,253] due to the increased degree of crystallinity of the carbon matrix [94]. For example, the dissolution rate constants of coal can vary in the range 0.0011 to $0.0036\,s^{-1}$ [253].

The kinetics of metal-oxides reduction can be investigated by using TGA coupled with an infrared analyzer at temperatures above $1500\,^{\circ}C$ [307,308]. A high graphitization degree of the carbon matrix is beneficial for the high reaction rate. However, reductants with a low degree of graphitization and large occupying volume fraction tend to have a higher reactivity towards manganese ores due to their large surface area [278]. Coke has a faster slag reactivity than charcoal and much faster than graphite, whereas the highest reactivity of charcoal is observed in alkali lean charcoal samples [309]. However, the kinetics for slag and MnO reduction can change by the presence of metal phases, such as reduced FeO and SiO_2 [253,309]. Silicon and ferrosilicon production can be accounted as slag free [30], and the main reduction reaction is given by the SiO-reactivity as discussed above. Thus, highest quartz-carbon reactivity can be observed for charcoal and coal [304]. The different tendencies in reactivity of gas-carbon and slag-carbon inhibit a correlation between the two properties [306].

5. Available Charcoal Technologies

Charcoal has been produced batch-wise in different kiln technologies since millenia, and some of the old inefficient technologies are still used in developing and newly industrializing countries [8]. In general, charcoal production can be divided into kiln technologies where the heat is provided directly by the partial combustion of additional feedstock, respectively [25]. A schematic of the different heat supply system for charcoal production is shown in Figure 2, whereas the production capacity and production rate are summarized in Table 7. In general, maximum heat treatment temperature for charcoal production in kilns and retorts is between 450 and $650\,^{\circ}C$ [68,161,220].

The pits and earthmound kilns are the simplest technologies for the charcoal production. The space capacity of the pits are in the range of 4 to $30\,m^3$ of fuel wood, while earthmound kilns can be enlarged up to a capacity of $150\,m^3$ [25]. Earthmound kilns have been improved in the second half of the 19th century by adding a chimney, and provided large amounts of charcoal for iron smelting industry [25]. The additional chimney enabled a precise control of the air supply by ensuring optimum draught conditions, and resulting in a greater charcoal yield with better properties [25,310]. The chimney can also act as an air cooler to condense heavy hydrocarbons, acids and water, which are value-added by-products from pyrolysis process and improve the control of the kiln. Chidumayo estimated that it is unlikely to futher increase the charcoal production rate of 23.3% in earthmound kilns [311]. Brick kilns have been used and improved over the last decades. The Missouri kilns provide a volume capacity of $180–800\,m^3$ and a charcoal yield of 20–30% [160,161], about twice than that of classical earthmound kilns. The cycle time of these kilns is typically in the range from 7 to more than $30\,days$ [11,160,226]. Largest kilns have a capacity of $2000\,m^3$ and can produce 10,000 tons of charcoal per year [161].

Most of the charcoal in the EU is produced in steel retorts [161]. Biomass carbonization is thereby carried out in continuously operated processes, such as the Lambiotte retort, which is used since 1942 with a production capacity of 2000–9000 tons of charcoal per year [68]. Heat treatment temperature in Lambiotte retort is 600 °C in the carbonization zone [68]. Other technologies, such as the Carbon engineering technology are operated at 500 °C that is similar to the maximum temperature in kilns [161,220]. A horizontal retort (also called Twin-retort) has been developed in the 1990s by Norit B.V., in which multiple retorts can be combined to increase production rate and reduce emissions [6]. Similar technologies are the VMR systems with two retorts and one central combustion chamber and a production rate of 6000–7000 tons per year [226]. Other retorts are Reichert retort, Lurgi/Degussa process, Nichols-Herreshoff Process and the rotative furnace [8,25]. Carbonex is the most recent process, combining the pyrolysis unit with a 1.4 MW power production [161]. The largest industrial charcoal plants produce about 25,000–27,000 tons of charcoal annually [207,208].

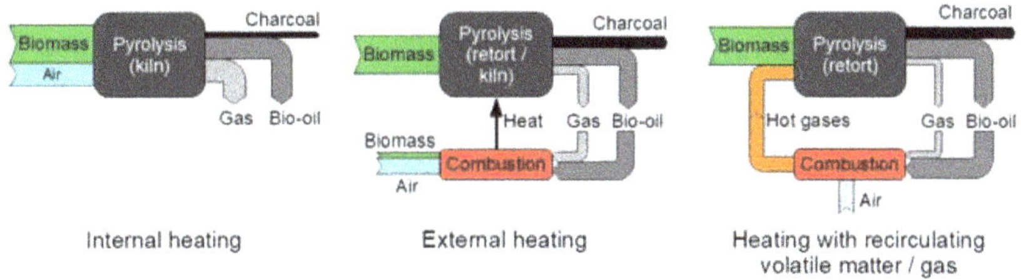

Figure 2. Possible heating systems in charcoal production (after [25,29]).

Table 7. Charcoal production technologies.

Brand or Technology	Size	Production Rate [mg/year]	Sources
Kilns			
Earthmound kilns	9.5 m^3	945 kg per cycle	[11]
Earthmound kilns	400–32,000 m^3		[182]
Egyptian process	3–7 tons per cycle (2–3 weeks duration)		[5]
	1–2 tons per cycle (3–5 weeks cycle)		[5]
Double wall kiln	1 m^3		[5]
Reactangular kiln	64–106	10 tons per cycle	[220]
Brazilian rectangular with tar recovery	80,000 m^3		[182]
Brazilian round brick (with chimney)	20,000 m^3		[182]
Brazilian hot-tail (no chimney)	4000 m^3		[182]
Rima Container Kiln (semi-continuous)		1 ton/hour	[160]
Retorts			
VMR		6000–7000	[226]
Twin-retort	3 m^3 of solid wood	900–11,000	[6]
CML (12 retorts)		2000–3000	[161]
DPC technology		3000–15,000	[161]
CK-1 EKKO (mobile kiln)		300–600	[161]
Lambiotte process	16.3 m height, 600 m^3 volume	2000–8000	[68]

5.1. Economics

Charcoal markets are established in Africa, Asia and South-America, where charcoal is the primary fuel source to urban and rural households. The largest wood charcoal markets in 2018 were Brazil, Ethiopia and Zambia, with a total market size of about 6.7 billion US dollars [312]. This charcoal was often produced by earthmound kilns at a low efficiency, but also at low costs [311]. Due to the low capital expenditures (CapEx) and operating expenses (OpEx), the economics of such charcoal production is mainly driven by the charcoal market prices and capacity factor for charcoal production [6]. The charcoal production in the EU countries is always highly affected by the investment costs, OpEx and biomass price, such as energy saving and production efficiency [166]. Charcoal price of 250 Euro per ton of charcoal in 2002 was required to cover investment costs of about 500,000 Euro [6]. However, the cost of the biomass feedstock is one of the major obstacles for charcoal or bio-oil production [172].

Sweden, Norway and Finland have large unused potential of forest areas available for wood production, e.g., 19.6 million ha in Finland [313] or about 10–15 million m^3 in Norway [200]. Cornifer roundwood, such as spruce and pine, are the main wood species in the northern countries, accounting for more than 75% of the annual harvest [313,314]. Some sawlogs may not provide the strength characteristics for a construction industry and can be considered as a possible feedstock for charcoal production [313]. There is a deficit of biomass mobilization and only limited integration of biomass resources in other industrial sectors [26], which is also valid for global roundwood markets with a total global trade volume of 135 million m^3 (7% of industrial roundwood production) [313]. The limited market integration can lead to the fluctuations in prices and can also inhibit investment in new technologies or methods to produce charcoal, whereas the metallurgical industry can be a large scale customer for renewable reductants over the long term.

Bio-refineries can improve the economics of charcoal production routes by enabling integrated value-added products at reduced fuel prices, and can be a possible key for future commercialization [315]. Bio-fuel production, which is well integrated in Europe, can establish the first stage of a bio-refinery [60], and the solid residues from the emerging bio-fuel markets can be used as feedstock for metallurgical grade charcoal production. The principles of bio-refinery can improve the overall economics by utilizing the complete feedstock or minimizing waste stream generation [316]. Charcoal can be extracted from bio-crude production, which would result in an increased number of decentralized pyrolysis plants to cover the bio-fuel product stream [317]. Bio-oil can be processed and specific compounds can be separated and used as value-added products [178]. For example, a natural smoke was extracted and commercialized by Ensyn as a food additive, whereas acetic acid can be used by chemical industry to produce vinyl acetate or ethyl acetate [178].

Charcoal production in developing countries is often carried out by unskilled operators in pits and earthmound kilns without a licence [221]. In some countries more than 75% of energy demand are covered by firewood and charcoal [318,319]. However, the income is especially important for households with loss agricultural capacity and limited stocks [318] and can account for more than 2% of GDP of the country [221]. Although the cost share of licensing is very low, sustainable wood sourcing requires the adequate governmental regulations to reflect the sustainability of a sustainable secondary forestry [7]. Certification of charcoal by Programme for the Endorsement of Forest Certification (PEFC), Forest Stewardship Council (FSC) or European Biochar Certificate (EBC) can help to distinguish between sustainable charcoal produced from secondary forestry and charcoal produced from wood made from primary forestry. Modern technologies can also increase the conversion efficiency and reduce the risk of local pollution, while long term contracts can ensure a stable price range for local charcoal producers and the ferroalloy industry.

6. Discussion

Charcoal has been used in the metallurgical industry over millenia and is partly used in the ferroalloy industry, such as silicon and silicomanganese production in Brazil. Typical reductants used for the silicon production are coal, petroleum coke and charcoal, while woodchips are used to ensure a good gas permeability in the burden [23]. Thus, fossil fuel reductants and renewable reductants are considered as carbon sources in silicon metal production [320]. Woodchips and charcoal may fully replace fossil-based reductants when their properties are comparable with the requirements for the use in open furnaces, e.g., reactivity, mechanical abrasion and mechanical strength. Minimum requirements vary by ferroalloy production and furnace type, in which reactivity toward CO_2 and dissolution rate into the metal are important parameters for ferromanganese and ferrochromium production, whereas SiO reactivity is important for ferrosilicon and ferrochromium reduction [94]. The high volatile matter content of classical charcoal is larger than 10 wt.% resulting in off-gas cleaning issues in closed hearth furnaces during reduction of manganese ferroalloys (both FeMn and SiMn).

Charcoal fines can cause two types of problem. Charcoal fines can reduce the gas permeability in the burden, resulting in an unstable operation of the furnace. Another challenge is that charcoal fines can be carried over into the gas cleaning system by the generated gases. Transport and handling can result in a fines generation up to 20%, which are not suitable in SAF. Charcoal pellets mixed with the bio-oil binder showed similar mechanical properties without an impact on the CO_2 reactivity to that of fossil-based coke. The use of organic binders i.e., molasses or starch resulted also in high mechanical stable pellets and briquettes [275]. However, the high ash content in molasses i.e., K, Na, Ca can interact with the organic compounds in other organic binders. The disadvantage of charcoal pelletization is the requirement to comminute particles. Briquettes can be produced from fines without additional milling and with satisfactory mechanical properties for the ferroalloy industry. The secondary heat treatment is required to ensure the high mechanical stability of charcoal-based pellets or briquettes.

The large volatile matter content and high CO_2 reactivity of charcoal may cause problems in the burden of SAF. The structural and chemical properties, mineral matter composition of the charcoal can be adjusted using post-treatment processes [291]. While volatile matter content decreases with increasing pyrolysis temperature, the ash content increases inversely proportional to the release of volatile matter. The increased ash content in charcoal may inhibit the use of alkali rich feedstock i.e., wheat straw, rice husks, bark, etc. In addition, the high alkali metal content of charcoal may cause problems in the upper part of the SAF where alkali metals tend to accumulate. In addition, different post-treatment processes, such as acid leaching and high temperature treatment, have shown that the ash content and reactivity of charcoal can be optimized approaching properties of fossil fuel reductants.

Currently, most of the charcoal is produced in developing countries and is locally used for heating and cooking. An increased global demand can be covered in a certain range, but a sustainable production may be limited due to the availability of energy efficient retorts or kilns. Earthmound kiln and pits can emit harmful emissions and pollute local environment by unskilled operators, such as local farmers involved in the unlicensed charcoal production. Deforestation and inefficient charcoal production can finally result in the net GHG emissions and should then be avoided as a substitute to fossil-based reductants. Certificated charcoal or charcoal from certified producers can reduce the risk of charcoal from primary forestry, e.g., by Forest Stewardship Council (FSC) or European Biochar Certificate (EBC). In the long term, modern charcoal retorts can produce high quality charcoal at high conversion efficiency, ensuring an income for farmers and operators at a stable price.

Charcoal production in Europe can ensure a sustainable biomass production and efficient charcoal manufacturing. However, the lack of market integration and fluctuation of feedstock prices may inhibit investments in new technologies. Bio-refineries can combine the production of chemicals, fuels and reductants. Recent research has shown that properties of charcoal were not negatively affected by recovering value-added compounds before pyrolysis, for example waxes, steroids or fatty acids. The liquid by-product from pyrolysis can be used as a binder to enlarge the charcoal yield with the limited influence on the CO_2 reactivity.

7. Summary

Charcoal can be considered as a renewable reducing agent for the ferroalloy industry. Retorts operated at 550–600 °C produce a charcoal with a fixed carbon content greater than 85%, in which pyrolysis by-products are utilized to provide the thermal heat for the process. Charcoal produced under these conditions can directly be used in open hearth furnaces for the ferrosilicon and silicon production. The high SiO reactivity of charcoal makes it a superior reductant compared to coal, semi-coke or coke. However, the mechanical properties of charcoal are generally inferior to fossil-based reductants, resulting in up to 20% losses of fines by handling and transport. In addition, the compressive strength of charcoal is half compared to metallurgical coke, but still sufficient to withstand the mechanical load by the burden in SAF.

Charcoal fines can be densified through pelletization or briquetting using organic binder such as bio-oil, starch or lignin to reduce waste streams. It was shown that the mechanical properties of densified charcoal are improved by a second heat treatment and would also withstand the load of the burden in SAF. The organic binders i.e., bio-oil are preferred over the inorganic binders due to the impact of alkali matter content on the reactivity and mechanical strength of charcoal samples. The high potassium content in charcoal may be a hurdle, since it catalyzes the Boudouard reaction. Acid leaching of biomass feedstock or charcoal can reduce the potassium content, and in combination with a high temperature treatment, concurrently adjust CO_2 reactivity that approaches properties of fossil-based reductants. A secondary heat treatment at temperatures above $1100\,^\circ$C is required to reduce the volatile matter content to below 1.5%.

Charcoal production in developing countries is only sustainable in a certain range. An additional charcoal demand is expected to increase local prices, possibly leading to a larger charcoal production using primary forestry. This can result in a short-term deforestation and charcoal production by inefficient technologies. Certification such as European Biochar Certificate (EBC) can reduce the risk of primary forestry and inefficient charcoal production. Using certified charcoal can substantially contribute to the reduction of anthropogenic CO_2 emissions from ferroalloy industry.

Harvesting cycles of 10 years can provide a metallurgical grade charcoal with the properties approaching that of currently used fossil-based reductants. Large potential of wood is available in Norway, Sweden and Finnland. However, the high CapEx and OpEx for classical charcoal production inhibit local charcoal development. Bio-refineries can provide high value-added chemicals for the energy and chemical sector with the concurrent production of charcoal used by metallurgical industries.

Author Contributions: Conceptualization, G.R.S., A.T. and M.T.; methodology, G.R.S.; validation, G.R.S., A.T. and M.T.; writing—original draft preparation, G.R.S.; writing—review and editing, G.R.S., A.T. and M.T.; visualization, G.R.S.; supervision, M.T.; project administration, M.T.; funding acquisition, M.T. All authors have read and agreed to the published version of the manuscript.

Funding: This research was funded by Research Council of Norway grant number 280968. The APC was funded by Norwegian University of Science and Technology.

Acknowledgments: The authors gratefully acknowledge the financial support from Research Council of Norway (Project Code: 280968) and our industrial partners Elkem ASA, TiZir Titanium & Iron AS, Eramet Norway AS, Finnfjord AS and Wacker Chemicals Norway AS for financing *KPN Reduced CO$_2$ emissions in metal production* project.

Conflicts of Interest: The authors declare no conflict of interest.

Abbreviations

The following abbreviations are used in this manuscript.

a	Ash
BF	Blast furnace
CapEx	Capital expenditures
CCS	Carbon capture and storage
CHL	Cellulose, hemicellulose and lignin
CRI	Coke reactivity index
CSR	Coke strength after reaction
EAF	Electric arc furnace
FC	Fixed carbon
FeMn	Ferromanganese
GDP	Gross domestic product
GHG	Green house gases
HC FeCr	High carbon ferrochrome
HC FeMn	High carbon ferromanganese
HTC	Hydrothermal carbonization
OpEx	Operating expenses
PAH	Polycyclic aromatic hydrocarbons
PM	Particulate matter
SAF	Submerged arc furnace
ScCO$_2$	Supercritical CO_2 extraction
SiMn	Silicomanganese
TGA	thermogravimetric analysis
VM	Volatile matter
VOC	Volatile organic compounds
γ	yield

References

1. Antal, M.J.; Grønli, M. The Art, Science, and Technology of Charcoal Production. *Ind. Eng. Chem. Res.* **2003**, *42*, 1619–1640. [CrossRef]
2. Hummel, R.E. The First Materials (Stone Age and Copper—Stone Age). In *Understanding Materials Science*; Springer: New York, NY, USA, 2004.
3. Gallagher, R.M.; Ingram, P. *New Coordinated Science: Chemistry Students' Book: For Higher Tier*; New Coordinated Science-3rd Edition for Higher Tier; Oxford University Press: Oxford, UK, 2001; Volume 3.
4. Klavina, K.; Karklina, K.; Blumberga, D. Charcoal production environmental performance. *Agron. Res.* **2015**, *13*, 511–519.
5. Gomaa, H.; Fathi, M. A simple charcoal kiln. In Proceedings of the ICEHM2000, Giza, Egypt, 9–12 September 2000; pp. 167–174.
6. Reumerman, P.J.; Frederiks, B. Charcoal production with reduced emissions. In Proceedings of the 12th European Conference on Biomass for Energy, Industry and Climate Protection, Amsterdam, The Netherlands 17–21 June 2002; pp. 1–4.
7. African Forestry and Wildlife Commission. *Sustainable Charcoal Production for Food Security and Forest Landscape Restoration*; Report Number FO:AFWC/2020/4.2; Food and Agriculture Organization of the United Nations: Rome, Italy, 2020; pp. 1–6.
8. Muylaert, M.S.; Sala, J.; de Freitas, M.A.V. The charcoal's production in Brazil—Process efficiency and environmental effects. *Renew. Energy* **1999**, *16*, 1037–1040. [CrossRef]
9. Nabukalu, C.; Gieré, R. Charcoal as an Energy Resource: Global Trade, Production and Socioeconoic Practices Observed in Uganda. *Resources* **2019**, *8*, 183. [CrossRef]
10. Mara dos Santos Barbosa, J.; Ré-Poppi, N.; Santiago-Silva, M. Polycyclic aromatic hydrocarbons from wood pyrolyis in charcoal production furnaces. *Environ. Res.* **2006**, *101*, 304–311. [CrossRef] [PubMed]
11. Coomes, O.T.; Burt, G.J. Peasant charcoal production in the Peruvian Amazon: rainforest use and economic reliance. *For. Ecol. Manag.* **2001**, *140*, 39–50. [CrossRef]
12. Carrari, E.; Ampoorter, E.; Bottalico, F.; Chirici, G.; Coppi, A.; Travaglini, D.; Verheyen, K.; Selvi, F. The old charcoal kiln sites in Central Italian forest landscapes. *Quat. Int.* **2017**, *458*, 214–223. [CrossRef]
13. Edenhofer, O.; Madruga, R.; Sokona, Y.; Minx, J.; Farahani, E.; Kadner, S.; Seyboth, K.; Adler, A.; Baum, I.; Brunner, S.; et al. *Climate Change 2014 Mitigation of Climate Change: Working Group III Contribution to the Fifth Assessment Report of the Intergovernmental Panel on Climate Change*; Pacific Northwest National Lab. (PNNL): Richland, WA, USA, 2014.
14. World Steel Association. *Steel's Contribution to a Low Carbon Future and Climate Resilient Societies-Worldsteel Position Paper*; World Steel Association: Brussels, Belgium, 2017.
15. Balomenos, E.; Panias, D.; Paspaliaris, I. Energy and Exergy Analysis of the Primary Aluminum Production Processes: A Review on Current and Future Sustainability. *Miner. Process. Extr. Metall. Rev.* **2011**, *32*, 69–89. [CrossRef]
16. Olsen, S.E.; Tangstad, M.; Lindstad, T. *Production of Manganese Ferroalloys*; Tapir Akademisk Forlag: Trondheim, Norway, 2007; p. 247.
17. Holappa, L. Basics of Ferroalloys. In *Handbook of Ferroalloys-Theory and Technology*; Butterworth-Heinemann: Oxford, UK, 2013; pp. 9–28.
18. Eidem, P.A. Electrical Resistivity of Coke Beds. Ph.D. Thesis, NTNU, Trondheim, Norway, 2008.
19. Dhainaut, M. Simulation of the electric field in a submerged arc furnace. In Proceedings of the Infacon X, Cape Town, South Africa, 1–4 February 2004; pp. 605–613.
20. Ng, K.W.; MacPhee, J.A.; Giroux, L.; Todoschuk, T. Reactivity of bio-coke with CO_2. *Fuel Process. Technol.* **2011**, *92*, 801–804. [CrossRef]
21. Seo, M.W.; Jeong, H.M.; Lee, W.J.; Yoon, S.J.; Ra, H.W.; Kim, Y.K.; Lee, D.; Han, S.W.; Kim, S.D.; Lee, J.G.; et al. Carbonization characteristics of biomass/coking coal blends for the application of bio-coke. *Chem. Eng. J.* **2020**, *394*, 124943. [CrossRef]
22. Monsen, B.; Grønli, M.; Nygaard, L.; Tveit, H. The Use of Biocarbon in Norwegian Ferroalloy Production. In Proceedings of the INFACON IX, Quebec City, ON, Canada, 3–6 June 2001; Volume 9, pp. 268–276.
23. Myrvågnes, V. Analyses and Characterization of Fossil Carbonaceous Materials for Silicon Production. Ph.D. Thesis, NTNU, Trondheim, Norway, 2008.

24. United States Environmental Protection Agency. Charcoal. Wood Products Industry. In *AP 42*, 5th ed.; Environmental Protection Agency: Washington, DC, USA, 1995; Volume I, Chapter 10. Available online: https://www.epa.gov/air-emissions-factors-and-quantification/ap-42-compilation-air-emissions-factors (accessed on 28 September 2020).

25. Emrich, W. *Handbook of Charcoal Making: The Traditional and Industrial Methods;* Solar Energy R&D in the Ec Series E; Springer: Berlin/Heidelberg, Germany, 1985.

26. Scarlat, N.; Dallemand, J.-F.; Skjelhaugen, O.J.; Asplund, D.; Nesheim, L. An overview of the biomass resource potential of Norway for bioenergy use. *Renew. Sustain. Energy Rev.* **2011**, *15*, 3388–3398. [CrossRef]

27. Norsk Industri. The Norwegian Process Industries' Roadmap-Combining Growth and Zero Emissions by 2050. 2016. Available online: https://www.norskindustri.no/siteassets/dokumenter/rapporter-og-brosjyrer/the-norwegian-process-industries-roadmap-summary.pdf (accessed on 7 December 2018).

28. Dalaker, H.; Ringdalen, E.; Kolbeinsen, L.; Mårdalen, J. Road-Map for Gas in the Norwegian Metallurgical Industry: Greater Value Creation and Reduced Emissions. 2017. Available online: https://sintef.brage.unit.no/sintef-xmlui/bitstream/handle/11250/2452079/Road-map%20for%20gas%20in%20the%20Norwegian%20metallurgical%20industry.pdf?sequence=6 (accessed on 11 September 2020).

29. Tangstad, M. *Ferrosilicon and Silicon Technology;* Butterworth-Heinemann: Oxford, UK, 2018.

30. Schei, A.; Tuset, J.K.; Tveit, H. *Production of High Silicon Ferroalloys;* Tapir Forlag: Trondheim, Norway, 1998.

31. Meng, J.; He, T.; Sanganyado, E.; Lan, Y.; Zhang, W.; Han, X.; Chen, W. Development of the straw biochar returning concept in China. *Biochar* **2019**, *1*, 139–149. [CrossRef]

32. Park, J.; Lee, Y.; Ryu, C.; Park, Y.K. Slow pyrolysis of rice straw: Analysis of products properties, carbon and energy yields. *Bioresour. Technol.* **2014**, *155*, 63–70. [CrossRef]

33. Carpenter, D.; Westover, T.L.; Czernik, S.; Jablonski, W. Biomass feedstocks for renewable fuel production: A review of the impacts of feedstock and pretreatment on the yield and product distribution of fast pyrolysis bio-oils and vapors. *Green Chem.* **2014**, *16*, 384–406. [CrossRef]

34. Henrich, E.; Dinjus, E.; Rumpel, S.; Stahl, R. A Two-Stage Pyrolysis/Gasification Process for Herbaceous Waste Biomass from Agriculture. In *Progress in Thermochemical Biomass Conversion;* John Wiley & Sons: Hoboken, NJ, USA, 2008; pp. 221–236.

35. Yang, C.; Li, R.; Zhang, B.; Qui, Q.; Wang, B.; Yang, H.; Ding, Y.; Wang, C. Pyrolysis of microalgae: A critical review. *Fuel Process. Technol.* **2019**, *186*, 53–72. [CrossRef]

36. Monsen, B.; Tangstad, M.; Midtgaard, H. Use of charcoal in silicomanganese production. In Proceedings of the Infacon X, Cape Town, South Africa, 1–4 February 2004; Volume 68, pp. 155–164.

37. Monsen, B.; Tangstad, M.; Solheim, I.; Syvertsen, M.; Ishak, R.; Midtgaard, H. Charcoal for manganese alloy production. In Proceedings of the INFACON XI, New Delhi, India, 18–21 February 2007; pp. 297–310.

38. ECN.TNO. Phyllis2, Database for (Treated) Biomass, Algae, Feedstocks for Biogas Production and Biochar. Available online: https://phyllis.nl/ (accessed on 12 October 2020).

39. Nygård, R.; Elfving, B. Stem basic density and bark proportion of 45 woody species in young savanna coppice forests in Burkina Faso. *Ann. For. Sci.* **2000**, *57*, 143–153.

40. Demirbaş, A.; Ahmad, W.; Alamoudi, R.; Sheikh, M. Sustainable charcoal production from biomass. *Energy Sources Part A* **2016**, *38*, 1882–1889. [CrossRef]

41. Bassam, N.E. *Handbook of Bioenergy Crops;* Earthscan Ltd.: London, UK, 2010.

42. Räisänen, T.; Athanassiadis, D. Basic Chemical Composition of the Biomass Components of Pine, Spruce and Birch. 2013. Available online: http://biofuelregion.se/wp-content/uploads/2017/01/1_2_IS_2013-01-31_Basic_chemical_composition.pdf (accessed on 15 September 2020).

43. Balat, M. Mechanisms of Thermochemical Biomass Conversion Processes. Part 1: Reactions of Pyrolysis. *Energy Sources Part A* **2008**, *30*, 620–635. [CrossRef]

44. Jones, J.R.; Chen, Q.; Ripberger, G.D. Secondary Reactions and the Heat of Pyrolysis of Wood. *Energy Technol.* **2020**, *8*, 2000130. [CrossRef]

45. Evans, R.J.; Milne, T.A. Molecular characterization of the pyrolysis of biomass. 1. Fundamentals. *Energy Fuels* **1987**, *1*, 123–137. [CrossRef]

46. Broido, A.; Kilzer, F.J. A Critique of the Present State of Knowledge of The Mechanism of Cellulose Pyrolysis. *Fire Res.* **1963**, *5*, 157–161.

47. Antal, M.J.; Gábor, V.; Jakab, E. Cellulose Pyrolysis Kinetics: Revisited. *Ind. Eng. Chem. Res.* **1998**, *37*, 1267–1275. [CrossRef]

48. Collard, F.-X.; Blin, J. A review on pyrolysis of biomass constituents: Mechanisms and composition of the products obtained from the conversion of cellulose, hemicelluloses and lignin. *Renew. Sustain. Energy Rev.* **2014**, *38*, 594–608. [CrossRef]

49. Peng, Y.; Wu, S. The structural and thermal characteristics of wheat straw hemicellulose. *J. Anal. Appl. Pyrolysis* **2010**, *88*, 134–139. [CrossRef]

50. Shen, D.K.; Gu, S.; Bridgwater, A.V. Study on the pyrolytic behaviour of xylan-based hemicellulose using TG–FTIR and Py–GC–FTIR. *J. Anal. Appl. Pyrolysis* **2010**, *87*, 199–206. [CrossRef]

51. Werner, K.; Pommer, L.; Broström, M. Thermal decomposition of hemicelluloses. *J. Anal. Appl. Pyrolysis* **2014**, *110*, 130–137. [CrossRef]

52. Branca, C.; Albano, A.; Di Blasi, C. Critical evaluation of global mechanisms of wood devolatilization. *Thermochim. Acta* **2005**, *429*, 133–141. [CrossRef]

53. Du, S.; Valla, J.A.; Bollas, G.M. Characteristics and origin of char and coke from fast and slow, catalytic and thermal pyrolysis of biomass and relevant model compounds. *Green Chem.* **2013**, *15*, 3214–3229. [CrossRef]

54. Hosoya, T.; Kawamoto, H.; Saka, S. Pyrolysis behaviors of wood and its constituent polymers at gasification temperature. *J. Anal. Appl. Pyrolysis* **2007**, *78*, 328–336. [CrossRef]

55. Yang, H.; Yan, R.; Chen, H.; Lee, D.H. Characteristics of hemicellulose, cellulose and lignin pyrolysis. *Fuel* **2007**, *86*, 1781–1788. [CrossRef]

56. Zaror, C.A.; Pyle, D.L. The pyrolysis of biomass: A general review. *Proc. Indian Acad. Sci. Sect. C Eng. Sci.* **1982**, *5*, 269–285.

57. Trubetskaya, A.; Jensen, P.A.; Jensen, A.D.; Glarborg, P.; Larsen, F.H.; Andersen, M.L. Characterization of free radicals by electron spin resonance spectroscopy in biochars from pyrolysis at high heating rates and at high temperatures. *Biomass Bioenergy* **2016**, *94*, 117–129. [CrossRef]

58. Trubetskaya, A.; Jensen, P.A.; Jensen, A.D.; Steibel, M.; Spliethoff, H.; Glarborg, P. Influence of fast pyrolysis conditions on yield and structural transformation of biomass chars. *Fuel Process. Technol.* **2015**, *140*, 205–214. [CrossRef]

59. Persson, H.; Kantarelis, E.; Evangelopoulos, P.; Yang, W. Wood-derived acid leaching of biomass for enhanced production of sugars and sugar derivatives during pyrolysis: Influence of acidity and treatment time. *J. Anal. Appl. Pyrolysis* **2017**, *127*, 329–334. [CrossRef]

60. Hoover, A.; Emerson, R.; Williams, C.L.; Ramirez-Corredores, M.M.; Ray, A.; Schaller, K.; Hernandez, S.; Li, C.; Walton, M. Grading Herbaceous Biomass for Biorefineries: A Case Study Based on Chemical Composition and Biochemical Conversion. *Bioenergy Res.* **2019**, *12*, 977–991. [CrossRef]

61. Demirbaş, A. Relationships between Carbonization Temperature and Pyrolysis Products from Biomass. *Energy Explor. Exploit.* **2004**, *22*, 411–420. [CrossRef]

62. Surup, G.R.; Foppe, M.; Schubert, D.; Deike, R.; Heidelmann, M.; Timko, M.T.; Trubetskaya, A. The effect of feedstock origin and temperature on the structure and reactivity of char from pyrolysis at 1300–2800 °C. *Fuel* **2019**, *235*, 306–316. [CrossRef]

63. Surup, G.; Vehus, T.; Eidem, P.A.; Trubetskaya, A.; Nielsen, H.K. Characterization of renewable reductants and charcoal-based pellets for the use in ferroalloy industries. *Energy* **2019**, *167*, 337–345. [CrossRef]

64. Pereira, B.L.C.; Carneiro, A.C.O.; Márcia, A.; Colodette, J.; Oliveira, A.; Fontes, M.P.F. Influence of Chemical Composition of Eucalyptus Wood on Gravimetric Yield and Charcoal Properties. *BioResources* **2013**, *8*, 4574–4592. [CrossRef]

65. Nielsen, H.K.; Lærke, P.E.; Liu, N.; Jørgensen, U. Sampling procedure in a willow plantation for estimation of moisture content. *Biomass Bioenergy* **2015**, *78*, 62–70. [CrossRef]

66. Da Silva, V.M.; Silva, L.A.; de Andrade, J.B.; da Cunha Veloso, M.C.; Santos, G.V. Determination of moisture content and water activity in algae and fish by thermoanalytical techniques. *Quim. Nova* **2008**, *31*, 901–905. [CrossRef]

67. Hosseinizand, H.; Sokhansanj, S.; Lim, C.J. Studying the drying mechanism of microalgae *Chlorella vulgaris* and the optimum drying temperature to preserve quality characteristics. *Dry. Technol.* **2018**, *36*, 1049–1060. [CrossRef]

68. Klavina, K.; Klavins, J.; Veidenbergs, I.; Blumberga, D. Charcoal production in a continuous operation retort. Experimental data processing. *Energy Procedia* **2016**, *95*, 208–215. [CrossRef]

69. Bergman, R. Drying and Control of Moisture Content and Dimensional Changes. In *Wood Handbook-Wood as an Engineering Material*; General Technical Report FPL-GTR-190; USDA Forest Service: Washington, DC, USA, 2010; p. 509.

70. Dong, J.; Chi, Y.; Tang, Y.; Ni, M.; Nzihou, A.; Weiss-Hortala, E.; Huang, Q. Effect of Operating Parameters and Moisture Content on Municipal Solid Waste Pyrolysis and Gasification. *Energy Fuels* **2016**, *30*, 3994–4001. [CrossRef]

71. Eke, J.; Onwudili, J.A.; Bridgwater, A.V. Influence of Moisture Contents on the Fast Pyrolysis of Trommel Fines in a Bubbling Fluidized Bed Reactor. *Waste Biomass Valoriz.* **2020**, *11*, 3711–3722. [CrossRef]

72. Canal, W.D.; Carvalho, A.M.; Figueiró, C.G.; de Cassia Oliveira Carneiro, A.; de Freitas Fialho, L.; Donato, D.B. Impact of Wood Moisture in Charcoal Production and Quality. *Floresta Ambiente* **2020**, *27*, e20170999. [CrossRef]

73. Emrich, W.; Booth, H.; Beaumont, E. *Industrial Charcoal Making*; Food and Agriculture Organization of the United Nations: Rome, Italy, 1985.

74. Bridgwater, T.; Meier, D.; Radlein, D. An Overview of Fast Pyrolysis of Biomass. *Org. Geochem.* **1999**, *30*, 1479–1493. [CrossRef]

75. Yamagishi, K.; Endo, K.; Saga, J. A comprehensive analysis of the furnace interior for high-carbon ferrochromium. In Proceedings of the INFACON I, Johannesburg, South Africa, 22–26 April 1974; pp. 143–147.

76. De Waal, A.; Barker, I.J.; Rennie, M.S.; Klopper, J.; Groeneveld, B.S. Electrical Factors Affecting the Economic Optimization of Submerged-arc Furnaces. In Proceedings of INFACON VI, Cape Town, South Africa, 8–11 March 1992; Volume 1, pp. 247–252.

77. Chahal, A.; Ciolkosz, D. A review of wood-bark adhesion: Methods and mechanics of debarking for woody biomass. *Wood Fiber Sci.* **2019**, *51*, 1–12. [CrossRef]

78. Heppelmann, J.B.; Labelle, E.R.; Wittkopf, S.; Seeling, U. In-stand debarking with the use of modifed harvesting heads: a potential solution for key challenges in European forestry. *Eur. J. For. Res.* **2019**, *138*, 1067–1081. [CrossRef]

79. Momeni, M.; Yin, C.; Kœr, S.K.; Hansen, T.B.; Jensen, P.A.; Glarborg, P. Experimental Study on Effects of Particle Shape and Operating Conditions on Combustion Characteristics of Single Biomass Particles. *Energy Fuels* **2012**, *27*, 507–514. [CrossRef]

80. Momeni, M. Fundamental Study of Single Biomass Particle Combustion. Ph.D. Thesis, Aalborg University, Aalborg, Denmark, 2012.

81. Himmel, M.; Tucker, M.; Baker, J.; Rivard, C.; Oh, K.; Grohmann, K. Comminution of Biomass: Hammer and Knife Mills. *Biotechnol. Bioeng. Symp.* **1985**, *15*, 39–58.

82. Paulrud, S.; Mattsson, J.E.; Nilsson, C. Particle and handling chracteristics of wood fuel powder: Effects of different mills. *Fuel Process. Technol.* **2002**, *76*, 23–39. [CrossRef]

83. Paulrud, S. Upgraded Biofuels-Effects of Quality on Processing, Handling Characteristics, Combustion and ASH Melting. Ph.D. Thesis, Umeå University, Umeå, Sweden, 2004.

84. Schell, D.J.; Harwood, C. Milling of Lignocellulosic Biomass. *Appl. Biochem. Biotechnol.* **1994**, *46*, 159–169. [CrossRef]

85. Mandø, M.; Rosendahl, L.; Yin, C.; Sørensen, H. Pulverized straw combustion in a low NO_x multifuel burner: Modeling the transition from coal to straw. *Fuel* **2010**, *89*, 3051–3062. [CrossRef]

86. Miao, Z.; Grift, T.E.; Hansen, A.C.; Ting, K.C. Energy requirement for comminution of biomass in relation to particle physical properties. *Ind. Crop. Prod.* **2011**, *33*, 504–513. [CrossRef]

87. Williams, O.; Newbolt, G.; Eastwick, C.; Kingman, S.; Giggings, D.; Lormor, S.; Lester, E. Influence of mill type on densified biomass comminution. *Appl. Energy* **2016**, *182*, 219–231. [CrossRef]

88. Masche, M.; Puig-Arnavat, M.; Wadenbäck, J.; Clausen, S.; Jensen, P.A.; Ahrendfeldt, J.; Henriksen, U. Wood pellet milling tests in a suspension-fired power plant. *Fuel Process. Technol.* **2018**, *173*, 89–102. [CrossRef]

89. Westover, T.L.; Phanphanich, M.; Clark, M.L.; Rowe, S.R.; Egan, S.E.; Zacher, A.H.; Santosa, D. Impact of thermal pretreatment on the fast pyrolysis conversion of southern pine. *Biofuels* **2013**, *4*, 45–61. [CrossRef]

90. Stelte, W.; Sanadi, A.R.; Shang, L.; Holm, J.K.; Ahrenfeldt, J.; Henriksen, U.B. Recent developments in biomass pelletization—A review. *BioResources* **2012**, *7*, 4451–4490.

91. Harun, N.Y.; Afzal, M.T. Effect of Particle Size on Mechanical Properties of Pellets Made from Biomass Blends. *Procedia Eng.* **2016**, *148*, 93–99. [CrossRef]

92. Trubetskaya, A.; Beckmann, G.; Poyraz, Y.; Weber, R. Secondary comminution of wood pellets in power plant and laboratory-scale mills. *Fuel* **2017**, *160*, 675–683. [CrossRef]

93. Jensen, A.; Dam-Johansen, K.; Wojtowicz, M.A.; Serio, M.A. TG-FTIR study of the influence of potassium chloride on wheat straw pyrolysis. *Energy Fuels* **1998**, *12*, 929–938. [CrossRef]

94. Pistorius, P.C. Reductant selection in ferro-alloy production: The case for the importance of dissolution in the metal. *J. S. Afr. Inst. Min. Metall.* **2002**, *1*, 33–36.

95. Kong, L.; Tian, S.H.; Li, Z.; Luo, R.; Chen, D.; Tu, Y.; Xiong, Y. Conversion of recycled sawdust into high HHV and low NO_x emission bio-char pellets using lignin and calcium hydroxide blended binders. *Renew. Energy* **2013**, *60*, 559–565. [CrossRef]

96. Jensen, P.A.; Sander, B.; Dam-Johansen, K. Removal of K and Cl by leaching of straw char. *Biomass Bioenergy* **2001**, *20*, 447–457. [CrossRef]

97. Trubetskaya, A.; Larsen, F.H.; Shchukarev, A.; Ståhl, K.; Umeki, K. Potassium and soot interaction in fast biomass pyrolysis at high temperatures. *Fuel* **2018**, *225*, 89–94. [CrossRef]

98. Xu, W.; Wei, J.; Chen, J.; Zhang, B.; Xu, P.; Ren, J.; Yu, Q. Comparative Study of Water-Leaching and Acid-Leaching Pretreatment on the Thermal Stability and Reactivity of Biomass Silica for Viability as a Pozzolanic Additive in Cement. *Materials* **2018**, *11*, 1697. [CrossRef] [PubMed]

99. Hussein, A.; Larachi, F.; Ziegler, D.; Alamdari, H. Effects of heat treatment and acid washing on properties and reactivity of charcoal. *Biomass Bioenergy* **2016**, *90*, 101–113. [CrossRef]

100. Trubetskaya, A.; Jensen, P.A.; Jensen, A.D.; Llamas, A.D.G.; Umeki, K.; Glarborg, P. Effect of fast pyrolysis conditions on biomass solid residues at high temperatures. *Fuel Process. Technol.* **2016**, *143*, 118–129. [CrossRef]

101. Bridgwater, A.V. Review of fast pyrolysis of biomass and product upgrading. *Biomass Bioenergy* **2012**, *38*, 68–94. [CrossRef]

102. Fahmi, R.; Bridgwater, A.V.; Donnison, I.S.; Yates, N.; Jones, J.M. The effect of lignin and inorganic species in biomass on pyrolysis oil yields, quality and stability. *Fuel* **2008**, *87*, 1230–1240. [CrossRef]

103. Das, P.; Ganesh, A.; Wangikar, P.P. Influence of pretreatment of deashing of sugarcane bagasse on pyrolysis products. *Biomass Bioenergy* **2004**, *27*, 445–457. [CrossRef]

104. Runge, T.; Wipperfurth, P.; Zhang, C. Improving biomass combustion quality using a liquid hot water treatment. *Biofuel* **2013**, *4*, 73–83. [CrossRef]

105. Surup, G.R. Renewable Reducing Agents for the Use in Ferroalloy Industries. Ph.D. Thesis, University of Agder, Kristiansand, Norway, 2019; p. 238.

106. Paul, M.; Seferinoğlu, M.; Ayçik, G.; Sandström, A.; Smith, M.L.; Paul, J. Acid leaching of ash and coal: Time dependence and trace element occurrences. *Int. J. Miner. Process.* **2006**, *79*, 27–41. [CrossRef]

107. Attard, T.M.; Arshadi, M.; Nilsson, C.; Budarin, V.L.; Valencia-Reyes, E.; Clark, J.H.; Hunt, A.J. Impact of supercritical extraction on solid fuel wood pellet properties and off-gassing during storage. *Green Chem.* **2016**, *18*, 2682–2690. [CrossRef]

108. Hill, K. Fats and oils as oleochemical raw materials. *Eur. J. Pharm. Sci.* **2000**, *72*, 1255–1264. [CrossRef]

109. Ruston, N.A. Commercial uses of fatty acids. *J. Am. Oil Chem. Soc.* **1952**, *29*, 495–498. [CrossRef]

110. Gill, I.; Valivety, R. Polyunsaturated fatty acids: 1. Occurrence, biological activities and applications. *Trends Biotechnol.* **1997**, *15*, 401–409. [CrossRef]

111. White, K.; Lorenz, N.; Potts, T.; Penney, W.R.; Babcock, R.; Hardison, A.; Canuel, E.A.; Hestekin, J.A. Production of biodiesel fuel from tall oil fatty acids via high temperature methanol reaction. *Fuel* **2011**, *90*, 3193–3199. [CrossRef]

112. Pereira, C.G.; Meireles, M.A.A. Economic analysis of rosemary, fennel and anise essential oils obtained by supercritical fluid extraction. *Flavour Fragr. J.* **2007**, *22*, 407–413. [CrossRef]

113. Chemat, F.; Abert Vian, M.; Cravotto, G. Green Extraction of Natural Products: Concept and Principles. *Int. J. Mol. Sci.* **2012**, *13*, 8615–8627. [CrossRef]

114. Ruiz, H.A.; Rodriquez-Jasso, R.M.; Fernandes, B.D.; Vicente, A.A.; Teixeira, J.A. Hydrothermal processing, as an alternative for upgrading agriculture residues and marine biomass according to the biorefinery concept: A review. *Renew. Sustain. Energy Rev.* **2013**, *21*, 35–51. [CrossRef]

115. Hunt, A.J.; Sin, E.H.K.; Marriott, R.; Clark, J.H. Generation, Capture, and Utilization of Industrial Carbon Dioxide. *ChemSusChem* **2010**, *3*, 306–322. [CrossRef]

116. Sin, E.H.K.; Marriott, R.; Hunt, A.J.; Clark, J.H. Identification, quantification and Chrastil modelling of wheat straw wax extraction using supercritical carbon dioxide. *C. R. Chim.* **2014**, *17*, 293–300. [CrossRef]

117. Brunner, G. *An Introduction to Fundamentals of Supercritical Fluids and the Application to Separation Processes*; Springer: Berlin/Heidelberg, Germany, 1994; p. 387.

118. Brunner, G. Supercritical fluids: Technology and application to food processing. *J. Food Eng.* **2005**, *67*, 21–33. [CrossRef]

119. Arshadi, M.; Hunt, A.J.; Clark, J.H. Supercritical fluid extraction (SFE) as an effective tool in reducing auto-oxidation of dried pine sawdust for power generation. *RSC Adv.* **2012**, *2*, 1806–1809. [CrossRef]

120. Budarin, V.L.; Shuttleworth, P.S.; Dodson, J.R.; Hunt, A.J.; Lanigan, B.; Marriott, R.; Milkowski, K.J.; Wilson, A.J.; Breeden, S.W.; Fan, J.; et al. Use of green chemical technologies in an integrated biorefinery. *Energy Environ. Sci.* **2011**, *4*, 471–479. [CrossRef]

121. Smith, S.M.; Sahle Demessie, E.; Morrell, J.J.; Levien, K.L.; Ng, H. Supercritical fluid (SCF) treatment: its effect on bending strength and stiffness of ponderosa pine sapwood. *Wood Fiber Sci.* **1993**, *25*, 119–123.

122. Surup, G.R.; Hunt, A.J.; Attard, T.; Budarin, V.L.; Forsberg, F.; Arshadi, M.; Abdelsayed, V.; Shekhawat, D.; Trubetskaya, A. The effect of wood composition and supercritical CO_2 extraction on charcoal production in ferroalloy industries. *Energy* **2020**, *193*, 116696. [CrossRef]

123. Singh, S.; Chakraborty, J.C.; Mondal, M.K. Torrefaction of woody biomass (*Acacia nilotica*): Investigation of fuel and flow properties to study its suitability as a good quality solid fuel. *Renew. Energy* **2020**, *153*, 711–724. [CrossRef]

124. Tumuluru, J.S.; Wright, C.T.; Hess, J.R.; Kenney, K.L. A review of biomass densification systems to develop uniform feedstock commodities for bioenergy application. *Biofuels Bioprod. Biorefin.* **2011**, *5*, 683–707. [CrossRef]

125. Trubetskaya, A.; Grams, J.; Leahy, J.J.; Kwapinska, M.; Monaghan, R.F.D.; Johnson, R.; Gallagher, P.; Monaghan, R. The effect of particle size, temperature and residence time on the yields and reactivity of olive stones from torrefaction. *Renew. Energy* **2020**, *160*, 998–1011. [CrossRef]

126. Medic, C.; Darr, M.; Shah, A.; Potter, B.; Zimmermann, J. Effects of torrefaction process parameters on biomass feedstock upgrading. *Fuel* **2012**, *91*, 147–154. [CrossRef]

127. Cholewinski, M.; Kaminski, M. Torrefaction as a Means to Increase the use of Solid Biomass for Power Generation and Heating. *Acta Energy* **2016**, *2*, 60–67. [CrossRef]

128. Sangines, P.; Dominguez, M.P.; Sanchez, F.; San Miguel, G. Slow pyrolysis of olive stones in a rotary kiln: Chemical and energy characterization of solid, gas, and condensable products. *J. Renew. Sustain. Energy* **2015**, *7*, 1–13. [CrossRef]

129. Surup, G.R.; Leahy, J.J.; Timko, M.T.; Trubetskaya, A. Hydrothermal carbonization of olive wastes to produce renewable, binder-free pellets for use as metallurgical reducing agents. *Renew. Energy* **2020**, *155*, 347–357. [CrossRef]

130. Hamid, M.F.; Idroas, M.Y.; Ishak, M.Z.; Zainal Alauddin, Z.A.; Miskam, M.A.; Abdullah, M.K. An Experimental Study of Briquetting Process of Torrefied Rubber Seed Kernel and Palm Oil Shell. *BioMed Res. Int.* **2016**, *1*, 1–11. [CrossRef] [PubMed]

131. Funke, A.; Ziegler, F. Heat of reaction measurements for hydrothermal carbonization of biomass. *Bioresour. Technol.* **2011**, *102*, 7595–7598. [CrossRef]

132. Saha, P.; McGaughy, K.; Hasan, R.; Reza, M.T. Pyrolysis and carbon dioxide gasification kinetics of hydrochar produced from cow manure. *Environ. Prog. Sustain. Energy* **2018**, *102*, 7595–7598. [CrossRef]

133. Reza, M.T.; Uddin, M.H.; Lynam, J.; Hoekman, S.K.; Coronella, C. Hydrothermal carbonization of loblolly pine: Reaction chemistry and water balance. *Biomass Convers. Biorefin.* **2014**, *4*, 311–321. [CrossRef]

134. Ahmad, F.; Silva, E.L.; Varesche, M.B.A. Hydrothermal processing of biomass for anaerobic digestion—A review. *Renew. Sustain. Energy Rev.* **2018**, *98*, 108–124. [CrossRef]

135. He, C.; Tang, C.; Li, C.; Yuan, J.; Tran, K.Q.; Bach, Q.V.; Qiu, R.; Yang, Y. Wet torrefaction of biomass for high quality solid fuel production: A review. *Renew. Sustain. Energy Rev.* **2018**, *91*, 259–271. [CrossRef]

136. Delahaye, L.; Hobson, J.T.; Rando, M.P.; Sweeney, B.; Brown, A.B.; Tompsett, G.A.; Ates, A.; Deskins, N.A.; Timko, M.T. Experimental and Computational Evaluation of Heavy Metal Cation Adsorption for Molecular Design of Hydrothermal Char. *Energies* **2020**, *13*, 4203. [CrossRef]

137. Volpe, M.; Fiori, L. From olive waste to solid biofuel through hydrothermal carbonization: The role of temperature and solid load on secondary char formation and hydrochar energy properties. *J. Anal. Appl. Pyrolysis* **2017**, *124*, 63–72. [CrossRef]

138. Lucian, M.; Fiori, L. Hydrothermal carbonization of waste biomass: Process design, modeling, energy efficiency and cost analysis. *Energies* **2017**, *10*, 211. [CrossRef]

139. Mandova, H.; Leduc, S.; Wang, C.; Wetterlund, E.; Patrizio, P.; Gale, W.; Kraxner, F. Possibilities for CO$_2$ emission reduction using biomass in European integrated steel plants. *Biomass Bioenergy* **2018**, *115*, 231–243. [CrossRef]

140. Shang, L.; Nielsen, N.P.K.; Stelte, W.; Dahl, J.; Ahrendfeldt, J.; Holm, J.K.; Arnavat, M.P.; Bach, L.S. Lab and Bench-Scale Pelletization of Torrefied Wood Chips–Process Optimization and Pellet Quality. *Bioenergy Res.* **2014**, *7*, 87–94. [CrossRef]

141. Dutta, S.K. Iron ore–coal composite pellets/briquettes as new feed material for iron and steelmaking. *Mater. Sci. Eng. Int. J.* **2017**, *1*, 10–13.

142. Surup, G.R.; Nielsen, H.K.; Großarth, M.; Deike, R.; Van den Bulcke, J.; Kibleur, P.; Müller, M.; Ziegner, M.; Yazhenskikh, E.; Beloshapkin, S.; et al. Effect of operating conditions and feedstock composition on the properties of manganese oxide or quartz charcoal pellets for the use in ferroalloy industries. *Energy* **2020**, *193*, 116736. [CrossRef]

143. Riva, L. Production and Application of Sustainable Metallurgical Biochar Pellets. Ph.D. Thesis, University of Agder, Kristiansand, Norway, 2020; p. 284.

144. Nascimento, R.C.; Mourao, M.B.; Capocchi, J.D.T. Reduction-swelling behaviour of pellets bearing iron ore and charcoal. *Can. Metall. Q.* **1998**, *37*, 441–448. [CrossRef]

145. Yoshikoshi, H.; Takeuchi, O.; Miyashita, T.; Kuwana, T.; Kishikawa, K. Development of Composite Cold Pellet for Silico-Manganese Production. *Trans. ISIJ* **1984**, *24*, 492–497. [CrossRef]

146. Tangstad, M.; Leroy, D.; Ringdalen, E. Behaviour of agglomerates in ferromanganese production. In Proceedings of the INFACON XII, Hensinki, Finland, 6–10 June 2010; pp. 457–466.

147. Abedi, A.; Cheng, H.; Dalai, A.K. Effects of natural additives on the properties of sawdust fuel pellets. *Energy Fuels* **2018**, *32*, 1863–1873. [CrossRef]

148. Bach, Q.V.; Skreiberg, O. Upgrading biomass fuels via wet torrefaction: A review and comparison with dry torrefaction. *Renew. Sustain. Energy Rev.* **2016**, *54*, 665–677. [CrossRef]

149. Wnukowski, M.; Owczarek, P.; Niedzwiecki, L. Wet torrefaction of miscanthus-Characterization of hydrochars in view of handling, storage and combustion properties. *J. Ecol. Eng.* **2015**, *16*, 161–167. [CrossRef]

150. Riva, L.; Surup, G.R.; Buø, T.V.; Nielsen, H.K. A study of densified biochar as carbon source in the silicon and ferrosilicon production. *Energy* **2019**, *181*, 985–996. [CrossRef]

151. Liu, Z.; Quek, A.; Balasubramanian, R. Preparation and characterization of fuel pellets from woody biomass, agro-residues and their corresponding hydrochars. *Appl. Energy* **2014**, *113*, 1315–1322. [CrossRef]

152. Antal, M.J.; Mok, W.S.L.; Varhegyi, G.; Szekely, T. Review of methods for improving the yield of charcoal from biomass. *Energy Fuels* **1990**, *4*, 221–225. [CrossRef]

153. Antal, M.J.; Croiset, E.; Dai, X.; DeAlmeida, C.; Mok, W.S.L.; Norberg, N.; Richard, J.-R.; Majthoub, A. High-Yield Biomass Charcoal. *Energy Fuels* **1996**, *10*, 652–658. [CrossRef]

154. Lehmann, J.; Joseph, S. *Biochar for Environmental Management: Science, Technology and Implementation*; Routledge: Abingdon, UK, 2015.

155. Weber, K.; Quicker, P. Properties of biochar. *Fuel* **2018**, *217*, 240–261. [CrossRef]

156. Kan, T.; Strezov, V.; Evans, T.J. Lignocellulosic biomass pyrolysis: A review of product properties and effects of pyrolysis parameters. *Renew. Sustain. Energy Rev.* **2016**, *57*, 1126–1140. [CrossRef]

157. Wang, L.; Trninic, M.; Skreiberg, Ø.; Grønli, M.; Considine, R.; Antal, M.J. Is elevated pressure required to achieve a high fixed-carbon yield of charcoal from biomass? Part 1: Roundrobin results for three different corncob materials. *Energy Fuels* **2011**, *25*, 3251–3265. [CrossRef]

158. Wang, L.; Skreiberg, Ø.; Grønli, M.; Specht, G.P.; Antal, M.J. Is elevated pressure required to achieve fixed-carbon yield of charcoal from biomass? Part 2. The importance of particle size. *Energy Fuels* **2013**, *27*, 2146–2156. [CrossRef]

159. Rousset, P.; Figueiredo, C.; De Souza, M.; Quirino, W. Pressure effect on the quality of eucalyptus wood charcoal for the steel industry: A statistical analysis approach. *Fuel Process. Technol.* **2011**, *92*, 1890–1897. [CrossRef]

160. De Oliveira Vilela, A.; Lora, E.S.; Quintero, Q.R.; Vicintin, R.A.; da Silva e Souza, T.P. A new technology for the combined production of charcoal and electricity through cogeneration. *Biomass Bioenergy* **2014**, *69*, 222–240. [CrossRef]

161. Kajina, W.; Junpen, A.; Garivait, S.; Kamnoet, O.; Keeratiisariyakul, P.; Rousset, P. Charcoal production processes: An overview. *J. Sustain. Energy Environ.* **2019**, *10*, 19–25.

162. Amen-Chen, C.; Pakdel, H.; Roy, C. Separation of phenols from *Eucalyptus* wood tar. *Biomass Bioenergy* **1997**, *13*, 25–37. [CrossRef]

163. Teplitz-Sembitzky, W. *The Malawi Charcoal Project-Experiences and Lessons*; Industry and Energy Department Working Paper, Energy Series Paper No. 20; The World Bank: Washington, DC, USA, 1990.

164. Trninić, M.; Jovović, A.; Stojiljković, D. A steady state model of agricultural waste pyrolysis: A mini review. *Waste Manag. Res.* **2016**, *34*, 851–865. [CrossRef]

165. Lam, K.-L.; Lee, C.-W.; Hui, C.-W. Multi-stage Waste Tyre Pyrolysis: An Optimisation Approach. *Chem. Eng. Trans.* **2010**, *21*, 853–858.

166. Oyedun, A.O.; Lam, K.L.; Hui, C.W. Charcoal Production via Multistage Pyrolysis. *Chin. J. Chem. Eng.* **2012**, *20*, 455–460. [CrossRef]

167. Cheung, K.-Y.; Lee, K.-L.; Lam, K.-L.; Chan, T.-Y.; Lee, C.-W.; Hui, C.-W. Operation strategy for multi-stage pyrolysis. *J. Anal. Appl. Pyrolysis* **2011**, *91*, 165–182. [CrossRef]

168. Dahmen, N.; Dinjus, E.; Kolb, T.; Arnold, U.; Leibold, H.; Stahl, R. State of the Art of the Bioliq® Process for Synthetic Biofuels Production. *Environ. Prog. Sustain.* **2012**, *31*, 176–181. [CrossRef]

169. Vehviläinen, J. Challenges in rendering co-rich gas from submerged arc furnaces sufficiently clean for co-generation application. In Proceedings of the International Ferro-Alloys Congress XIII, Almaty, Kazakhstan, 9–13 June 2013; pp. 1009–1014.

170. Kero, I.T.; Eidem, P.A.; Ma, Y.; Indresand, H.; Aarhaug, T.A.; Grådahl, S. Airborne Emissions from Mn Ferroalloy Production. *JOM* **2019**, *71*, 349–365. [CrossRef]

171. Trubetskaya, A.; Timko, M.T.; Umeki, K. Prediction of fast pyrolysis products yields using lignocellulosic compounds and ash contents. *Appl. Energy* **2020**, *257*, 113897. [CrossRef]

172. Funke, A.; Demus, T.; Willms, T.; Schenke, L.; Echterhof, T.; Niebel, A.; Pfeifer, H.; Dahmen, N. Application of fast pyrolysis char in an electric arc furnace. *Fuel Process. Technol.* **2018**, *174*, 61–68. [CrossRef]

173. Septien, S.; Valin, S.; Dupont, C.; Peyrot, M.; Salvador, S. Effect of particle size and temperature on woody biomass fast pyrolysis at high temperature (1000–1400 °C). *Fuel* **2012**, *97*, 202–210. [CrossRef]

174. Liaw, S.B.; Wu, H. A New Method for Direct Determination of Char Yield during Solid Fuel Pyrolysis in Drop-Tube Furnace at High Temperature and Its Comparison with Ash Tracer Method. *Energy Fuels* **2019**, *33*, 1509–1517. [CrossRef]

175. Zhang, Y.; Kajitani, S.; Ashizawa, M.; Oki, Y. Tar destruction and coke formation during rapid pyrolysis and gasification of biomass in a drop-tube furnace. *Fuel* **2010**, *89*, 302–309. [CrossRef]

176. Trubetskaya, A.; Souihi, N.; Umeki, K. Categorization of tars from fast pyrolysis of pure lignocellulosic compounds at high temperature. *Renew. Energy* **2019**, *141*, 751–759. [CrossRef]

177. Radlein, D.; Quignard, A. A Short Historical Review of Fast Pyrolysis of Biomass. *Oil Gas Sci. Technol.* **2013**, *68*, 765–783. [CrossRef]

178. Pires, A.P.P.; Arauzo, J.; Fonts, I.; Domine, M.E.; Arroyo, A.F.; Garcia-Perez, M.E.; Montoya, J.; Chejne, F.; Pfromm, P.; Garcia-Perez, M. Challenges and Opportunities for Bio-oil Refining: A Review. *Energy Fuels* **2019**, *33*, 4683–4720. [CrossRef]

179. Mostafazadeh, A.K.; Solomatnikova, O.; Drogui, P.; Tyagi, R.D. A review of recent research and developments in fast pyrolysis and bio-oil upgrading. *Biomass Convers. Biorefin.* **2018**, *8*, 739–773. [CrossRef]

180. Riva, L.; Nielsen, H.K.; Skreiberg, Ø.; Wang, L.; Bartocci, P.; Barbanera, M.; Bidini, G.; Fantozzi, F. Analysis of optimal temperature, pressure and binder quantity for the production of biocarbon pellet to be used as a substitute for coke. *Appl. Energy* **2019**, *256*, 113933. [CrossRef]

181. Elkasabi, Y.; Boateng, A.A.; Jackson, M.A. Upgrading of bio-oil distillation bottoms into biorenewable calcined coke. *Biomass Bioenergy* **2015**, *81*, 415–423. [CrossRef]

182. Pennise, D.M.; Smith, K.R.; Kithinji, J.P.; Rezende, M.E.; Raad, T.J.; Zhang, J.; Fan, C. Emissions of greenhouse gases and other airborne pollutants from charcoal making in Kenya and Brazil. *J. Geophys. Res.* **2001**, *106*, 143–155. [CrossRef]

183. Simoneit, B.R.T.; Rogge, W.F.; Lang, Q.; Jaffé, R. Molecular characterization of smoke from campfire burning of pine wood (textitPinus elliottii). *Chemosph. Glob. Chang. Sci.* **2000**, *2*, 107–122. [CrossRef]

184. Oanh, N.T.K.; Nghiem, L.H.; Phyu, Y.L. Emission of Polycyclic Aromatic Hydrocarbons, Toxicity, and Mutagenicity from Domestic Cooking Using Sawdust Briquettes, Wood, and Kerosene. *Environ. Sci. Technol.* **2002**, *36*, 833–839. [CrossRef] [PubMed]

185. Andersson, J.T.; Achten, C. Time to Say Goodbye to the 16 EPA PAHs? Toward an Up-to-Date Use of PACs for Environmental Purposes. *Polycyl. Aromat. Compd.* **2015**, *35*, 330–354. [CrossRef] [PubMed]

186. National Institute for Occupational Safety and Health (NIOSH). *Criteria for a Recommended Standard: Occupational Exposure to Carbon Black*; DHHS (NIOSH) Publication Number 78-204; NIOSH: Washington, DC, USA, 1978.

187. Chen, H.; Si, Y.; Chen, Y.; Yang, H.; Chen, D.; Chen, W. NO$_x$ precursors from biomass pyrolysis: Distribution of amino acids in biomass and Tar-N during devolatilization using model compounds. *Fuel* **2017**, *187*, 367–375. [CrossRef]

188. Cheah, S.; Malone, S.C.; Feik, C.J. Speciation of Sulfur in Biochar Produced from Pyrolysis and Gasification of Oak and Corn Stover. *Environ. Sci. Technol.* **2014**, *48*, 8474–8480. [CrossRef]

189. Van Dam, J. *The Charcoal Transition: Greening the Charcoal Value Chain to Mitigate Climate Change and Improve Local Livelihoods*; FAO: Rome, Italy, 2017.

190. Brocksiepe, H.-G. Charcoal. In *Ullmann's Encyclopedia of Industrial Chemistry*; Wiley-VCH: Weinheim, Germany, 2012; Volume 8, pp. 93–98.

191. Sparrevik, M.; Adam, C.; Martinsen, V.; Jubaedah, V.; Cornelissen, G. Emissions of gases and particles from charcoal/biochar production in rural areas using medium.sized traditional and improved "retort" kilns. *Biomass Bioenergy* **2015**, *72*, 65–73. [CrossRef]

192. Cornelissen, G.; Pandit, N.R.; Taylor, P.; Pandit, B.H.; Sparrevik, M.; Schmidt, H.P. Emissions and Char Quality of Flame-Curtain "Kon Tiki" Kilns for Farmer-Scale Charcoal/Biochar Production. *PLoS ONE* **2016**, *11*, e0154617. [CrossRef]

193. Chen, J.; Li, S.; Liang, C.; Xu, Q.; Li, Y.; Fuhrmann, J.J. Response of microbial community structure and function to short-term biochar amendment in an intensively managed bamboo (*Phyllostachys praecox*). *Sci. Total Environ.* **2017**, *574*, 24–33. [CrossRef]

194. Zhu, W.; Song, W.; Lin, W. Catalytic gasification of char from co-pyrolysis of coal and biomass. *Fuel Process. Technol.* **2008**, *89*, 890–896. [CrossRef]

195. Demirbaş, A. Sustainable Charcoal Production and Charcoal Briquetting. *Energy Sources Part A* **2009**, *31*, 1694–1699. [CrossRef]

196. Akowuah, J.O.; Kemausuor, F.; Mitchual, S.J. Physico-chemical characteristics and market potential of sawdust charcoal briquette. *IJEE* **2012**, *3*, 1–6. [CrossRef]

197. Li, F.; Tangstad, M.; Solheim, I. Quartz and carbon black pellets for silicon production. In Proceedings of the INFACON XIV, Kiev, Ukraine, 31 May–4 June 2015; pp. 390–401.

198. Hardianto, T.; Pambudi, F.F.; Irhamna, A.R. A study on lignin characteristic as internal binder in hot briquetting process of organic municipal solid waste. In Proceedings of the AIP Conference Proceedings 1984, Bali, Indonesia, 17–19 November 2017; p. 030013.

199. Lumadue, M.R.; Cannon, F.; Brown, N.R. Lignin as both fuel and fusing binder in briquetted anthracite fines for foundry coke substitute. *Fuel* **2012**, *97*, 869–875. [CrossRef]

200. Surup, G.R.; Heidelmann, M.; Nielsen, H.K.; Trubetskaya, A. Characterization and reactivity of charcoal from high temperature pyrolysis (800–1600 °C). *Fuel* **2019**, *235*, 1544–1554. [CrossRef]

201. Surup, G.R.; Pedersen, T.A.; Chaldien, A.; Beukes, J.P.; Tangstad, M. Electrical Resistivity of Carbonaceous Bed Material at High Temperature. *Processes* **2020**, *8*, 933. [CrossRef]

202. Chen, J.; Zhou, F.; Si, T.; Zhou, J.; Cen, K. Mechanical strength and combustion properties of biomass pellets prepared with coal tar residue as a binder. *Fuel Process. Technol.* **2018**, *179*, 229–237. [CrossRef]

203. Coulon, F.; Orsi, R.; Turner, C.; Walton, C.; Daly, P.; Pollard, S.J. Understanding the fate and transport of petroleum hydrocarbons from coal tar within gasholders. *Environ. Int.* **2009**, *35*, 248–252. [CrossRef]

204. Wang, X.; Shen, J.; Niu, Y.; Sheng, Q.; Liu, G.; Wang, Y. Solvent extracting coal gasification tar residue and the extracts characterization. *J. Clean. Prod.* **2016**, *133*, 965–970. [CrossRef]
205. Craven, J.M.; Swithenbank, J.; Sharifi, V.N.; Peralta-Solorio, D.; Kelsall, G.; Sage, P. Hydrophobic coatings for moisture stable wood pellets. *Biomass Bioenergy* **2015**, *80*, 278–285. [CrossRef]
206. Asadullah, M.; Zhang, S.; Li, C.-Z. Evaluation of structural features of chars from pyrolysis of biomass of different particle sizes. *Fuel Process. Technol.* **2010**, *91*, 877–881. [CrossRef]
207. Ronsse, F. Commercial biochar production and its certification. Presented at the Interreg Conference, Groningen, The Netherlands, 10 December 2013.
208. Quicker, P.; Weber, K. *Biokohle-Herstellung, Eigenschaften und Verwendung von Biomassekarbonisaten*; Springer Vieweg: Medford, MA, USA, 2016.
209. Antal, M.J.; Allen, S.G.; Dai, X.; Shimizu, B.; Tam, M.S.; Grønli, M. Attainment of the Theoretical Yield of Carbon from Biomass. *Ind. Eng. Chem. Res.* **2000**, *39*, 4024–4031. [CrossRef]
210. Amini, E.; Safdari, M.-S.; DeYoung, J.T.; Weise, D.R.; Fletcher, T.H. Characterization of pyrolysis products from slow pyrolysis of live and dead vegetation native to the southern United States. *Fuel* **2019**, *235*, 1475–1491. [CrossRef]
211. Wesenbeeck, S.V.; Higashi, C.; Legarra, M.; Wang, L.; Antal, M.J. Biomass Pyrolysis in Sealed Vessels. Fixed-Carbon Yields from Avicel Cellulose That Realize the Theoretical Limit. *Energy Fuels* **2016**, *30*, 480–491. [CrossRef]
212. Soto Veiga, J.P.; Romanelli, T.L. Mitigation of greenhouse gas emissions using exergy. *J. Clean. Prod.* **2020**, *260*, 121092. [CrossRef]
213. Yoshida, T.; Turn, S.Q.; Yost, R.S.; Antal, M.J. Banagrass vs Eucalyptus Wood as Feedstocks for Metallurgical Biocarbon Production. *Ind. Eng. Chem. Res.* **2008**, *47*, 9882–9888. [CrossRef]
214. Kimura, L.M.; Santos, L.C.; Vieira, P.F.; Parreira, P.M.; Henrique, H.M. Biomass Pyrolysis: Use of Some Agricultural Wastes for Alternative Fuel Production. In Proceedings of the Seventh International Latin American Conference on Powder Technology, Atibaia, Brazil, 8–10 November 2009; pp. 274–279.
215. Islam, M.N.; Ali, M.H.; Ahman, I. Fixed bed slow pyrolysis of biomass solid waste for bio-char. *IOP Conf. Ser. Mater. Sci. Eng.* **2017**, *206*, 012014. [CrossRef]
216. Czajczyńska, D.; Anguilano, L.; Ghazal, H.; Krzyżyńska, R.; Reynolds, A.J.; Spencer, N.; Jouhara, H. Potential of pyrolysis processes in the waste management sector. *Therm. Sci. Eng. Prog.* **2017**, *3*, 171–197. [CrossRef]
217. Roy, P.; Dias, G. Prospects for pyrolysis technologies in the bioenergy sector: A review. *Renew. Sustain. Energy Rev.* **2017**, *77*, 59–69. [CrossRef]
218. Antal, M.J.; Mochidzuki, K.; Paredes, L.S. Flash Carbonization of Biomass. *Ind. Eng. Chem. Res.* **2003**, *42*, 3690–3699. [CrossRef]
219. Ringer, M.; Putsche, V.; Scahill, J. *Large-Scale Pyrolysis Oil Production: A Technology Assessment and Economic Analysis*; Report Number: NREL/TP-510-37779; National Renewable Energy Laboratory: Golden, CO, USA, 2006.
220. Tintner, J.; Preimesberger, C.; Pfeifer, C.; Theiner, J.; Ottner, F.; Wriessnig, K.; Puchberger, M.; Smidt, E. Pyrolysis profile of a rectangular kiln—Natural scientific investigation of a traditional charcoal production process. *J. Anal. Appl. Pyrolysis* **2020**, *146*, 104757. [CrossRef]
221. Chidumayo, E.N. Is charcoal production in *Brachystegia-Julbernardia* woodlands of Zambia sustainable? *Biomass Bioenergy* **2019**, *125*, 1–7. [CrossRef]
222. Duarte, S.J.; Lin, J.; Alviso, D.; Rolón, J.C. Effect of Temperature and Particle Size on the Yield of Bio-oil, Produced from Conventional Coconut Core Pyrolysis. *Int. J. Chem. Eng. Appl.* **2016**, *7*, 102–108. [CrossRef]
223. Neves, D.; Matos, A.; Tarelho, L.; Thunman, H.; Larsson, A.; Seemann, M. Volatile gases from biomass pyrolysis under conditions relevant for fluidized bed gasifiers. *J. Anal. Appl. Pyrolysis* **2017**, *127*, 57–67. [CrossRef]
224. Chen, H. *Lignocellulose Biorefinery Conversion Engineering*; Woodhead Publishing: Cambridge, UK, 2015; pp. 87–124.
225. Bridgwater, T.; Carson, P.; Coulson, M. A comparison of fast and slow pyrolysis liquids from mallee. *Int. J. Glob. Energy* **2007**, *27*, 204–216. [CrossRef]
226. Grønli, M. *Industrial Production of Charcoal*; SINTEF Energy Research: Trondheim, Norway, 1999.
227. Scholze, B.; Meier, D. Characterization of the water-insoluble fraction from pyrolysis oil (pyrolytic lignin). Part I. PY–GC/MS, FTIR, and functional groups. *J. Anal. Appl. Pyrolysis* **2001**, *60*, 41–54. [CrossRef]

228. Mullen, C.A.; Boateng, A.A. Chemical Composition of Bio-oils Produced by Fast Pyrolysis of Two Energy Crops. *Energy Fuels* **2008**, *22*, 2104–2109. [CrossRef]

229. Meng, J.; Moore, A.; Tilotta, D.C.; Kelley, S.S.; Adhikar, S.; Park, S. Thermal and Storage Stability of Bio-Oil from Pyrolysis of Torrefied Wood. *Energy Fuels* **2015**, *29*, 5117–5126. [CrossRef]

230. Diebold, J.P. *A Review of the Chemical and Physical Mechanisms of the Storage Stability of Fast Pyrolysis Bio-Oils; Report N. NREL/SR-570-27613; Contract No.: DE-AC36-99-GO10337*; National Renewable Energy Laboratory: Golden, CO, USA, 1999.

231. Lawal, A.T. Polycyclic aromatic hydrocarbons. A review. *Cogent Environ. Sci.* **2017**, *31*, 1339841. [CrossRef]

232. Chen, D.; Zhou, J.; Zhang, Q.; Zhu, X. Evaluation methods and research progresses in bio-oil storage stability. *Renew. Sustain. Energy Rev.* **2014**, *40*, 69–79. [CrossRef]

233. Sarkar, J.K.; Wang, Q. Different Pyrolysis Process Conditions of South Asian Waste Coconut Shell and Characterization of Gas, Bio-Char, and Bio-Oil. *Energies* **2020**, *13*, 1970. [CrossRef]

234. Crombie, K.; Mašek, O. Investigating the potential for a self-sustaining slow pyrolysis system under varying operating conditions. *Bioresour. Technol.* **2014**, *162*, 148–156. [CrossRef]

235. Panwara, N.L.; Kothari, R.; Tyagi, V.V. Thermo chemical conversion of biomass – Eco friendly energy routes. *Renew. Sustain. Energy Rev.* **2012**, *16*, 1801–1816. [CrossRef]

236. Jin, W.; Pastor-Pérez, L.; Yu, J.; Odriozola, J.A.; Gu, S.; Reina, T.R. Cost-effective routes for catalytic biomass upgrading. *Curr. Opin. Green Sustain. Chem.* **2020**, *23*, 1–9. [CrossRef]

237. Zheng, J.-L.; Zhu, Y.-H.; Zhu, M.-Q.; Kang, K.; Sun, R.-C. A review of gasification of bio-oil for gas production. *Sustain. Energy Fuels* **2019**, *7*, 1600–1622. [CrossRef]

238. Dahmen, N.; Henrich, E.; Dinjus, E.; Weirich, F. The bioliq® bioslurry gasification process for the production of biosynfuels, organic chemicals, and energy. *Energy Sustain. Soc.* **2012**, *2*, 1–44. [CrossRef]

239. Subramanian, M.; Harman, C.N. Problems and Prospects of carbonaceous reducing agents in ferro alloys production. In Proceedings of the Seminar on Problems and Prospects of Ferro-Alloy Industry in India, Jamshedpur, India, 24–26 October 1983; pp. 133–140.

240. Lindstad, T.; Olsen, S.; Tranell, G.; Færden, T.; Lubetsky, J. Greenhouse gas emissions from ferroalloy production. In Proceedings of the INFACON XI, New Delhi, India, 18–21 February 2007; pp. 457–466.

241. Díez, M.A.; Alvarez, R.; Barriocanal, C. Coal for metallurgical coke production: Predictions of coke quality and future requirements for cokemaking. *Int. J. Coal Geol.* **2002**, *50*, 389–412. [CrossRef]

242. Valia, H.S. *Coke Production for Blast Furnace Ironmaking*; Ispat Inland Inc.: Chicago, IN, USA, 2015.

243. Babich, A.; Senk, D. Chapter 12-Coal use in iron and steel metallurgy. In *The Coal Handbook: Towards Cleaner Production*; Osborne, D., Ed.; Woodhead Publishing Series in Energy; Woodhead Publishing: Cambridge, UK, 2013; Volume 2, pp. 267–311.

244. United States International Trade Commission. *Metallurgical Coke: Baseline Analysis of the U.S. Industry and Imports: [Investigation No. 332–342]*; USITC Publication; U.S. International Trade Commission: Washington, DC, USA, 1994.

245. Schobert, H.; Schobert, N. Comparative Carbon Footprints of Metallurgical Coke and Anthracite for Blast Furnace and Electric Arc Furnace Use. 2015. Available online: http://www.blaschakcoal.com/wp-content/uploads/Carbon-Footprint-Archival-Report-v-4-September-2015.pdf (accessed on 17 September 2020).

246. Eidem, P.A.; Runde, M.; Tangstad, M.; Bakken, J.A.; Zhou, Z.Y.; Yu, A.B. Effect of Contact Resistance on Bulk Resistivity of Dry Coke Beds. *Metall. Mater. Trans. B* **2009**, *40B*, 388–396. [CrossRef]

247. Khanna, R.; Li, K.; Wang, Z.; Sun, M.; Zhang, J.; Mukherjee, P.S. Biochars in iron and steel industries. In *Char and Carbon Materials Derived from Biomass: Production, Characterization and Applications*; Elsevier: Amsterdam, The Netherlands, 2019; pp. 429–446.

248. Yunos, N.F.M. Recycling Agricultural Waste from Palm Shells during Electric Arc Furnace Steelmaking. *Energy Fuels* **2011**, *26*, 278–286. [CrossRef]

249. Wang, C.; Mellin, P.; Lövgren, J.; Nilsson, L.; Yang, W.; Salam, H.; Hultgren, A.; Larsson, M. Biomass as blast furnace injectant—Considering availability pretreatment and deployment in the Swedish steel industry. *Energy Convers. Manag.* **2015**, *102*, 217–226. [CrossRef]

250. Júnior, A.F.D.; de Oliveira, R.N.; Deglise, X.; de Souzam, N.D.; Brito, J.O. Infrared spectroscopy analysis on charcoal generated by the pyrolysis of Corymbia. citiodora wood. *Matéria (Rio de Janeiro)* **2019**, *24*, 1–7.

251. Labbé, N.; Harper, D.; Rials, T. Chemical Structure of Wood Charcoal by Infrared Spectroscopy and Multivariate Analysis. *J. Agric. Food Chem.* **2006**, *54*, 3492–3497. [CrossRef]

252. Chen, Z.; Ma, W.; Wu, J.; Wei, K.; Lei, Y.; Ly, G. A Study of the Performance of Submerged Arc Furnace Smelting of Industrial Silicon. *Silicon* **2018**, *10*, 1121–1127. [CrossRef]

253. Sahajwalla, V.; Dubikova, M.; Khanna, R. Reductant characterisation and selection: Implications for ferroalloys processing. In Proceedings of the INFACON X, Cape Town, South Africa, 1–4 February 2004; pp. 351–362.

254. Chen, Y.; Aanjaneya, K.; Atreya, A. A study to investigate pyrolysis of wood particles of various shapes and sizes. *Fire Saf. J.* **2017**, *91*, 820–827. [CrossRef]

255. Kenney, W.A.; Sennerby-Forsse, L.; Layton, P. A Review of Biomass Quality Research Relevant to the Use of Poplar and Willow for Energy Conversion. *Biomass* **1990**, *21*, 163–188. [CrossRef]

256. Plötze, M.; Niemz, P. Porosity and pore size distribution of different wood types as determined by mercury intrusion porosimetry. *Eur. J. Wood Wood Prod.* **2011**, *69*, 649–657. [CrossRef]

257. Hu, Q.; Yang, H.; Yao, D.; Zhu, D.; Wang, X.; Shao, J.; Chen, H. The densification of bio-char: Effect of pyrolysis temperature on the qualities of pellets. *Bioresource* **2016**, *200*, 521–527. [CrossRef]

258. Brewer, C.E.; Chuang, V.J.; Masiello, C.A.; Gonnermann, H.; Gao, X.; Dugan, B.; Driver, L.E.; Panzacchi, P.; Zygourakis, K.; Davies, C.A. New approaches to measuring biochar density and porosity. *Biomass Bioenergy* **2014**, *66*, 176–185. [CrossRef]

259. Somerville, M.; Jahanshahi, S. The effect of temperature and compression during pyrolysis on the density of charcoal made from Australian eucalypt wood. *Renew. Energy* **2015**, *80*, 471–478. [CrossRef]

260. Florentino-Madiedo, L.; Díaz-Faes, E.; Barriocanal, C. Mechanical strength of bio-coke from briquettes. *Renew. Energy* **2020**, *146*, 1717–1724. [CrossRef]

261. Sato, H.; Patrick, J.W.; Walker, A. Effect of coal properties and porous structure on tensile strength of metallurgical coke. *Fuel* **1998**, *77*, 1203–1208. [CrossRef]

262. Montiano, M.G.; Díaz-Faes, E.; Barriocanal, C.; Alvarez, R. Influence of biomass on metallurgical coke quality. *Fuel* **2014**, *116*, 175–182. [CrossRef]

263. Newalkar, G.; Iisa, K.; D'Amico, A.D.; Sievers, C.; Agrawal, P. Effect of Temperature, Pressure, and Residence Time on Pyrolysis of Pine in an Entrained Flow Reactor. *Energy Fuels* **2014**, *28*, 5144–5157. [CrossRef]

264. Lee, Y.; Park, J.; Ryu, C.; Gang, K.S.; Yang, W.; Park, Y.; Jung, J.; Hyun, S. Comparison of biochar properties from biomass residues produced by slow pyrolysis at 500 °C. *Bioresour. Technol.* **2013**, *148*, 196–201. [CrossRef]

265. Yan, M.; Zhang, S.; Wibowo, H.; Grisdanurak, N.; Cai, Y.; Zhou, X.; Kanchanatip, E.; Antoni. Biochar and pyrolytic gas properties from pyrolysis of simulated municipal solid waste (SMSW) under pyrolytic gas atmosphere. *Waste Dispos. Sustain. Energy* **2020**, *2*, 37–46. [CrossRef]

266. Ronsse, F.; van Hecke, S.; Dickinson, D.; Prins, W. Production and characterization of slow pyrolysis biochar: influence of feedstock type and pyrolysis conditions. *GCB Bioenergy* **2013**, *5*, 104–115. [CrossRef]

267. Wang, L.; Hovd, B.; Bui, H.-H.; Valderhaug, A.; Buø, T.V.; Birkeland, R.G.; Skreiberg, Ø.; Tran, K.-Q. CO_2 Reactivity Assessment of Woody Biomass Biocarbons for Metallurgical Purposes. *Chem. Eng. Trans.* **2016**, *50*, 55–60.

268. Mukherjee, A.; Zimmerman, A.R.; Harris, W. Surface chemistry variations among a series of laboratory-produced biochars. *Geoderma* **2011**, *163*, 247–255. [CrossRef]

269. Forest Products Laboratory. *Charcoal Production, Marketing, and Use*; Technical Report No. 2213; US Department of Agriculture: Madison, WI, USA, 1961.

270. Foley, G. *Charcoal Making in Developing Countries*; Energy Information Programme Technical Report No. 5; Earthscan: London, UK, 1986.

271. Food and Agriculture Organization of the United Nations. *Simple Technologies for Charcoal Making*; FAO Forestry Paper 41, Bd 4; Food and Agriculture Organization of the United Nations: Rome, Italy, 1987.

272. Mizuno, S.; Ida, T.; Fuchihata, M.; Sanchez, E.; Yoshikuni, K. Formation characteristics of bio-coke produced from waste agricultural biomass. In Proceedings of the ASME 2015 InterPACK2015, San Francisco, CA, USA, 6–9 July 2015; pp. 1–5.

273. Byrne, C.E.; Nagle, D.C. Carbonization of wood for advanced materials applications. *Carbon* **1997**, *35*, 259–266. [CrossRef]

274. Riva, L.; Cardarelli, A.; Andersen, G.J.; Buø, T.V.; Barbanera, M.; Bartocci, P.; Fantozzi, F.; Nielsen, H.K. On the self-heating behavior of upgraded biochar pellets blended with pyrolysis oil: Effects of process parameters. *Fuel* **2020**, *278*, 118395. [CrossRef]

275. Safana, A.A.; Abdullah, N.; Sulaiman, F. Bio-char and bio-oil mixtures derived from the pyrolysis of mesocarp fibre for for briquettes production. *J. Oil Palm Res.* **2018**, *30*, 130–140.

276. Huang, H.; Fox, J.T.; Cannon, F.S.; Komarneni, S. In situ growth of silicon carbide nanowires from anthracite surfaces. *Ceram Int.* **2011**, *37*, 1063–1072. [CrossRef]

277. Wang, N.; Low, M.J.D. Spectroscopic studies of carbons. XV. The pyrolysis of a lignin. *Mater. Chem. Phys.* **1990**, *26*, 67–80. [CrossRef]

278. Dijs, H.; Smith, D. Factors affecting the resistivity and reactivity of carbonaceous reducing agents for the electric-smelting industry. *J. S. Afr. Inst. Min. Metall.* **1980**, *80*, 286–296.

279. Eidem, P.A.; Tangstad, M.; Bakken, J.A. Measurement of material resistivity and contact resistance of metallurgical coke. In Proceedings of the INFACON XI, New Delhi, India, 18–21 February 2007; pp. 561–571.

280. Eidem, P.A.; Tangstad, M.; Bakken, J.A.; Ishak, R. Influence of coke particle size on the electrical resistivity of coke beds. In Proceedings of the INFACON XII, Hensinki, Finland, 6–9 June 2010; pp. 349–358.

281. Feng, B.; Bhatia, S.K.; Barry, J.C. Structural ordering of coal char during heat treatment and its impact on reactivity. *Carbon* **2002**, *40*, 481–496. [CrossRef]

282. Tangstad, M.; Beukes, J.P.; Steenkamp, J.; Ringdalen, E. 14—Coal-based reducing agents in ferroalloys and silicon production. In *New Trends in Coal Conversion Combustion, Gasification, Emissions, and Coking*; Woodhead Publishing: Cambridge, UK, 2019; pp. 405–438.

283. Adinaveen, T.; Vijaya, J.J.; Kennedy, L. Comparative Study of Electrical Conductivity on Activated Carbons Prepared from Various Cellulose Materials. *Arab. J. Sci. Eng.* **2014**, *41*, 55–65. [CrossRef]

284. Behazin, E.; Ogunsona, E.; Rodriguez-Uribe, A.; Mohanty, A.; Misra, M.; Anyia, A.O. Mechanical, Chemical, and Physical Properties of Wood and Perennial Grass Biochars for Possible Composite Application. *Bioresources* **2016**, *11*, 1334–1348. [CrossRef]

285. Mrozowski, S. Semiconductivity and Diamagnetism of Polycrystalline Graphite and Condensed Ring Systems. *Phys. Rev.* **1952**, *85*, 609–620. [CrossRef]

286. Golden, T.C.; Jenkins, R.G.; Otake, Y.; Scaroni, A.W. Oxygen Complexes on Carbon Surfaces. *Proc. Electrochem. Soc.* **1983**, *84*, 61–78.

287. Sánchez-González, J.; Stoeckli, F.; Centeno, T.A. The role of the electric conductivity of carbons in the electrochemical capacitor performance. *J. Electroanal. Chem.* **2011**, *657*, 176–180. [CrossRef]

288. Mochidzuki, K.; Soutric, F.; Takokoro, K.; Antal, M.J.; Toth, M.; Zelei, B.; Várhegyi, G. Electrical and Physical Properties of Carbonized Charcoals. *Ind. Eng. Chem. Res.* **2003**, *42*, 5140–5151. [CrossRef]

289. Koursaris, A.; See, J.B. The resistivity of mixtures of Mamatwan manganese ore and reducing agents. *J. S. Afr. Inst. Min. Metall.* **1980**, *80*, 229–238.

290. Lindstad, T.; Monsen, B.; Osen, K.S. How the ferroalloys industry can meet greenhouse gas regulations. In Proceedings of the INFACON XII, Helsinki, Finland, 6–9 June 2010; pp. 63–70.

291. Bouraoui, Z.; Jeguirim, M.; Guizani, C.; Limousy, L.; Dupint, C.; Gadiou, R. Thermogravimetric study on the influence of structural, textural and chemical properties of biomass chars on CO_2 gasification reactivity. *Energy* **2015**, *88*, 703–710. [CrossRef]

292. Bui, H.-H.; Wang, L.; Tran, K.-Q.; Skreiberg, O.; Luengnaruemitchai, A. CO_2 Gasification of Charcoals in the Context of Metallurgical Application. *Energy Procedia* **2017**, *105*, 316–321. [CrossRef]

293. Rollinson, A.N.; Karmakar, M.K. On there activity of various biomass species with CO_2 using a standardised methodology for fixed-bed gasification. *Chem. Eng. Sci.* **2015**, *128*, 82–91. [CrossRef]

294. Bui, H.-H.; Wang, L.; Tran, K.-Q.; Skreiberg, Ø. CO_2 gasificiation of charcoals produced at various pressures. *Fuel Process. Technol.* **2016**, *152*, 207–214. [CrossRef]

295. Mani, T.; Mahinpey, N.; Murugan, P. Reaction kinetics and mass transfer studies of biomass char gasification with CO_2. *Chem. Eng. Sci.* **2011**, *66*, 36–41. [CrossRef]

296. Calo, J.M.; Perkins, M.T. A heterogeneous surface model for the "steady-state" kinetics of the Boudouard reaction. *Carbon* **1987**, *25*, 395–407. [CrossRef]

297. Cetin, E.; Moghtaderi, B.; Gupta, R.; Wall, T.F. Biomass gasification kinetics: Influences of pressure and char structure. *Compos. Sci. Technol.* **2005**, *177*, 765–791. [CrossRef]

298. Gomez, A.; Silbermann, R.; Mahinpey, N. A comprehensive experimental procedure for CO_2 coal gasification: Is there really a maximum reaction rate? *Appl. Energy* **2014**, *124*, 73–81. [CrossRef]

299. Gomez, A.; Mahinpey, N. A new model to estimate CO_2 coal gasification kinetics based only on parent coal characterization properties. *Appl. Energy* **2015**, *137*, 126–133. [CrossRef]

300. Olsen, S.E.; Ding, W.; Kossyreva, O.A.; Tangstad, M. Equilibrium in production of high carbon ferromanganese. In Proceedings of the INFACON VII, Trondheim, Norway, 11–14 June 1995; pp. 591–600.

301. Kapure, G.; Kari, C.; Rao, S.M.; Raju, K.S. Use of chemical energy in submerged arc furnace to produce ferrochrome: Prospects and limitations. In Proceedings of the INFACON XI, New Delhi, India, 18–21 February 2007; pp. 165–170.

302. Lindstad, T.; Gaal, S.; Hansen, S.; Prytz, S. Improved SINTEF SiO-Reactivity Test. In Proceedings of the INFACON XI, New Delhi, India, 18–21 February 2007; pp. 414–423.

303. Raaness, O.; Kolbeinsen, L.; Byberg, J.A. Statistical Analysis of Properties for Coals Used in the Production of Silicon Rich Alloys. In Proceedings of the INFACON VIII, Beijing, China, 7–10 June 1998; pp. 116–120.

304. Li, F.; Tangstad, M.; Ringdalen, E. Carbothermal Reduction of Quartz and Carbon Pellets at Elevated Temperatures. *Metall. Mater. Trans. B* **2018**, *49*, 1078–1088. [CrossRef]

305. Chen, Z.; Ma, W.; Li, S.; Wu, J.; Wei, K.; Yu, Z.; Ding, W. Influence of carbon material on the production of different electric arc furnaces. *J. Clean. Prod.* **2018**, *174*, 17–25. [CrossRef]

306. Safarian, J.; Tangstad, M. Slag-carbon reactivity. In Proceedings of the INFACON XII, Helsinki, Finland, 6–9 June 2010; pp. 327–338.

307. Kim, P.P.; Tangstad, M. Kinetic Investigations of SiMn Slags From Different Mn Sources. *Metall. Mater. Trans. B* **2018**, *49*, 1185–1196. [CrossRef]

308. Kang, T.W.; Gupta, S.; Saha-Chaudhury, N.; Sahajwalla, V. Wetting and Interfacial Reaction Investigations of Coke/Slag Systems and Associated Liquid Permeability of Blast Furnaces. *ISIJ Int.* **2005**, *45*, 1526–1535. [CrossRef]

309. Safarian, J.; Kolbeinsen, L.; Tangstad, M.; Tranell, G. Kinetics and Mechanism of the Simultaneous Carbothermic Reduction of FeO and MnO from High-Carbon Ferromanganese Slag. *Metall. Mater. Trans. B* **2009**, *40*, 929–939. [CrossRef]

310. Hall, D.O.; Kaale, B.K.; Karekezi, S.; Ravindranath, N.K. *Application of Biomass-Energy Technologies*; United Nations Centre for Human Settlements (Habitat): Nairobi, Kenya, 1993.

311. Chidumayo, E.N. Woody biomass structure and utilisation for charcoal production in a Zambian Miombo woodland. *Bioresour. Technol.* **1991**, *37*, 43–52. [CrossRef]

312. INDEXBOX. Global Wood Charcoal Market Reached \$24B, Buoyed By Robust Demand in Africa. Available online: https://www.globaltrademag.com/global-wood-charcoal-market-reached-24b-buoyed-by-robust-demand-in-africa/ (accessed on 18 September 2020).

313. Chudy, R.R.; Hagler, R.W. Dynamics of global roundwood prices-Cointegration analysis. *For. Policy Econ.* **2020**, *115*, 1–14. [CrossRef]

314. Dibdiakova, J.; Gjølsjø, S.; Wang, L. *Report on Solid Biofuels from Forest-Fuel Specification and Quality Assurance*; Report 8/2014; Norwegian Forest and Landscape Institute: Ås, Norway, 2014.

315. Iisa, K.; Robichaud, D.J.; Watson, M.J.; Dam, J.; Dutta, A.; Mukarakate, C.; Kim, S.; Nimlos, M.R.; Baldwin, R. Improving biomass pyrolysis economics by integrating vapor and liquid phase upgrading. *Green Chem.* **2018**, *20*, 567–582. [CrossRef]

316. Awasthi, M.K.; Sarsaiya, S.; Patel, A.; Juneja, A.; Singh, R.P.; Yan, B.; Awasthi, S.K.; Jain, A.; Liu, T.; Duan, Y.; et al. Refining biomass residues for sustainable energy and bio-products: An assessment of technology, its importance, and strategic applications in circular bio-economy. *Renew. Sustain. Energy Rev.* **2020**, *127*, 109876. [CrossRef]

317. Funke, A.; Niebel, A.; Richter, D.; Abbas, M.M.; Müller, A.-K.; Radloff, S.; Paneru, M.; Maier, J.; Dahmen, N.; Sauer, J. Fast pyrolysis char—Assessment of alternative uses within the bioliq® concept. *Bioresour. Technol.* **2016**, *200*, 905–913. [CrossRef]

318. Khundi, F.; Jagger, P.; Shively, G.; Sserunkuuma, D. Income, poverty and charcoal production in Uganda. *For. Policy Econ.* **2011**, *13*, 199–205. [CrossRef]

319. Adanguidi, J.; Padonou, E.A.; Zannou, A.; Houngbo, S.B.E.; Saliou, I.O.; Agbahoungba, S. Fuelwood consumption and supply strategies in mangrove forests—Insights from RAMSAR sites in Benin. *For. Policy Econ.* **2020**, *116*, 102192. [CrossRef]
320. Myrhaug, E.H. Non-Fossil Reduction Materials in the Silicon Process-Properties and Behaviour. Ph.D. Thesis, NTNU, Trondheim, Norway, 2003.

Publisher's Note: MDPI stays neutral with regard to jurisdictional claims in published maps and institutional affiliations.